AF688353

Introductory Nuclear Physics

Introductory Nuclear Physics

Leonel Ware

NY Research Press
New York

Published by NY Research Press
118-35 Queens Blvd., Suite 400,
Forest Hills, NY 11375, USA
www.nyresearchpress.com

Introductory Nuclear Physics
Leonel Ware

© 2022 NY Research Press

International Standard Book Number: 978-1-63238-893-3 (Hardback)

This book contains information obtained from authentic and highly regarded sources. All chapters are published with permission under the Creative Commons Attribution Share Alike License or equivalent. A wide variety of references are listed. Permissions and sources are indicated; for detailed attributions, please refer to the permissions page. Reasonable efforts have been made to publish reliable data and information, but the authors, editors and publisher cannot assume any responsibility for the validity of all materials or the consequences of their use.

Trademark Notice: Registered trademark of products or corporate names are used only for explanation and identification without intent to infringe.

Cataloging-in-Publication Data

Introductory nuclear physics / Leonel Ware.
p. cm.
Includes bibliographical references and index.
ISBN 978-1-63238-893-3
1. Nuclear physics. 2. Physics. 3. Nuclear physicists. I. Ware, Leonel.
QC173 .I58 2022
539.7--dc23

Contents

Preface

The field of physics which studies atomic nuclei, and their constituents and interactions is known as nuclear physics. Advancements in the field of nuclear physics have led to the development of many new fields such as nuclear power, nuclear weapons, nuclear medicine and radiocarbon dating. Evolution in nuclear physics has also led to the conception of the field of particle physics. There are several aspects which are studied in this discipline such as nuclear decay, nuclear fusion, nuclear fission and the production of heavy elements. The topics covered in this extensive book deal with the core subjects of nuclear physics. It presents this complex subject in the most comprehensible and easy to understand language. The book is appropriate for students seeking detailed information in this area as well as for experts.

To facilitate a deeper understanding of the contents of this book a short introduction of every chapter is written below:

Chapter 1- The domain of physics which is involved in the study of atomic nuclei along with their constituents and interactions is referred to as nuclear physics. Some of the major areas of interest within this field are nuclear size, nuclear density and nuclear binding energy. This is an introductory chapter which will provide a brief introduction to the different aspects of nuclear physics.

Chapter 2- A nuclear model is a theoretical description which is used to predict the properties of nuclei. Some of the major nuclear models are nuclear shell model, liquid-drop model and collective model. This chapter closely examines these nuclear models to provide an extensive understanding of the subject.

Chapter 3- The process wherein two nuclei collide and produce one or more different nuclides is termed as a nuclear reaction. They are classified into a few major types such as direct reactions and compound nucleus reactions. The topics elaborated in this chapter will help in gaining a better perspective about these types of nuclear reactions.

Chapter 4- The process through which an unstable nucleus loses energy through radiation is termed as nuclear decay. It is also known as radioactivity. The different particles which can undergo nuclear decay are alpha particles, beta particles and gamma particles. This chapter has been carefully written to provide an easy understanding of the varied facets of the nuclear decay of these particles.

Chapter 5- Usage of nuclear reactions for the purpose of producing nuclear energy is termed as nuclear power. The thermal power station where heat is produced using nuclear power is called a nuclear power plant. This chapter closely examines nuclear power as well as the key components of nuclear power plants such as nuclear fuel, nuclear reactor and nuclear waste to provide an extensive understanding of the subject.

Finally, I would like to thank the entire team involved in the inception of this book for their valuable time and contribution. This book would not have been possible without their efforts. I would also like to thank my friends and family for their constant support.

Leonel Ware

1

Introduction to Nuclear Physics

The domain of physics which is involved in the study of atomic nuclei along with their constituents and interactions is referred to as nuclear physics. Some of the major areas of interest within this field are nuclear size, nuclear density and nuclear binding energy. This is an introductory chapter which will provide a brief introduction to the different aspects of nuclear physics.

Nuclear physics is a field of physics that involves investigation of the building blocks and interactions of atomic nuclei. It includes studies of nuclear components such as protons and neutrons, forces such as the strong force (or strong interaction), and phenomena such as radioactive decay, nuclear fission, and nuclear fusion.

Nuclear power and nuclear weapons are the most commonly known applications of nuclear physics, but the research field is also the basis for a far wider range of less common applictions, such as in medicine (nuclear medicine, magnetic resonance imaging), materials engineering (ion implantation), and archaeology (radiocarbon dating).

Related Fields

Nuclear physics is sometimes used synonymously with atomic physics, but physicists usually differentiate between the two. Atomic physics studies the combined system of the atomic nucleus and the arrangement of electrons around the nucleus.

Particle physics involves study of the elementary constituents of matter and radiation, and the interactions between them. Particle physics evolved out of nuclear physics and, for this reason, has been included under the same term in earlier times.

Rutherford's Team Discovers the Nucleus

In 1906, Ernest Rutherford published "Radiation of the α Particle from Radium in passing through Matter." Hans Geiger expanded on this work in a communication to the Royal Society with experiments he and Rutherford had done passing α particles through air, aluminum foil and gold leaf. More work was published in 1909, by Geiger and Ernest Marsden, and was greatly expanded in 1910 by Geiger. In 1911-12, Rutherford went before the Royal Society to explain the experiments and propound the new theory of the atomic nucleus as we now understand it.

The key experiment behind this announcement was performed in 1909, when Hans Geiger and Ernest Marsden, under Rutherford's supervision, fired alpha particles (helium nuclei) at a thin film of gold foil. The plum pudding model predicted that the alpha particles should come out of the foil with their trajectories being at most slightly bent. Their actual observations, however, were shocking: A few particles were scattered through large angles, with some bouncing completely backwards.

Those observations, upon analysis, led to the Rutherford model of the atom, in which the atom has a very small, dense nucleus containing most of its mass, and consisting of heavy positively charged particles with several embedded electrons that would (at least partially) balance out the charge (since the neutron was unknown). As an example of this model (which is not the modern one), nitrogen-14 was thought to consist of a nucleus with 14 protons and 7 electrons (21 total particles), and the nucleus was surrounded by 7 more orbiting electrons.

The Rutherford model worked quite well until studies of nuclear spin were carried out by Franco Rasetti at the California Institute of Technology in 1929. By 1925, it was known that protons and electrons had a spin of 1/2, and in the Rutherford model of nitrogen-14, 20 of the 21 particles should have paired up to cancel each other's spin, and the final odd particle should have left the nucleus with a spin of 1/2. Rasetti discovered, however, that nitrogen-14 has a spin of 1.

Chadwick Discovers the Final Necessary Particle

In 1932, Chadwick realized that radiation that had been observed by Walther Bothe, Herbert L. Becker, and Irène and Frédéric Joliot-Curie was actually due to a neutral particle of about the same mass as the proton, that he called the neutron (following a suggestion about the need for such a particle, by Rutherford). That same year, Dmitri Ivanenko suggested that neutrons were in fact spin 1/2 particles and that the nucleus contained neutrons to explain the mass not due to protons, and that there were no electrons in the nucleus—only protons and neutrons. The neutron spin immediately solved the problem of the spin of nitrogen-14, as the one unpaired proton and one unpaired neutron in this model, each contribute a spin of 1/2 in the same direction, for a final total spin of 1.

With the discovery of the neutron, scientists could at last calculate what fraction of binding energy each nucleus had, from comparing the nuclear mass with that of the protons and neutrons that composed it. Differences between nuclear masses calculated in this way, and when nuclear reactions were measured, where found to agree with Einstein's calculation of the equivalence of mass and energy to high accuracy (within 1 percent, when calculated).

Yukawa's Meson Postulated to Bind Nuclei

In 1935, Hideki Yukawa proposed the first significant theory of the strong force to explain how neutrons and protons are held together in the nucleus. In the Yukawa interaction, a virtual particle, later called a meson, mediated a force between all nucleons, including protons and neutrons. This force explained why nuclei did not disintegrate under the influence of proton repulsion, and it also gave an explanation of why the attractive strong force had a more limited range than the electromagnetic repulsion between protons. Later, the discovery of the pi meson showed it to have the properties of Yukawa's particle.

With Yukawa's papers, the modern model of the atom was nearing completion. The center of the atom contains a tight ball of neutrons and protons, which is held together by the strong nuclear force, unless the nucleus is too large. Unstable nuclei may undergo alpha decay, when they emit an energetic helium nucleus, or beta decay, when they eject an electron (or positron). After one of these decays, the resultant nucleus may be left in an excited state, and in this case it decays to its ground state by emitting high energy photons (gamma decay).

The study of the strong and weak nuclear forces (the latter explained by Enrico Fermi via Fermi's interaction in 1934) led physicists to collide nuclei and electrons at ever higher energies. This research became the science of particle physics, the crown jewel of which is the standard model of particle physics, which unifies the strong, weak, and electromagnetic forces.

Modern Nuclear Physics

A heavy nucleus can contain hundreds of nucleons, which means that with some approximation it can be treated as a classical system, rather than a quantum-mechanical one. In the resulting liquid-drop model, the nucleus has an energy that arises partly from surface tension and partly from electrical repulsion of the protons. The liquid-drop model is able to reproduce many features of nuclei, including the general trend of binding energy with respect to mass number, as well as the phenomenon of nuclear fission.

Superimposed on this classical picture, however, are quantum mechanical effects, which can be described using the nuclear shell model, developed in large part by Maria Goeppert-Mayer. Nuclei with certain numbers of neutrons and protons (the magic numbers 2, 8, 20, 50, 82, 126, ...) are particularly stable, because their shells are filled.

Other, more complicated models for the nucleus have also been proposed, such as the interacting boson model, in which pairs of neutrons and protons interact as bosons, analogously to Cooper pairs of electrons.

Much of current research in nuclear physics relates to the study of nuclei under extreme conditions, such as high spin and excitation energy. Nuclei may also have extreme shapes (similar to that of American footballs) or extreme neutron-to-proton ratios. Experimenters can create such nuclei using artificially induced fusion or nucleon transfer reactions, employing ion beams from a particle accelerator. Beams with even higher energies can be used to create nuclei at very high temperatures, and there are signs that these experiments have produced a phase transition from normal nuclear matter to a new state, the quark-gluon plasma, in which quarks mingle with one another, rather than being segregated in triplets as they are in neutrons and protons.

Modern Topics in Nuclear Physics

Nuclear Decay: Spontaneous Changes from One Nuclide to Another

There are 80 elements that have at least one stable isotope, and 250 such stable isotopes. However, there are thousands more well-characterized isotopes that are unstable. These radioisotopes may be unstable and decay along timescales ranging from fractions of a second to weeks, years, or even many millions of years.

If a nucleus has too few or too many neutrons it may be unstable, and will decay after some period of time. For example, in a process called beta decay, a nitrogen-16 atom (7 protons, 9 neutrons) is converted to an oxygen-16 atom (8 protons, 8 neutrons) within a few seconds of being created. In this decay, a neutron in the nitrogen nucleus is turned into a proton and an electron and anti-neutrino, by the weak nuclear force. The element is transmuted to another element in the process because, while it previously had seven protons (which makes it nitrogen), it now has eight (which makes it oxygen).

In alpha decay, the radioactive element decays by emitting a helium nucleus (2 protons and 2 neutrons), giving another element, plus helium-4. In many cases this process continues through several steps of this kind, including other types of decays, until a stable element is formed.

In gamma decay, a nucleus decays from an excited state into a lower energy state by emitting a gamma ray. The element is not changed in the process.

Other, more exotic decays, are also possible. For example, in internal conversion decay, the energy from an excited nucleus may be used to eject one of the inner orbital electrons from the atom. This process produces high speed electrons, but it is not beta decay, and (unlike beta decay) does not transmute one element to another.

Nuclear Fusion

When two light nuclei come into very close contact with each other, it is possible for the strong force to fuse the two together. It takes a great deal of energy to push the nuclei close enough together for the strong or nuclear forces to have an effect, so the process of nuclear fusion can take place only at very high temperatures or high densities. Once the nuclei are close enough together, the strong force overcomes their electromagnetic repulsion and squishes them into a new nucleus. A very large amount of energy is released when light nuclei fuse together because the binding energy per nucleon increases with mass number up until nickel-62.

Stars like the Sun are powered by the fusion of four protons into a helium nucleus, two positrons, and two neutrinos. The uncontrolled fusion of hydrogen into helium is known as thermonuclear runaway. Research to find an economically viable method of using energy from a controlled fusion reaction is currently being undertaken by various research establishments.

Nuclear Fission

For nuclei heavier than nickel-62, the binding energy per nucleon decreases with mass number. It is therefore possible for energy to be released if a heavy nucleus breaks apart into two lighter ones. This splitting of atomic nuclei is known as nuclear fission.

The process of alpha decay may be thought of as a special type of spontaneous nuclear fission. This process produces a highly asymmetrical fission because the four particles that make up the alpha particle are especially tightly bound to each other, making production of this nucleus in fission particularly likely.

For some of the heaviest nuclei that produce neutrons on fission, and which also easily absorb neutrons to initiate fission, a self-igniting type of neutron-initiated fission can be obtained, in a so-called chain reaction. [Chain reactions were known in chemistry before physics, and in fact many familiar processes like fires and chemical explosions are chemical chain reactions].

The fission or "nuclear" chain-reaction, using fission-produced neutrons, is the source of energy for nuclear power plants and fission type nuclear bombs such as the two that the United States used against Hiroshima and Nagasaki to bring an end to World War II in the Pacific theater. Heavy nuclei such as uranium and thorium may undergo spontaneous fission, but they are much more likely to undergo alpha decay.

For a neutron-initiated chain-reaction to occur, there must be a critical mass of the element present in a certain space under certain conditions (these conditions slow and conserve neutrons for the reactions). There is one known example of a natural nuclear fission reactor, which was active in two regions of Oklo, Gabon, Africa, over 1.5 billion years ago. Measurements of natural neutrino emission have demonstrated that around half of the heat emanating from the Earth's core results from radioactive decay. However, it is not known if any of this results from fission chain-reactions.

Production of Heavy Elements

As the Universe cooled after the Big Bang, it eventually became possible for particles as we know them to exist. The most common particles created in the Big Bang that are still easily observable to us were protons (hydrogen) and electrons (in equal numbers). Some heavier elements were created as the protons collided with each other, but most of the heavy elements we see today were created inside of stars during a series of fusion stages, such as the proton-proton chain, the CNO cycle, and the triple-alpha process.

Progressively heavier elements are created during the evolution of a star. Since the binding energy per nucleon peaks around iron, energy is released only through fusion processes occurring below this point. Since the creation of heavier nuclei by fusion costs energy, nature resorts to the process of neutron capture. Neutrons (due to their lack of charge) are readily absorbed by a nucleus. The heavy elements are created by either a slow neutron capture process (the so-called s process) or by the rapid (or r) process. The s process occurs in thermally pulsing stars (called AGB, or asymptotic giant branch stars) and takes hundreds to thousands of years to reach the heaviest elements of lead and bismuth. The r process is thought to occur in supernova explosions because the conditions of high temperature, high neutron flux and ejected matter are present. These stellar conditions make the successive neutron captures very fast, involving very neutron-rich species, which then undergo beta decay to heavier elements, especially at the so-called waiting points that correspond to more stable nuclides with closed neutron shells (magic numbers). The r process duration is typically in the range of a few seconds.

Nucleus

The atomic nucleus is a tiny massive entity at the center of an atom. Occupying a volume whose radius is 1/100,000 the size of the atom, the nucleus contains most (99.9%) of the mass of the atom. In describing the nucleus, we shall describe its composition, size, density, and the forces that hold it together. After describing the structure of the nucleus, we shall go on to describe some of the limits of nuclear stability.

The nucleus is composed of protons (charge = +1; mass = 1.007 atomic mass units (μ) and neutrons. The number of protons in the nucleus is called the atomic number Z and defines which chemical element the nucleus represents. The number of neutrons in the nucleus is called the neutron number N, whereas the total number of neutrons and protons in the nucleus is referred to as the mass number A, where A = N + Z. The neutrons and protons are referred to collectively as nucleons. A nucleus with a given N and Z is referred to as a nuclide. Nuclides with the same atomic number are isotopes, such as ^{12}C and ^{14}C, whereas nuclides with the same N, such as ^{14}C and ^{16}O, are called isotones. Nuclei such as ^{14}N and ^{14}C, which have the same mass number, are isobars. Nuclides are designated by a shorthand notation in which one writes, that is, for a nucleus with 6 protons and 8 neutrons, one writes, or just ^{14}C. The size of a nucleus is approximately 1 to 10 × 10^{-15} m, with the nuclear radius being represented more precisely as 1.2 × $A^{1/3}$ × 10^{-15} m. We can roughly approximate the nucleus as a sphere and thus we can calculate its density, where 1.66 × 10^{-27} kg is the mass of the nucleon. Thus the nuclear density is about 200,000 tonnes/ mm^3 and is independent of A. Imagine a cube that is 1 mm on a side. If filled with nuclear matter, it would have a mass of about 200,000 tonnes. This calculation demonstrates the enormous matter/energy density of nuclei and gives some idea as to why nuclear phenomena lead to large energy releases.

Of the 6,000 species of nuclei that can exist in the universe, about 2,700 are known, but only 270 of these are stable. The rest are radioactive, that is, they spontaneously decay. The driving force behind all radioactive decay is the ability to produce products of greater stability than one had initially. In other words, radioactive decay releases energy and because of the high energy density of nuclei, that energy release is substantial. Qualitatively we describe radioactive decay as occurring in three general ways: α-, β-, and γ-decay. Alpha-decay occurs in the heavy elements, and consists of the emission of a ^{4}He nucleus. Beta-decay occurs in nuclei whose N/Z ratio is different from that of a stable nucleus and consists of a transformation of neutrons into protons or vice versa to make the nucleus more stable. Gamma-decay occurs when excited nuclei get rid of some or all of their excitation energy via the emission of electromagnetic radiation, or via the radiationless transfer of energy to orbital electrons.

The force responsible for holding the neutrons and protons together within the very small nuclear volume must be unusually strong. The nuclear force, or strong interaction, is one of the four fundamental forces of nature (namely, the gravitational, electromagnetic, strong, and weak forces). The nuclear force is charge-independent, meaning that the nuclear force between two protons, or two neutrons, or a neutron and a proton, is the same. The nuclear force is short-ranged, meaning it acts over a distance of 10^{-15} to 10^{-14} m, that is, the size of nuclei. Of course the nuclear force is attractive, as it binds the nucleons in a nucleus. But some experiments have shown the nuclear force has a "repulsive core," meaning that at very short distances, the force switches from attractive to repulsive, preventing the nucleus from collapsing in on itself. The nuclear force is an "exchange" force, resulting from the virtual exchange of pions (short-lived particles with integral spin, produced normally in nuclear reactions) between interacting nucleons. More recently we have come to understand the nuclear force using a theory called quantum chromodynamics (QCD), which describes protons and neutrons as being made up of quarks. In particular, the proton is thought of as a combination of two up quarks (uu) and a down quark (d), whereas the neutron is thought to consist of one up quark (u) and two down quarks (dd). The up and down quarks are small particles with charges of +2/3 and −1/3, respectively. The quarks account for approximately 2 percent of the mass of the proton. The rest of the mass consists of gluons, which are the particles exchanged between the quarks to bind them together. The force acting between the quarks has the unusual property of being small when they are close together, and increasing as the distance between them grows. Because of this, no isolated quarks have been observed in nature.

In close analogy to atomic structure, we speak of the structure of various nuclei. Many nuclear properties can be described using a nuclear shell model in which the nucleons are placed in orbitals like electrons in atoms. These orbitals and their properties are predicted by applying quantum mechanics to the problem of defining the states of the nucleons, which move under the influence of the average force in the nucleus. Like atoms, there are certain configurations of nucleons that have special stability, for example, the so-called magic numbers akin to the inert gas structures in chemistry. In addition to those nuclear properties that are best described in terms of a shell model, there are other properties that seem to be best explained by the large-scale collective motion of a number of nucleons. These motions lead to nuclear rotations and vibrations, which are described by a nuclear collective model.

Current research on nuclei, their properties, and the forces that hold them together focuses on studying nuclei at the limits of stability. The basic idea is that when one studies nuclei under extreme conditions, one then has a unique ability to test theories and models that were designed to describe the "normal" properties of nuclei. One limit of nuclear stability is that of high Z, that is, as the atomic number of the nucleus increases, the repulsion between the nuclear protons becomes so large as to cause the nucleus to spontaneously fission. The competition between this repulsive Coulomb force and the cohesive nuclear force is what defines the size of the Periodic Table and the number of chemical elements. At present there are 112 known chemical elements, and evidence for the successful synthesis of elements having the atomic numbers 114 and 116 has been presented.

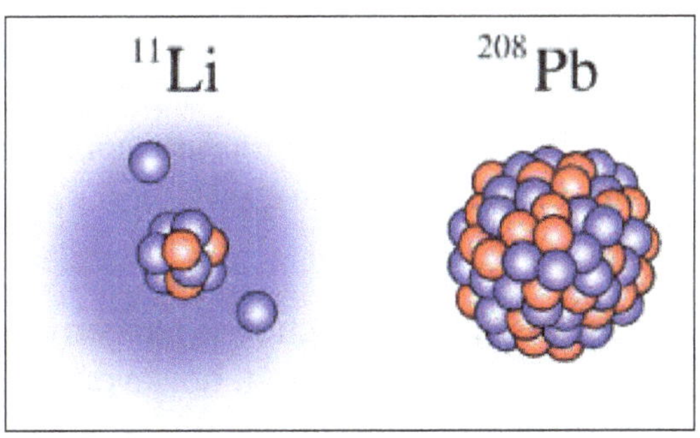

Schematic views of the nuclear halo nuclei ^{11}Li that compares it to ^{208}Pb and the Borromean rings of medieval times.

Another limit of nuclear stability is the extreme of the neutron to proton ratio, N/Z. For certain very neutron-rich nuclei, such as ^{11}Li, an unusual halo structure has been observed. In halo nuclei, a "core" of nucleons is surrounded by a "misty cloud, a halo" of valence nucleons that are weakly bound and extend out to great distances, analogous to electrons surrounding the nucleus in an atom. Halo nuclei are fragile objects, are relatively large, and interact easily with other nuclei (have enhanced reaction cross sections). The halo nucleus ^{11}Li, which has a ^{9}Li core surrounded by a two-neutron halo is shown in figure Li is as large as ^{208}Pb. ^{11}Li and other two-neutron halo nuclei are three-body systems (2 neutrons and a ^{9}Li core), which pose a special challenge to nuclear theorists. They are also examples of Borromean systems, in which the nucleus is no longer bound if any one of the three components is removed.

Proton

In physics, the proton is a subatomic particle with an electric charge of one positive fundamental unit. The proton is observed to be stable, although some theories predict that the proton may decay. The proton has a density of about 2.31×10^{17} kg m^{-3}.

Protons are spin -1/2 fermions and are composed of three quarks, making them baryons. The two up quarks and one down quark of the proton are also held together by the strong nuclear force, mediated by gluons. Protons may be transmuted into neutrons by inverse beta decay (that is, by capturing an electron); since neutrons are heavier than protons, this process does not occur spontaneously but only when energy is supplied. The proton's antimatter equivalent is the antiproton, which has the *same* magnitude charge as the proton but the opposite sign.

Protons and neutrons are both nucleons, which may be bound by the nuclear force into atomic nuclei. The most common isotope of the hydrogen atom is a single proton. The nuclei of other atoms are composed of various numbers of protons and neutrons. The number of protons in the nucleus determines the chemical properties of the atom and which chemical element it is.

In chemistry and biochemistry, the proton is thought of as the hydrogen ion, denoted H^+. In this context, a proton donor is an acid and a proton acceptor a base.

The proton is the simplest composite particle. It contains three fermions, called quarks, that have shed all their color charge into a halo of gluons as dictated by the quantum probability wavefunction of the color interaction.

A gluon is similar to a photon of light, except that a gluon goes both backwards and forwards in time, it has a quantum 'color' charge at one end, and a quantum 'anticolor' at the other end. In the middle, the gluon is colorless. The gluon halo places all the color and anticolor at the surface of the proton, while the center of the proton is colorless. This colorless center is where the now-essentially colorless quarks spend most of the time. The quarks are 'confined' to this colorless center but have plenty of space to move around in as size of the quarks compared to the extended gluon halo is as three dust particles are to New York City.

The surface of the proton, as far as the color interaction is concerned, can be compared to a color computer screen displaying white. On close inspection, however, it is composed of red, blue and green pixels. This is why the color surface of the proton with its pixels of gluon ends appears colorless. The energy of this relatively vast gluon halo is responsible for 99.9 percent of the mass-energy of the proton.

The quarks, while shedding their color charge, have no way to shed their electric charge. As the charged quarks are confined to the colorless central regions of the proton, the electric diameter of the proton—the region where all the electric charge is concentrated—is significantly smaller than the color-charge diameter of the proton.

There are two types of quarks in regular matter. The three quarks in a proton are two U-quarks each with +2/3 electric charge, and a D-quark with -1/3 charge. The composite proton has an overall charge of +1. In a neutron, the other composite particle found in atomic nuclei, there is one U and 2 Ds, and the neutron has zero overall charge. An aspect of the color interaction akin to polarization in light photons makes the neutron combination of quarks generates a few more gluons than the proton combination, and this gives the neutron slightly more mass-energy than the proton.

This is a diagram of a proton, with one of the gluons magnified. This particular gluon—one of the eight possible combinations of color-anticolor, has quantum red going forward in time and quantum blue (as its anticolor complement, yellow) going backwards in time. As long as the (sand-grain sized) quarks stay in the colorless center, the experience what is called 'asymptotic freedom' and are free of the color influence on them. In this state, they align themselves according to the quantum waveform of their electromagnetic interaction as they freely couple photons.

A representation of a proton.

For the quark to leave the colorless center and pick up color from the periphery, however, takes a lot of energy. So much energy, in fact, that [virtual particle]-pairs of quarks and antiquarks become real when a quark gets kicked out of the center (perhaps by a very energetic electron) and new combinations of quarks swathed in gluons appear, such as [pions] and other such combinations. At no time in any such process is there a "bare quark" which can be observed. Quarks are confined by what is called "infrared slavery" (they cannot escape their low energy colorless state) to always be observed in composite, colorless combinations.

Antiproton

The antiproton is the antiparticle of the proton. It was discovered in 1955 by Emilio Segre and Owen Chamberlain, for which they were awarded the 1959 Nobel Prize in Physics.

CPT-symmetry puts strong constraints on the relative properties of particles and antiparticles and, therefore, is open to stringent tests. For example, the charges of the proton and antiproton must sum to exactly zero. This equality has been tested to one part in 10^8. The equality of their masses is also tested to better than one part in 10^8. By holding antiprotons in a Penning trap, the equality of the charge to mass ratio of the proton and the antiproton has been tested to 1 part in 9×10^{11}. The magnetic moment of the antiproton has been measured with error of 8×10^{-3} nuclear Bohr magnetons, and is found to be equal and opposite to that of the proton.

High-energy Physics

Due to their stability and large mass (compared to electrons), protons are well suited to use in particle colliders such as the Large Hadron Collider at CERN and the Tevatron at Fermilab. Protons also make up a large majority of the cosmic rays, which impinge on the Earth's atmosphere. Such high-energy proton collisions are more complicated to study than electron collisions, due to the composite nature of the proton. Understanding the details of proton structure requires quantum chromodynamics.

Neutron

A neutron is a subatomic particle that normally is part of the nucleus of a chemical element. When free (not bound to a nucleus), a neutron can have important physical, chemical, and biological effects.

The mass m_n of a neutron is close to, but not equal to, the mass of a proton:

$$m_n = 1.674\,927\,211 \times 10^{-27}\ \text{kg}.$$

The neutron consists of three quarks. A free neutron shows beta decay, breaking down into a proton, an electron, and an antineutrino. Because it disintegrates, the free neutron does not exist in nature. Neutrons do not carry electric charge, they pass unhindered through the electrical fields within liquids and solids.

Properties

The quark structure of the neutron: There are two down quark and one up quark. The strong force is mediated by gluons. The strong force has three types of charges, the so called red, green and the blue. The color assignment of individual quarks is arbitrary, but all three colors must be present.

Outside the nucleus, neutrons are unstable and have a mean lifetime of 886 seconds (about 15 minutes), decaying by emitting an electron and antineutrino to become a proton. Neutrons in this unstable form are known as free neutrons. The same decay method (beta decay) occurs in some nuclei. Particles inside the nucleus are typically resonances between neutrons and protons, which transform into one another by the emission and absorption of pions. A neutron is classified as a baryon, and consists of two down quarks and one up quark. The neutron's antimatter equivalent is the antineutron.

Though most atoms have approximately equal numbers of protons and neutrons the number of neutrons can vary. Atoms of the same element, same number of protons, with different numbers of neutrons are called isotopes. The number of neutrons determines the isotope; for example, the carbon-12 isotope has six protons and six neutrons, while the carbon-14 isotope has six protons and eight neutrons.

Neutron Interactions

The neutron interacts through all four of the common classifications of physical interaction. These four are the electromagnetic, weak nuclear, strong nuclear, and gravitational interactions.

Although it is true that the neutron has zero net charge, it is nonetheless composed of electrically charged quarks, in the same way that a neutral atom is nonetheless composed of protons and electrons. As such, the neutron experiences the electromagnetic interaction. The net charge is zero, so if you are far enough away from the neutron that it appears to occupy no volume, then the total effect of the electric force will add up to zero. The movement of the charges inside the neutrons do not cancel however, and this is what gives the neutron its nonzero magnetic moment.

Gravity is often not discussed when talking about neutrons. This is because neutrons are usually studied in terms of subatomic interactions. In the subatomic world, gravity is undetectable relative to the other forces, which are much stronger. This having been said, a neutron accelerates at the same rate in the earth's gravitational field as a lead brick.

Charged particles (such as protons, electrons, or alpha particles) and electromagnetic radiation (such as gamma rays) lose energy in passing through matter. They exert electric forces that ionize atoms of the material through which they pass. The energy taken up in ionization equals the energy lost by the charged particle, which slows down, or by the gamma ray, which is absorbed or scattered. The neutron, by contrast, is seen by atoms it passes as containing no electric charge, and so does not create any ionization.

As far as the nuclear forces are concerned, it is a different story. Nuclear forces play the leading role when neutrons pass through regular matter. Consequently, a free neutron goes on its way unchecked until it makes a "head-on" collision with an atomic nucleus.

When this happens, the neutrons and target nuclei can be scattered (deflected or slowed down), absorbed, or transformed into something different. In the case of the reaction $n + {}^{3}He \rightarrow {}^{1}H + {}^{3}H$ (n:neutron; ${}^{3}He$: nucleus consisting of two protons and one neutron; ${}^{1}H$: nucleus consisting of a only proton; ${}^{3}H$: nucleus consisting of one proton and two neutrons) for example, the proton and the neutron appear to have exchanged places, and kinetic energy is released. In many cases, secondary particles are created and energy can be used up or released.

Neutrons, like other particles, can undergo elastic collisions. A collision is elastic under the special case where kinetic energy is conserved. Billiard balls for example typically undergo elastic collisions. The law of conservation of momentum also applies as it does for any collision. If the nucleus that is struck in an elastic collision is heavy, it acquires relatively little speed, but if it is a proton, which is approximately equal in mass to the neutron, it is projected forward with a large fraction of the original speed of the neutron, which is itself correspondingly slowed.

Neutron Detection

The common means of detecting a charged particle by looking for a track of ionization does not work for neutrons. The means that we have for detecting neutrons consists of allowing them to interact with atomic nuclei.

A common method for detecting neutrons involves converting the energy released from such reactions into electrical signals. The isotopes ^{3}He, ^{6}Li, ^{10}B, ^{233}U, ^{235}U and ^{239}Pu are useful for this purpose. These substances are mostly rare or tightly controlled, except for ^{10}B which is readily available.

Neutron Uses

The neutron plays an important role in many nuclear reactions. For example, neutron capture often results in neutron activation, inducing nuclear fission. In particular, knowledge of neutrons and their behavior has been important in the development of nuclear reactors and nuclear weapons.

The development of "neutron lenses" based on total internal reflection within hollow glass capillary tubes or by reflection from dimpled aluminum plates has driven ongoing research into neutron microscopy and neutron/gamma ray tomography.

One use of neutron emitters is the detection of light nuclei, particularly the hydrogen found in water molecules. When a fast neutron collides with a light nucleus, it loses a large fraction of its energy. By measuring the rate at which slow neutrons return to the probe after reflecting off of hydrogen nuclei, a neutron probe may determine the water content in soil.

Neutron Sources

Due to the fact that free neutrons are unstable, they (neutron radiation) can be obtained only from nuclear disintegrations or high-energy reactions (such as cosmic radiation or in accelerators). Free neutron beams are obtained from neutron sources by neutron transport.

Neutrons' lack of electric charge prevents engineers or experimentalists from being able to steer or accelerate them. Charged particles can be accelerated, decelerated, or deflected by electric or magnetic fields. However, these methods have almost no effect on neutrons (there is a small effect of a magnetic field on the free neutron because of its magnetic moment).

Discovery

In 1930 Walther Bothe and H. Becker in Germany found that if the very energetic alpha particles emitted from polonium fell on certain of the light elements, specifically beryllium, boron, or lithium, an unusually penetrating radiation was produced. At first this radiation was thought to be gamma radiation although it was more penetrating than any gamma rays known, and the details of experimental results were very difficult to interpret on this basis. The next important contribution was reported in 1932 by Irène Joliot-Curie and Frédéric Joliot in Paris. They showed that if this unknown radiation fell on paraffin or any other hydrogen-containing compound it ejected protons of very high energy. This was not in itself inconsistent with the assumed gamma ray nature of the new radiation, but detailed quantitative analysis of the data became increasingly difficult to reconcile with such an hypothesis. Finally (later in 1932) the physicist James Chadwick in England performed a series of experiments showing that the gamma ray hypothesis was untenable. He suggested that in fact the new radiation consisted of uncharged particles of approximately the mass of the proton, and he performed a series of experiments verifying his suggestion. Such uncharged particles were eventually called neutrons, apparently from the Latin root for neutral and the Greek ending -on (by imitation of electron and proton).

Antineutron

The antineutron is the antiparticle of the neutron. It was discovered by Bruce Cork in the year 1956, a year after the antiproton was discovered.

CPT-symmetry puts strong constraints on the relative properties of particles and antiparticles, and therefore is open to stringent tests. The masses of the neutron and antineutron are equal to one part in $(9\pm5)\times10^{-5}$.

Nuclear Size

The actual dimensions of the nucleus present in the atom are called nuclear size. Various experiments have been done to determine the size of nucleus (Rutherford's experiment). The nuclear size can be determined as:

$$r = R_0 A^{1/3}$$

Where r is the radius of nucleus and A is the mass no. of atom and R_0 is constant whose value is 1.5×10^{-13} cm or 1.5×10^{-15} m. The nuclear radius is of order of 10^{-14}, therefore very small. It is measured in fermi units.

Nuclear Radius

Closest Distance of Approach of an Alpha Particle

We can estimate the radius of a nucleus by firing an alpha particle directly at the nucleus. The alpha particle will lose kinetic energy as it approaches the nucleus and gain potential energy due to the electric field because of the electrostatic repulsion between the positively charged nucleus and the positively charged alpha particle.

We can then work out the closest distance of approach of the alpha particle to the nucleus.

Loss in kinetic energy E_k = gain in potential energy by the field E_p,

$$E_p = \frac{1}{4\pi\varepsilon_0}\frac{Q_\alpha Q_n}{r}$$

$$E_k = E_p$$

$$E_k = \frac{1}{4\pi\varepsilon_0}\frac{Q_\alpha Q_n}{r}$$

$$r = \frac{1}{4\pi\varepsilon_0}\frac{Q_\alpha Q_n}{E_k}$$

These experiments show the nucleus must be smaller than about 10^{-14} m.

Electron Diffraction

We can achieve a better estimate of the size of a nucleus by looking at the angle θ_m to the first minimum when electrons are diffracted by a foil target.

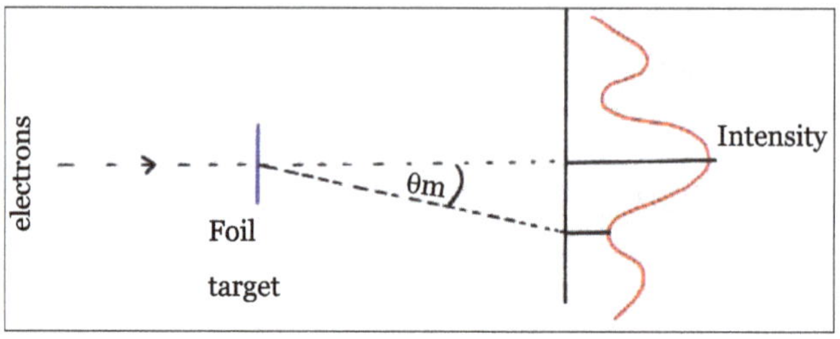

$$r = \frac{0.61\lambda}{\sin\theta_m}$$

where λ is the wavelength of the electrons.

This technique shows the nucleus to have a radius of about 10^{-15} m.

Nuclear Radius and Nucleon Number

From experimental data the equation below has been derived.

$$R = r_0 A^{1/3},$$

R = nuclear radius in metres (m),

r_o = is the radius of a nucleon approx 1.3 fm,

A = nucleon number.

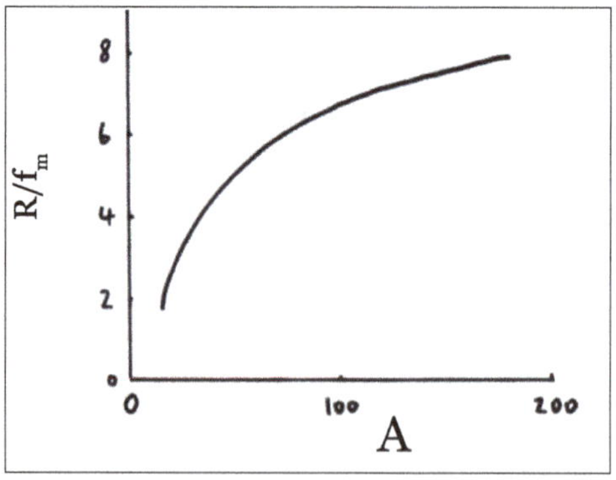

Nuclear Density

Nuclear density is the density of the nucleus of an atom. It is the ratio of mass per unit volume inside the nucleus. Since atomic nucleus carries most of atom's mass and atomic nucleus is very small in comparison to entire atom, the nuclear density is very high.

The nuclear density for a typical nucleus can be approximately calculated from the size of the nucleus and from its mass. Typical nuclear radii are of the order 10^{-14} m. Assuming spherical shape, nuclear radii can be calculated according to following formula:

$$r = r_0 . A^{1/3}$$

where $r_0 = 1.2 \times 10^{-15}$ m = 1.2 fm.

For example, natural uranium consists primarily of isotope ^{238}U (99.28%), therefore the atomic mass of uranium element is close to the atomic mass of ^{238}U isotope (238.03 u). Its radius of this nucleus will be:

$$r = r_0 . A^{1/3} = 7.44 \text{ fm}.$$

Assuming it is spherical, its volume will be:

$$V = 4\pi r^3/3 = 1.73 \times 10^{-42} \text{ m}^3.$$

The usual definition of nuclear density gives for its density:

$$\rho_{nucleus} = m / V = 238 \times 1.66 \times 10^{-27} / (1.73 \times 10^{-42}) = 2.3 \times 10^{17} \text{ kg/m}^3.$$

Thus, the density of nuclear material is more than 2.1014 times greater than that of water. It is an immense density. The descriptive term nuclear density is also applied to situations where similarly high densities occur, such as within neutron stars. Such immense densities are also found in neutron stars.

Density of Neutron Star

The densest material found on earth is the metal osmium, but its density pales by comparison to the densities of exotic astronomical objects such as white dwarf stars and neutron stars.

A neutron star is the collapsed core of a large star (usually of a red giant). Neutron stars are the smallest and densest stars known to exist and they are rotating extremely rapidly. A neutron star is basically a giant atomic nucleus about 11 km in diameter made especially of neutrons. It is believed that under the immense pressures of a collapsing massive stars going supernova it is possible for the electrons and protons to combine to form neutrons via electron capture, releasing a huge amount of neutrinos.

They are so dense that one teaspoon of its material would have a mass over 5.5×10^{12} kg. It is assumed they have densities of 3.7×10^{17} to 6×10^{17} kg/m^3, which is comparable to the approximate density of an atomic nucleus of 2.3×10^{17} kg/m^3.

Nuclear Binding Energy and Mass Defect

- Nuclear binding energy is the energy required to split a nucleus of an atom into its components.

- Nuclear binding energy is used to determine whether fission or fusion will be a favorable process.
- The mass defect of a nucleus represents the mass of the energy binding the nucleus, and is the difference between the mass of a nucleus and the sum of the masses of the nucleons of which it is composed.

Terms

- Nucleon One: Of the subatomic particles of the atomic nucleus, i.e. a proton or a neutron.
- Mass defect: The difference between the calculated mass of the unbound system and the experimentally measured mass of the nucleus.
- Strong force: The nuclear force, a residual force responsible for the interactions between nucleons, deriving from the color force.

Binding Energy

Nuclear binding energy is the energy required to split a nucleus of an atom into its component parts: protons and neutrons or collectively, the nucleons. The binding energy of nuclei is always a positive number, since all nuclei require net energy to separate them into individual protons and neutrons.

Mass Defect

Nuclear binding energy accounts for a noticeable difference between the actual mass of an atom's nucleus and its expected mass based on the sum of the masses of its non-bound components.

Recall that energy (E) and mass (m) are related by the equation:

$$E = mc^2$$

Here, *c* is the speed of light. In the case of nuclei, the binding energy is so great that it accounts for a significant amount of mass.

The actual mass is always less than the sum of the individual masses of the constituent protons and neutrons because energy is removed when the nucleus is formed. This energy has mass, which is removed from the total mass of the original particles. This mass, known as the mass defect, is missing in the resulting nucleus and represents the energy released when the nucleus is formed.

Mass defect (M_d) can be calculated as the difference between observed atomic mass (m_o) and that expected from the combined masses of its protons (m_p, each proton having a mass of 1.00728 amu) and neutrons (m_n, 1.00867 amu):

$$M_d = (m_n + m_p) - m_o$$

Nuclear Binding Energy

Once mass defect is known, nuclear binding energy can be calculated by converting that mass to energy by using $E=mc^2$. Mass must be in units of kg.

Once this energy, which is a quantity of joules for one nucleus, is known, it can be scaled into per-nucleon and per-mole quantities. To convert to joules/mole, simply multiply by Avogadro's number. To convert to joules per nucleon, simply divide by the number of nucleons.

Nuclear binding energy can also apply to situations when the nucleus splits into fragments composed of more than one nucleon; in these cases, the binding energies for the fragments, as compared to the whole, may be either positive or negative, depending on where the parent nucleus and the daughter fragments fall on the nuclear binding energy curve. If new binding energy is available when light nuclei fuse, or when heavy nuclei split, either of these processes result in the release of the binding energy. This energy—available as nuclear energy—can be used to produce nuclear power or build nuclear weapons. When a large nucleus splits into pieces, excess energy is emitted as photons, or gamma rays, and as kinetic energy, as a number of different particles are ejected.

Nuclear binding energy is also used to determine whether fission or fusion will be a favorable process. For elements lighter than iron-56, fusion will release energy because the nuclear binding energy increases with increasing mass. Elements heavier than iron-56 will generally release energy upon fission, as the lighter elements produced contain greater nuclear binding energy. As such, there is a peak at iron-56 on the nuclear binding energy curve.

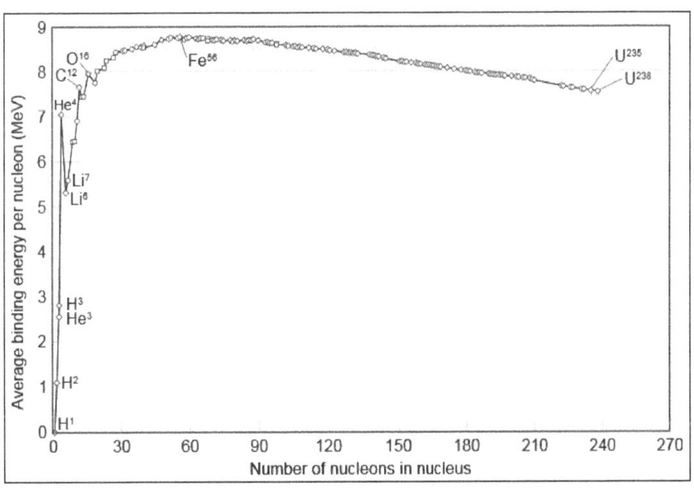

Nuclear binding energy curve: This graph shows the nuclear binding energy (in MeV) per nucleon as a function of the number of nucleons in the nucleus. Notice that iron-56 has the most binding energy per nucleon, making it the most stable nucleus.

The rationale for this peak in binding energy is the interplay between the coulombic repulsion of the protons in the nucleus, because like charges repel each other, and the strong nuclear force, or strong force. The strong force is what holds protons and neutrons together at short distances. As the size of the nucleus increases, the strong nuclear force is only felt between nucleons that are close together, while the coulombic repulsion continues to be felt throughout the nucleus; this leads to instability and hence the radioactivity and fissile nature of the heavier elements.

Binding Energy per Nucleon

A more useful measure of the binding between protons and neutrons is the binding energy per nucleon or E_{bn}. It is the ratio of the binding energy of a nucleus to the number of nucleons in the nucleus:

$$E_{bn} = E_b/A$$

where Eb is the binding energy of the nucleus and A is the number of nucleons in it. So, the binding energy per nucleon is the average energy per nucleon needed to separate a nucleus into its individual nucleons. Let's look at a plot of the binding energy per nucleon versus the mass number for a large number of nuclei:

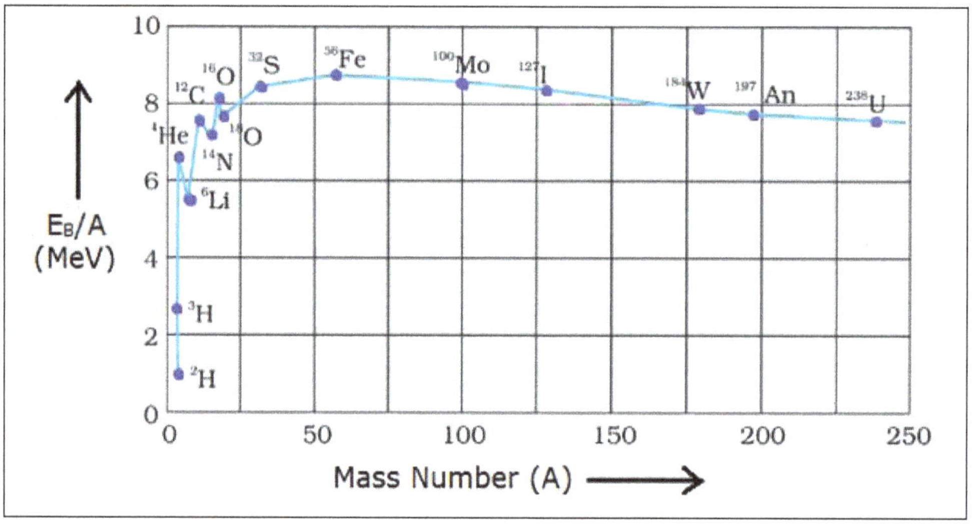

Here is what we can observe:

- The maximum binding energy per nucleon is around 8.75 MeV for mass number (A) = 56.
- The minimum binding energy per nucleon is around 7.6 MeV for mass number (A) = 238.
- For $30 < A < 170$, E_{bn} is nearly constant.
- Ebn is low for both light nuclei ($A < 30$) and heavy nuclei ($A > 170$).

Conclusion 1: The force is attractive in nature and very strong producing a binding energy of around a few MeV per nucleon.

Conclusion 2:

- Why is E_{bn} nearly constant in the range $30 < A < 170$? The answer is simple – the nuclear force is short-ranged. Now, imagine a really large nucleus. A nucleon (N_A) in this nucleus will be influenced only by some of its neighbours. These neighbours are those which lie in the short-range of the nuclear force.

- This means that all nucleons beyond the range of the nuclear force form N_A will have no influence on the binding energy of N_A. So, we can conclude that if a nucleon has 'p' neighbours within the range of the nuclear force, then its binding energy is proportional to 'p'.
- In the same large nucleus, if we increase the mass number (A) by adding nucleons, it will not change the binding energy of N_A. Why? Because, in a large nucleus, most of the nucleons lie inside it and not on the surface. Hence, the change in binding energy, if any, would be negligibly small.
- Remember, the binding energy per nucleon is a constant and is equal to pk, where k is a constant having the dimensions of energy. Also, the property that a given nucleon influences only nucleons close to it is also referred to as saturation property of the nuclear force.

Conclusion 3: Next, imagine a very heavy nucleus having A = 240. This has a low binding energy. Therefore, if a nucleus A = 240 breaks down into two A = 120 nuclei, then the nucleons get bound more tightly. Also, in the process energy is released. This concept is used in nuclear fission.

Conclusion 4: On the other hand, imagine two very light nuclei with A < 10. If these two nuclei were to join to form a heavier nucleus, then the binding energy per nucleon of the fused and heavier nucleus is more than the E_{bn} of the lighter nuclei. So, the nucleons are more tightly bound post fusion. And, energy is released in the process. This is how the Sun works.

Example: Obtain the binding energy (in MeV) of a nitrogen nucleus ($^{14}_{7}N$), given m ($^{14}_{7}N$) = 14.00307 u.

Solution: A nucleus of Nitrogen contains 7 protons and 7 neutrons. Therefore, the mass defect of this nucleus is,

$$\Delta M = 7m_p + 7m_n - m\,(^{14}_{7}N)$$

Now,

Mass of a proton: m_p = 1.007825 u

Mass of a neutron: m_n = 1.008665 u

Therefore,

$$\Delta M = (7 \times 1.007825\text{ u}) + (7 \times 1.008665\text{ u}) - 14.00307\text{ u} = 0.11236\text{ u}$$

Now, we know that 1u = 931.5 MeV/c_2.

Therefore,

$$\Delta M = 0.11236 \times 931.5\text{ MeV}/c^2$$

The binding energy of the nucleus is,

$$E_b = \Delta Mc^2$$

$$= 0.11236 \times 931.5\,(\text{MeV}/c^2) \times c^2$$

= 0.11236 x 931.5 MeV

= 104.66334 MeV.

Nuclear Forces

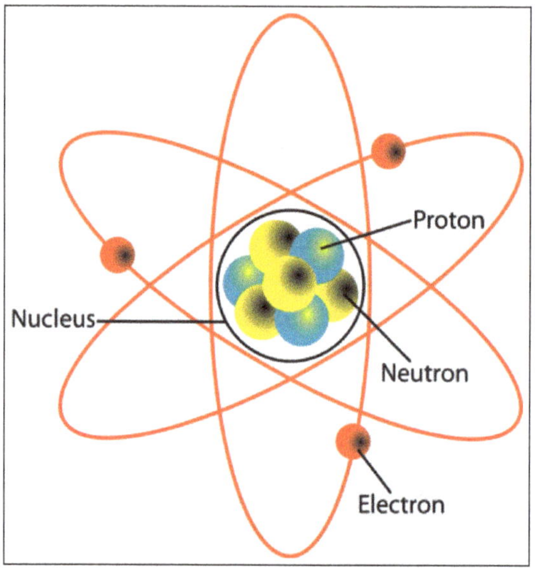

The atomic nucleus made up from protons (blue) and neutrons (yellow) and held together by nuclear forces.

Nuclear forces (also known as nuclear interactions or strong forces) are the forces that act between two or more nucleons. They bind protons and neutrons ("nucleons") into atomic nuclei. The nuclear force is about 10 millions times stronger than the chemical binding that holds atoms together in molecules. This is the reason why nuclear reactors produce about a million times more energy per kilogram fuel as compared to chemical fuel like oil or coal. However, the range of the nuclear force is short, only a few femtometer (1 fm $=10^{-15}$ m), beyond which it decreases rapidly. That is why, in spite of its enormous strength, we do not feel anything of this force on the atomic scale or in everyday life. The development of a proper theory of nuclear forces has occupied the minds of some of the brightest physicists for seven decades and has been one of the main topics of physics research in the 20th century. The original idea was that the force is caused by the exchange of particles lighter than nucleons known as mesons, and this idea gave rise to the birth of a new sub-field of modern physics, namely, (elementary) particle physics. The modern perception of the nuclear force is that it is a residual interaction (similar to the van der Waals force between neutral atoms) of the even stronger force between quarks, which is mediated by the exchange of gluons and holds the quarks together inside a nucleon.

Properties of the Nuclear Force

Some properties of nuclear interactions can be deduced from the properties of nuclei. Nuclei exhibit a phenomenon known as saturation: the volume of nuclei increases proportionally to the number of nucleons. This property suggests that the nuclear (central) force is of short range (a few fm) and strongly attractive at that range, which explains nuclear binding. But the nuclear force has also a very complex spin-dependence. Evidence of this property first came from the observation that the deuteron (the proton-neutron bound state, the smallest atomic nucleus) deviates slightly from a spherical shape: it has a non-vanishing quadrupole moment. This suggests a force that depends on the orientation of the spins of the nucleons with regard to the vector joining the two nucleons (a tensor force). In heavier nuclei, a shell structure has been observed which, can be explained by a strong force between the spin of the nucleon and its orbital motion (the spin-orbit force). More clear-cut evidence for the spin-dependence is extracted from scattering experiments where one nucleon is scattered off another nucleon, with distinct spin orientations. In such experiments, the existence of the nuclear spin-orbit and tensor forces has clearly been established. Scattering experiments at higher energies (more than 200 MeV) provide evidence that the nucleon-nucleon interaction turns repulsive at short inter-nucleon distances (smaller than 0.5 fm, the hard core).

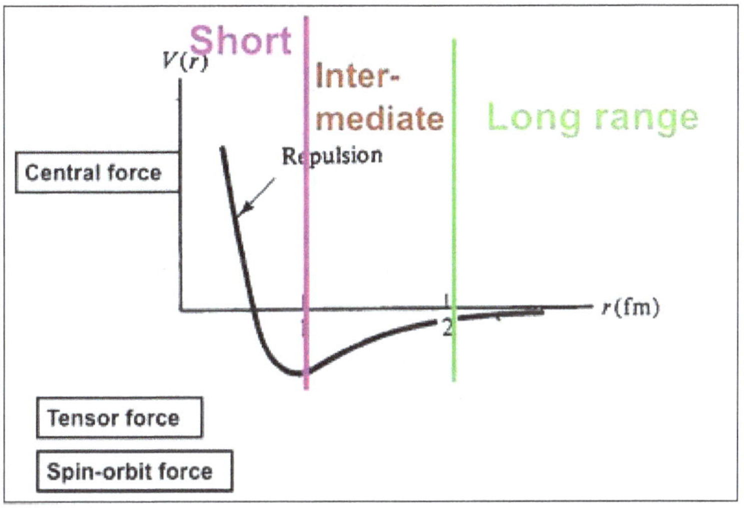

The major components of the nuclear force. The central force is best analyzed in terms of the three ranges indicated.

Besides the force between two nucleons ($_2$NF), there are also three-nucleon forces ($_3$NFs), four-nucleon forces ($_4$NFs), and so on. However, the $_2$NF is much stronger than the $_3$NF, which in turn is much stronger than the $_4$NF, and so on. In exact calculations of the properties of light nuclei based upon the "elementary" nuclear forces, it has been shown that 3NFs are important. Their contribution is small, but crucial. The need for 4NFs for explaining nuclear properties has not (yet) been established.

Nuclear forces are approximately charge-independent meaning that the force between two protons, two neutrons, and a proton and a neutron are nearly the same (in the same quantum mechanical state) when electromagnetic forces are ignored.

Phenomenological Approaches

Traditionally, research on the nuclear force has proceeded along two lines: phenomenology and theory. We start with phenomenology. At very low energy, nucleon-nucleon scattering can be described by the so-called effective range expansion which, in its simplest version, has only two parameters. In modern times, this expansion has been recovered from an EFT which does not include the pion ('pionless EFT'). At intermediate energies, the spin-dependence of the nuclear force becomes important. To produce a general expression for the *NN* potential (that includes spin and momentum dependences), Okuba and Marshak imposed the following symmetries:

- Translational invariance,
- Galilean invariance,
- Rotational invariance,
- Space reflection invariance,
- Time reversal invariance,
- Invariance under the interchange of particle 1 and 2,
- Isospin symmetry,
- Hermiticity.

And obtained:

$$\begin{aligned} V = {} & Vc + \tau_1 \cdot \tau_2\, Wc && \textit{central} \\ & + [Vs + \tau_1 \cdot \tau_2\, Ws]\vec{\sigma}_1 \cdot \vec{\sigma}_2 && \textit{spin-spin} \\ & + [V_{LS} + \tau_1 \cdot \tau_2\, W_{LS}]\vec{L} \cdot \vec{S} && \textit{spin-orbit} \\ & + [V_T + \tau_1 \cdot \tau_2\, W_T] S_{12}(\hat{r}) && \textit{tensor} \\ & + [V_{\sigma L} + \tau_1 \cdot \tau_2\, W_{\sigma L}] Q_{12} && \sigma - L \\ & + [V_{\sigma p} + \tau_1 \cdot \tau_2\, W_{\sigma p}]\vec{\sigma}_1 \cdot \vec{p}\,\vec{\sigma}_2 \cdot \vec{p} && \sigma - p \end{aligned}$$

With,

$$S_{12}(\hat{r}) = 3\vec{\sigma}_1 . \hat{r}\ \vec{\sigma}_2 \cdot \hat{r} - \vec{\sigma}_1 \cdot \sigma_2,$$

$$Q_{12} \equiv \frac{1}{2}\left[\vec{\sigma}_1 \cdot \vec{L}\ \vec{\sigma}_2 \cdot \vec{L} + \vec{\sigma}_2 \cdot \vec{L}\,\vec{\sigma}_1 \cdot \vec{L}\right],$$

and

$\vec{r} \equiv \vec{r}_1 - \vec{r}_2$ relative coordinate,

$\hat{r} \equiv \vec{r} / \vec{r}$ unit vector for relative coordinate,

$\vec{p} \equiv \frac{1}{2}(\vec{p}_1 - \vec{p}_2)$ relative momentum,

$\vec{L} \equiv \vec{L}_1 + \vec{L}_2 = \vec{r} \times \vec{p} = -i\vec{r} \times \vec{\nabla}$ total orbital angular momentum in position space,

$\vec{S} \equiv \frac{1}{2}(\vec{\sigma}_1 + \vec{\sigma}_2)$ total spin, where $\vec{r}_{1,2}, \vec{p}_{1,2}, \vec{L}_{1,2}, \vec{\sigma}_{1,2}$ and $\tau_{1,2}$ denote position, momentum, angular momentum, spin, and isospin, respectively, of nucleon 1 and 2. The V_i and W_i, with i=C, S, $L\,S$, T, σL, σp, are functions of r^2, p^2, and L^2 only, i.e.

$$V_i = V_i(r^2, p^2, L^2),$$
$$W_i = W_i(r^2, p^2, L^2).$$

Charge-independence or isospin invariance requires that the potential is a scalar in the isospin space of the two nucleons. The only such scalars are 1 and $\tau_1 \cdot \tau_2$, which explains the isospin structures in,

$$\begin{aligned}
V = {} & Vc + \tau_1 \cdot \tau_2\, Wc && \textit{central} \\
& + [Vs + \tau_1 \cdot \tau_2\, Ws]\vec{\sigma}_1 \cdot \vec{\sigma}_2 && \textit{spin-spin} \\
& + [V_{LS} + \tau_1 \cdot \tau_2\, W_{LS}]\vec{L} \cdot \vec{S} && \textit{spin-orbit} \\
& + [V_T + \tau_1 \cdot \tau_2\, W_T] S_{12}(\hat{r}) && \textit{tensor} \\
& + [V_{\sigma L} + \tau_1 \cdot \tau_2\, W_{\sigma L}] Q_{12} && \sigma - L \\
& + [V_{\sigma p} + \tau_1 \cdot \tau_2\, W_{\sigma p}]\vec{\sigma}_1 \cdot \vec{p}\, \vec{\sigma}_2 \cdot \vec{p} && \sigma - p
\end{aligned}$$

If energy is conserved in the scattering process ("on shell"), then there are only five independent terms and the σp term can be expressed as a combination of the other five terms. Note, however, that when a potential is applied in a scattering equation (Schrödinger or Lippmann-Schwinger equation) the potential goes off shell.

Potentials which are based upon the operator structure equation,

$$\begin{aligned}
V = {} & Vc + \tau_1 \cdot \tau_2\, Wc && \textit{central} \\
& + [Vs + \tau_1 \cdot \tau_2\, Ws]\vec{\sigma}_1 \cdot \vec{\sigma}_2 && \textit{spin-spin} \\
& + [V_{LS} + \tau_1 \cdot \tau_2\, W_{LS}]\vec{L} \cdot \vec{S} && \textit{spin-orbit} \\
& + [V_T + \tau_1 \cdot \tau_2\, W_T] S_{12}(\hat{r}) && \textit{tensor} \\
& + [V_{\sigma L} + \tau_1 \cdot \tau_2\, W_{\sigma L}] Q_{12} && \sigma - L \\
& + [V_{\sigma p} + \tau_1 \cdot \tau_2\, W_{\sigma p}]\vec{\sigma}_1 \cdot \vec{p}\, \vec{\sigma}_2 \cdot \vec{p} && \sigma - p
\end{aligned}$$

with functions Vi and Wi chosen such as to fit the *NN* data or phase shifts are called phenomenological potentials. To keep things simple, most phenomenological potentials do not include all six terms. A minimal set for a realistic potential is the central, spin-spin, spin-orbit and tensor term.

Phenomenological *NN* potentials are:

- Gammel and Thaler, first semi-quantitative NN potential, hard-core.
- Hamada and Johnston, first quantitative NN potential, hard core.

- Reid first quantitative soft core potential, very popular in the 1970s.
- Argonne V14 potential, based upon a set of 14 operators.
- Argonne V18 potential, based upon a set of 18 operators, charge-dependent, high-precision (low χ^2 for fit of NN data).

Meson Theory of Nuclear Forces

Yukawa's Idea of 1935

It introduced the concept of massive particle exchange to explain the nuclear force. He constructed an analogy to classical electrodynamics. In electrodynamics, the Coulomb potential,

$$\phi(r) = \frac{q}{4\pi}\frac{1}{r}$$

is the solution of Poisson's equation,

$$\nabla^2\phi(\vec{r}) = -q\delta^{(3)}(\vec{r}).$$

When adding a mass term to this equation (and flipping the sign on the r.h.s. to adjust for scalar coupling),

$$(\nabla^2 - m^2)\varphi(\vec{r}) = g\delta^{(3)}(\vec{r})$$

the solution becomes,

$$\varphi(r) = -\frac{g}{4\pi}\frac{e^{-mr}}{r},$$

which is the scalar field generated by one nucleon. A second nucleon, with also coupling g, at a distance r from the first one will be exposed to the interaction energy,

$$V(r) = -\frac{g^2}{4\pi}\frac{e^{-mr}}{r}$$

which is known as the Yukawa potential. The exponential in this expression, that is due to the mass m of the meson, restricts the potential to a finite range, which is the essential point. For $m \to 0$ we are back to the form of the Coulomb potential.

One Boson Exchange Model

Yukawa's original derivation was done for scalar bosons. When finally a real meson was discovered in 1947/48, it turned out to be pseudo-scalar with mass around 138 MeV and was dubbed the **π**-meson or pion. Consequently, in the 1950s, the attempts to derive the nuclear force focused on theories that included only pions. These 'pion theories' had many problems and little success--for reasons we understand today: pion dynamics is constrained by chiral symmetry, a concept that was unknown in the 1950s. In the early 1960s, heavier (non-strange) mesons were found in experiment, notably the vector (spin-1) mesons $\rho(770)$ and $\omega(782)$. Because of the problems with the pion theories, theoreticians were now happy to extend meson theory by including more and different species of mesons. This led to the one-boson-exchange (OBE) models, which were started in the 1960s and turned out to be very successful for the two-nucleon interaction.

Let's first address the question of which mesons to consider. When deriving the nuclear force, one has generally more confidence in the predictions for the longer ranged parts. Since the range, R_α, of the force created by a meson is inversely proportional to the meson mass, m_α, i.e.,

$$R_\alpha \sim \frac{1}{m_\alpha},$$

one starts with the lightest mesons and moves up to mesons with masses in the order of the nucleon mass. This includes essentially six mesons namely, π(138), η(548), σ(500), ρ(770), ω(782), a_0(980), where the numbers in parentheses are the masses in MeV. As it turns out, η and ao are not very important and so, we will focus here on:

- The pseudo-scalar isovector pion (0^-, 1),
- The scalar isoscalar sigma (0^+, 0),
- The vector isoscalar omega (1^-, 0),
- The vector isovector rho (1^-, 1),

where the parenthetical information, (J^P, I), summarizes spin J, parity P, and isospin I for each particle.

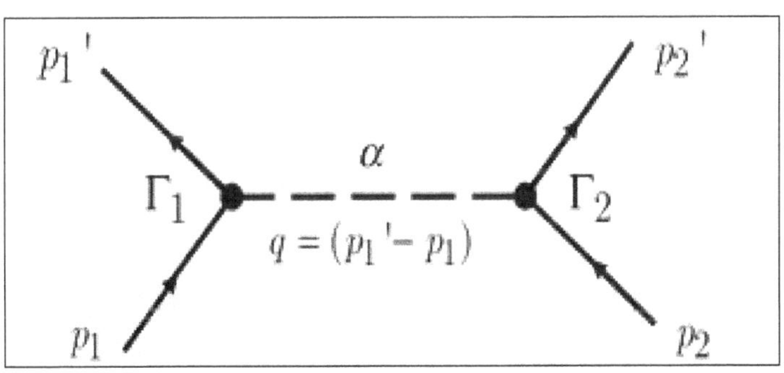

$$\text{Amplitude}: F_\alpha(p', p) = \frac{\bar{u}_1{}'\Gamma_1 u_1 P_\alpha \bar{u}_2{}'\Gamma_2 u_2}{q^2 - m_\alpha^2},$$

$$\text{with Dirac spinor } u(p,s) = \sqrt{\frac{E+M}{2M}}\begin{pmatrix} x_s \\ \frac{\vec{\sigma}\cdot\vec{p}}{E+M}x_s \end{pmatrix} = \begin{pmatrix} x_s \\ \frac{\vec{\sigma}\cdot\vec{p}}{E+M}x_s \end{pmatrix} = \begin{pmatrix} x_s \\ 0 \end{pmatrix},$$

where $E = \sqrt{\vec{p}^2 + M^2}$ and x_s is a two - components pauli spinor.

Feynman describing the exchange of a boson α between two nucleons. The amplitude F_α that results from this is stated, too. The $P\alpha$ and the Γi for the various meson exchanges.

Lagrangian: $\mathcal{L}_{\delta NN} = \frac{f_{\pi NN}}{m_\delta} \bar{\psi} \gamma^u \gamma_5 \vec{\tau} \psi \cdot \partial_u \vec{\varphi}^{(\pi)}$

Vertex: i times the Lagrangian stripped off the fields (for an incoming pion),

$$\Gamma \pi NN = (i)^2 \frac{f \pi NN}{m_\pi} \gamma^u \gamma_5 \vec{\tau} q_u = \frac{f \pi NN}{m_\pi} (\vec{\sigma} \cdot \vec{q}) \vec{\tau}$$

Potential: i times the amplitude $\left(p_\pi = i, q^2 \approx -\vec{q}^2\right)$,

$$V_\pi = iF_\pi \approx -\frac{f_\pi^2 NN}{m_\pi^2} \frac{(\vec{\sigma}_1 \cdot \vec{q})(\vec{\sigma}_2 \cdot \vec{q})}{\vec{q}^2 + m_\pi^2} \vec{\tau}_1 \cdot \vec{\tau}_2$$

Using the operator identity,

$$(\vec{\sigma}_1 \cdot \vec{q})(\vec{\sigma}_2 \cdot \vec{q}) = \frac{\vec{q}^2}{3} \left[\vec{\sigma}_1 \cdot \vec{\sigma}_2 + s_{12}(\vec{q})\right]$$

with $s_{12}(\vec{q}) = 3(\vec{\sigma}_1 \cdot \hat{q})(\vec{\sigma}_2 \cdot \hat{q}) - \vec{\sigma}_1 \cdot \vec{\sigma}_2$ ("Tensor operator")

The one-pion exchange potential (OPEP) can be written as,

$$v_\pi = \frac{f_\pi^2 NN}{3m_\pi^2} \frac{\vec{q}^2}{\vec{q}^2 + m_\pi^2} \left[-\vec{\sigma}_1 \cdot \vec{\sigma}_2 - s_{12}(\vec{q})\right] \vec{\tau}_1 \cdot \vec{\tau}_2$$

Derivation of the one-pion-exchange potential in non-relativistic approximation.

Lagrangian: $\mathcal{L}_{\sigma NN} = -g_\sigma \bar{\psi} \psi \varphi^{(\sigma)}$

Vertex: $\bar{u}(p')\Gamma_{\sigma NN}\, u(p) = -ig_\sigma \bar{u}(p')\, u(p) = -ig_\sigma \left(1 - \frac{(\vec{\sigma} \cdot \vec{p}\,')(\vec{\sigma} \cdot \vec{p})}{(E'+M)(E+M)}\right)$

$$= -ig_\sigma \left(1 - \frac{\vec{p}\,' \cdot \vec{p} + i\vec{\sigma} \cdot (\vec{p}\,' \times \vec{p})}{(E'+M)(E+M)}\right) = -ig_\sigma \left(1 - \frac{\vec{k}^2 - \frac{1}{4}\vec{q}^2 - \vec{\sigma} \cdot \vec{L}}{4M^2}\right)$$

Potential: Keeping all terms to $Q^2 / M^2 \left[P_\sigma = i, \vec{k} = \frac{1}{2}(\vec{p}\,' + \vec{p}), \vec{L} \cdot \vec{S} = -\frac{i}{2}(\vec{\sigma}_1 + \vec{\sigma}_2) \cdot (\vec{q} \times \vec{k})\right]$

$$V_\sigma = iF_\sigma \approx \sim \frac{g_\sigma^2}{\vec{q}^2 + m_\sigma^2} \left[-1 + \frac{\vec{k}^2}{2M^2} - \frac{\vec{q}^2}{8M^2} - \frac{\vec{L} \cdot \vec{S}}{2M^2}\right]$$

Derivation of the one-sigma-exchange potential up to terms in Q^2/M^2.

Lagrangian: $L\omega NN = -g\omega\bar{\psi}^{\mu}\,\psi\,\varphi_{\mu}^{(\omega)}$

Vertex:

$$\mu = 0 : \bar{u}(p')\Gamma_{\omega}^{0}NN^{u(p)} = -ig_{\omega}\bar{u}(p')\gamma^{0}u(p) = -ig_{\omega}\left(1 + \frac{(\vec{\sigma}\cdot\vec{p}\,')(\vec{\sigma}\cdot\vec{p})}{(E'+M)(E+M)}\right)$$

$$= -ig\omega\left(1 - \frac{\vec{\sigma}\cdot\vec{L}}{4M^{2}}\right), \textit{keeping all the } \vec{\sigma}\cdot\vec{L} \textit{ terms.}$$

Potential: Including also the $\vec{\gamma}$ terms : $\left[p_{\omega} = -ig_{\mu\nu} + \ldots\right]$

$$V_{\omega} = iF_{\omega} \approx \frac{g_{\omega}^{2}}{\vec{q}^{2} + m_{\omega}^{2}}\left[1 - 3\frac{\vec{L}\cdot\vec{S}}{2M^{2}}\right]$$

Derivation of the one-omega-exchange potential keeping only the central and spin-orbit term.

Yukawa's original considerations used classical field theory. A more proper derivation should be based upon quantum field theory, as we will use here. In the one-boson-exchange (OBE) model, the mesons are exchanged singly. In these derivations, we always start from an appropriate interaction Lagrangian for meson-nucleon coupling, which is designed with guidance from symmetry principles (the Lagrangian must be a Lorentz scalar). Concerning the Lagrangians for the vector mesons ω and ρ, we note that they may have both a vector and a tensor coupling (with coupling constants gv and fv, respectively):

$$\mathcal{L}_{vNN} = -g_{v}\bar{\psi}\gamma^{\mu}\psi\varphi_{\mu}^{(v)} - \frac{fv}{4M}\bar{\psi}\sigma^{\mu\nu}\psi(\partial_{\mu}\varphi_{\mu}^{(v)} - \partial_{v}\varphi_{\mu}^{(v)}).$$

These two couplings are similar to the interaction of a photon with a nucleon. The first is analogous to the coupling of the Dirac current to the electromagnetic vector potential, while the second one corresponds to the Pauli coupling of the anomalous magnetic moment. The analogy is not accidental; the vector-meson dominance model (VDM) for the electromagnetic form factor of the nucleon explains the close relationship. In the VDM one assumes that the photon couples to the nucleon through a vector boson, which explains the extended structure of the nucleon electromagnetic form factor. In the strict interpretation of this model, the ρ coupling constants ratio, $f\rho/g\rho$, should be 3.7, and from dispersion analysis one obtains even $f\rho/g\rho \approx 6$. In any case, the tensor coupling of the ρ is much larger than its vector coupling, which is why we omitted the vector coupling. For the ω meson, it is the other way around: the vector dominance model suggests a ω coupling constants ratio $f\omega/g\omega = -0.12$. Since this is close to zero, the ω is given no tensor coupling in most meson models.

The full propagator for vector bosons is,

$$P_{v} = i\frac{-g_{\mu\nu} + q_{\mu}q_{\nu}/m_{v}^{2}}{q^{2} - m_{v}^{2}}$$

where we dropped the $q_\mu q_\nu/m^2_\upsilon$-term. Due to nuclear current conservation, this term vanishes on-shell, but not off-shell. The off-shell effect of this term was examined by Holinde and Machleidt and was found to be unimportant.

It is customary to multiply the vertices with cutoffs, which suppress high-momentum components to ensure the convergence of the scattering equation. A simple form for these cutoffs is,

$$\left(\frac{\Lambda_\alpha^2 - m_\alpha^2}{\Lambda_\alpha^2 + \vec{q}^2}\right)^{n_\alpha},$$

where the cutoff mass Λ_α is typically chosen in the range 1.3 - 2.0 GeV. The multiplication by these form factors is not explicitly shown in our derivations. The calculation is performed in momentum space and in the center-of-mass (CMS) system of the two interacting nucleons, where $p_1 = (E, \vec{p})$ and $p_2 = (E, -\vec{p})$ in the initial states; and $p_1' = (E', \vec{p}\,')$ and $p_2' = (E', -\vec{p}\,')$ in the final states. Moreover,

$\vec{q} \equiv \vec{p}\,' - \vec{p}$ is the momentum transfer,

$\vec{k} \equiv \frac{1}{2}(\vec{p}\,' + \vec{p})$ the average momentum,

$\vec{L} = -i(\hat{q} \times \vec{k}) = -i(\vec{p}\,' - \vec{p})$ the total orbital angular momentum in momentum space,

$S_{12}(\vec{q}) \equiv 3\vec{\sigma}_1 \cdot \hat{q} \cdot \vec{\sigma}_2 \cdot \hat{q} - \vec{\sigma}_1 \cdot \vec{\sigma}_2$ the tensor operator in momentum space.

Lagrangian: $\mathcal{L}_{\rho NN}^{(tensor)} = -\frac{f\rho}{4M}\bar{\psi}\sigma^{\mu\nu}\bar{\tau}\psi \cdot \left(\partial_\mu \bar{\varphi}_\nu^{(\rho)} - \partial\nu\bar{\varphi}_\mu^{(\rho)}\right)$

Vertex: (incoming meson) $\Gamma_{\rho NN}^{(tensor)} = -\frac{f\rho}{4M}\sigma^{\mu\nu}(q_\mu - q_\nu)\vec{\tau} = -\frac{f\rho}{2M}\sigma^{\mu\nu} q_\mu\ \vec{\tau} = -\frac{f_\rho}{2M}(\vec{\sigma} \times \vec{q})\vec{\tau}$

Potential: $\left[P_\rho = -ig_{\mu\nu} + \ldots\right]$

$$V_\rho^{(\text{tensor})} = iF_\rho^{(\text{tensor})} = -\frac{f_\rho^2}{4M^2}\frac{(\vec{\sigma}_1 \times \vec{q})}{\vec{q}^2 + m_\rho^2}\vec{\tau}_1 \cdot \vec{\tau}_2$$

$$= -\frac{f_\rho^2}{4M^2}\frac{\vec{\sigma}_1 \cdot \vec{\sigma}_2 \vec{q}^2 - (\vec{\sigma}_1 \cdot \vec{q})(\vec{\sigma}_2 \cdot \vec{q})}{\vec{q}^2 - m_\rho^2}\vec{\tau}_1 \cdot \vec{\tau}_2$$

$$= \frac{f_\rho^2}{12M^2}\frac{\vec{q}^2}{\vec{q}^2 + m_\rho^2}\left[-2\vec{\sigma}_1 \cdot \vec{\sigma}_2 + s_{12}(\hat{q})\right]\vec{\tau}_1 \cdot \vec{\tau}_2$$

Derivation of the one-rho-exchange potential in non-relativistic approximation using only the tensor coupling of the ρ to the nucleon are derived.

The characteristic properties of the contributions from the various mesons derived in table.

The OBE *NN* potential (OBEP) is defined as the sum over the contributions from the four mesons discussed.

$$V_{OBE} = \sum_{\alpha=\pi,\sigma,\omega,\rho,\eta,a_0} V\alpha$$

Table: The four most important mesons and the main characteristics of their contributions to components of the nuclear force.

Meson	Central	Spin-Spin	Tensor	Spin-Orbit
π(138)	-	weak, long-ranged	strong, long-ranged	-
σ(500)	strong, attractive, intermediate-ranged	-	-	moderate, intermediate-ranged
ω(782)	strong, repulsive, short-ranged	-	-	strong, short-ranged, coherent with σ
ρ(770)	-	weak, short-ranged, coherent with π	moderate, short-ranged, opposite to π	-

OBEP in Position Space

The momentum-space potentials can be Fourier transformed,

$$V_\alpha(\vec{r}) = \frac{1}{(2\pi)^3}\int d^3q e^{i\vec{q}\cdot\vec{r}}\, V\alpha(\vec{q}),$$

to obtain the following equivalent position space potentials:

$$V_\pi(\vec{r}) = \frac{1}{3}\frac{f_{\pi NN}^2}{4\pi} m_\pi \left\{\left[Y(m_\pi r) - \frac{4\pi}{m_\pi^3}\delta^{(3)}(\vec{r})\right]\vec{\sigma}_1\cdot\vec{\sigma}_2 + \left(1+\frac{3}{m_\pi r}+\frac{3}{(m_\pi r)^2}\right)Y(m_\pi r)s_{12}(\hat{r})\right\}\tau_1\cdot\tau_2,$$

$$V_\sigma(\vec{r}) = -\frac{g_\sigma^2}{4\pi} m_\sigma \left\{\left[1-\frac{1}{4}\left(\frac{m_\sigma}{M}\right)\right]^2 Y(m_\sigma r) + \frac{1}{4M^2}\left[\vec{\nabla}^2 Y(m_\sigma r) + Y(m_\sigma r)\vec{\nabla}^2\right] + \frac{1}{2}\left(\frac{m_\sigma}{M}\right)^2\left(\frac{1}{m_\sigma r}+\frac{1}{(m_\sigma r)^2}\right)Y(m_\sigma r)\vec{L}\cdot\vec{S}\right\}$$

$$V_\omega(\vec{r})\frac{g_\omega^2}{4\pi} m_v \left[Y(m_\omega r) - \frac{3}{2}\left(\frac{m_\omega}{M}\right)^2\left(\frac{1}{m_\omega r}+\frac{1}{(m_\omega r)^2}\right)Y(m_\omega r)\vec{L}\cdot\vec{S}\right],$$

$$V_\rho(\vec{r})\frac{1}{12}\frac{f_\rho^2}{4\pi}\left(\frac{m_\rho}{M}\right)^2 m_\rho \left\{2\left[Y(m_\rho r) - \frac{4\pi}{m_\rho^3}\delta^{(3)}(\vec{r})\right]\vec{\sigma}_1\cdot\vec{\sigma}_2 - \left(1+\frac{3}{m_\rho r}+\frac{3}{(m_\rho r)^2}\right)Y(m_\rho r)s_{12}(\hat{r})\right\}\tau_1\cdot\tau_2,$$

with the Yukawa function,

$$Y(x) = e^{-x}/x$$

and $S_{12}(\hat{r})$ given in equation $S_{12}(\hat{r}) = 3\vec{\sigma}_1.\hat{r}\ \vec{\sigma}_2\cdot\hat{r} - \vec{\sigma}_1\cdot\sigma_2$.

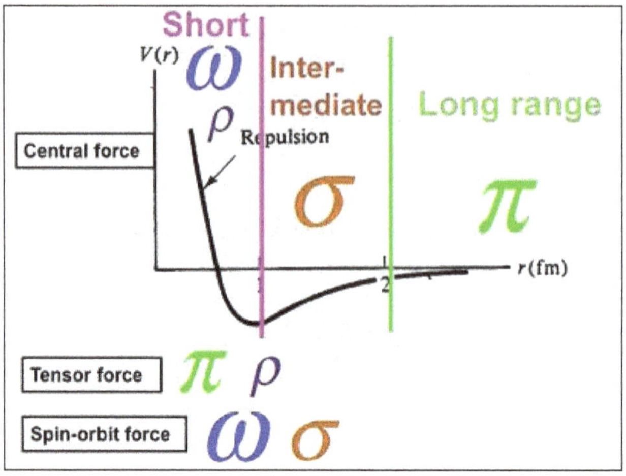

Nuclear force in the meson picture.

We can now better understand the success of the OBE model by summarizing its properties:

- The pseudo-scalar pion with a mass of about 138 MeV is the lightest meson and provides the longest-ranged part of the *NN* potential and the essential part of the tensor force.
- The vector meson ρ (770 MeV) cuts down the tensor force created by the pion at short distances to arrive at a tensor force of realistic strength.
- The scalar σ boson of about 500 MeV provides the intermediate-range attraction necessary for nuclear binding.
- The vector meson ω (782 MeV) produces a strong repulsive central force of short range (repulsive core) and the essential part of the nuclear spin-orbit force.

This takes care of all the empirical properties of the nuclear force and, therefore, allows for a quantitative description of the *NN* system.

Relativistic OBEPs

The first OBEPs ever developed were derived along the lines we followed above: one starts from the Feynman amplitude of an OBE and then performs a Q/M expansion up to terms of order $(Q/M)^2$. The motivation for this procedure was twofold. First, one wanted to see in a simple way what force components (e.g., central, spin-spin, tensor, spin-orbit) were created by the exchange of different mesons. Second, early researchers preferred a local potential in position space, i.e., an analytic expression for the potential that is a function of the relative distance between the two nucleons, $\vec{r}$. For this it is necessary that the Fourier transform of the momentum space expressions can be performed analytically.

However, there is no need to perform calculations in position space. Equally well and often in a more elegant way, calculations of *NN* scattering and nuclear bound states can be carried out in momentum space. Locality is not an issue and presents no advantages in momentum space. The original relativistic Feynman amplitudes of OBE are functions of p' and p and the relativistic OBEP is defined as,

$$V^{\text{rel}}_{\text{OBE}}(p',p) = i\sum_{\alpha} F_{\alpha}(p',p)$$

with $F_a(p', p)$ as evaluated in full beauty and without approximations. Two-nucleon scattering is described covariantly by the Bethe-Salpeter equation for which V_{OBE}^{rel} is input. Since the four-dimensional Bethe-Salpeter (BS) equation is difficult to solve, relativistic three-dimensional (3D) reductions of the BS equation are frequently used, which are more amenable to numerical solution. It is common to the derivation of all relativistic three-dimensional equations that the time component of the relative momentum is fixed in some covariant way, so that it no longer appears as a separate variable. Thus,

$$V_{OBE}^{rel}(p', p) \rightarrow V_{OBE}^{rel3D}(\vec{p}', \vec{p}).$$

Beyond the OBE Approximation

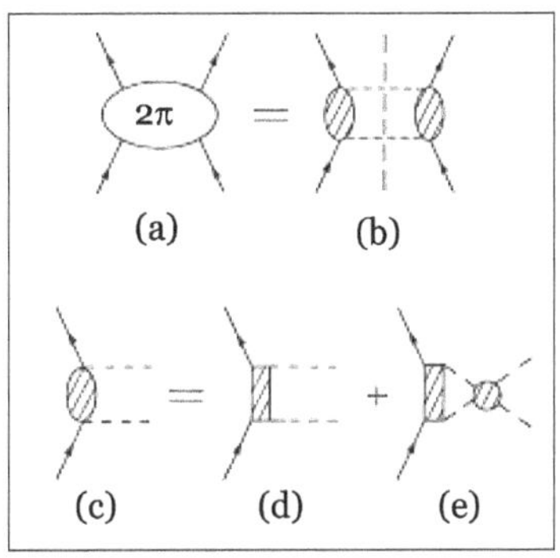

The 2π exchange contribution to the *NN* interaction as viewed by dispersion theory. Solid lines represent nucleons, dashed lines pions.

The OBE model is a great simplification of the complicated scenario of a meson theory for the *NN* interaction. Therefore, in spite of the quantitative success of the OBEPs, we should be concerned about the approximations involved in the model. Major concerns include:

- The scalar isoscalar σ 'meson' of about 500 MeV.
- The neglect of all non-iterative diagrams.
- The role of meson-nucleon resonances.

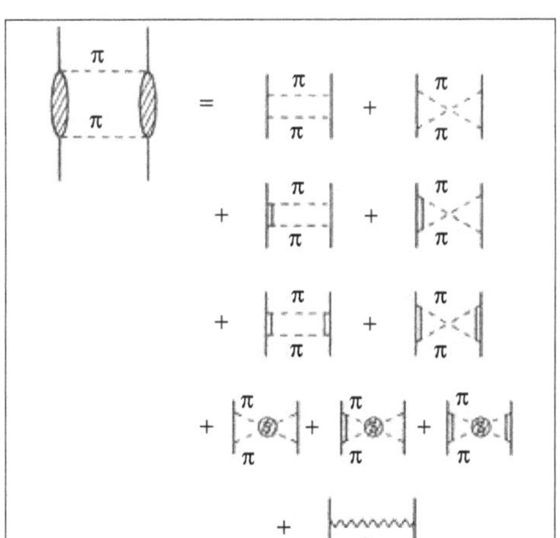

Field-theoretic model for the 2π exchange. Solid lines represent nucleons, double lines isobars, and dashed lines pions. The hatched circles are $\pi\pi$ correlations.

Two pions, when 'in the air', can interact strongly. When in a relative *P*-wave (L=1), they form a proper resonance, the ρ meson. They can also interact in a relative *S*-wave (L=0), which gives rise to the σ boson. Whether the σ is a proper resonance is controversial, even though the Particle Data Group lists an f_0 (500) or σ (500) mason, but with a width 400-700 MeV. What's for sure is that two pions have correlations, and if one doesn't believe in the σ as a two pion resonance, then one has to take these correlations into account. There are essentially two approaches to take care of the two-pion exchange contribution to the *NN* interaction (which generates the intermediate range attraction): dispersion theory and field theory.

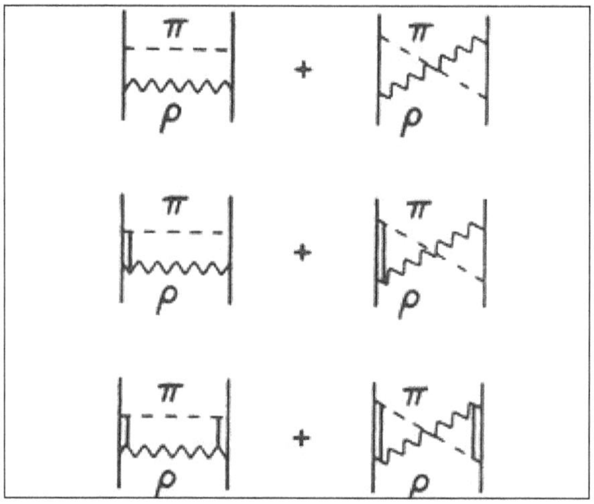

$\pi\rho$ contributions to the *NN* interaction.

The dispersion-theoretic picture is described schematically in figure above. In this approach one assumes that the total diagram: (a) can be analysed in terms of two 'halves', (b) The hatched ovals stand for all possible processes which two pions and a nucleon can undergo. This is made more explicit in, (d) and (e). The hatched boxes represent all possible baryon intermediate states including the nucleon. The shaded circle stands for $\pi\pi$ scattering. Quantitatively, these processes are taken into account by using empirical information from πN and $\pi\pi$ scattering (e.g. phase shifts) which represents the input for such a calculation. Dispersion relations then provide on-shell *NN* amplitude, which with some kind of plausible prescription is represented as a potential. The Paris potential is constructed along these lines complemented by one-pion-exchange and ω-exchange. For further details, we refer the interested reader to a pedagogical article written by Vinh Mau.

The model includes contributions from isobars as well as from $\pi\pi$ correlations. This can be understood in analogy to the dispersion relations picture. In general, only the lowest-lying πN resonance, the so-called Δ isobar (spin 3/2, isospin 3/2, mass 1232 MeV), is taken into account. The contributions from other resonances have proven to be small for the low-energy *NN* processes under consideration. A field-theoretic model treats the Δ isobar as an elementary (Rarita-Schwinger) particle. The crossed (non-iterative) two-particle exchanges (second diagram in each row) are important. They guarantee the proper (very weak) isospin dependence due to characteristic cancellations in the isospin dependent parts of box and crossed box diagrams. Furthermore, their contribution is about as large as the one from the corresponding box diagrams (iterative diagrams); therefore, they are not negligible. In addition to the processes discussed, also correlated 2π exchange has to be included. Quantitatively, these contributions are about as large as those from the uncorrelated processes.

Besides the contributions from two pions, there are also contributions from the combination of other mesons. The combination of π and ρ is particularly significant. This contribution is repulsive and important to suppress the 2π exchange contribution at high momenta (or small distances), which is too strong by itself.

The most developed meson-theoretic model for the *NN* interaction is the Bonn Full Model. Besides this, the Bonn Full Model also contains non-iterative graphs of π and σ and π and ω exchanges. However, the combined contribution of the latter two groups of diagrams is small and not significant.

Having highly sophisticated models at hand, like the Paris and the Bonn potentials, allows to check the approximations made in the simple OBE model. As it turns out, the highly complicated 2π exchange contributions to the *NN* interaction tamed by the $\pi\rho$ diagrams can well be simulated by the single scalar isoscalar boson, the σ, with a mass around 550 MeV. Retroactively, this fact provides justification for the simple OBE model.

QCD and the Nuclear Force

Quantum chromodynamics (QCD) is the theory of strong interactions. It deals with quarks, gluons and their interactions and is part of the Standard Model of Particle Physics. QCD is a non-Abelian gauge field theory with color $SU(3)$ the underlying gauge group. The non-Abelian nature of the theory has dramatic consequences. While the interaction between colored objects is weak at short distances or high momentum transfer (asymptotic freedom), it is strong at long distances (≥1fm) or low energies, leading to the confinement of quarks into colorless objects, the hadrons.

Consequently, QCD allows for a perturbative analysis at large energies, whereas it is highly non-perturbative in the low-energy regime. Nuclear physics resides at low energies and the force between nucleons is a residual color interaction similar to the van der Waals force between neutral molecules. Therefore, in terms of quarks and gluons, the nuclear force is a very complicated problem that, nevertheless, can be attacked with brute computing power on a discretized, Euclidean space-time lattice (known as lattice QCD). In a recent study, the neutron-proton scattering lengths in the singlet and triplet *S*-waves have been determined in fully dynamical lattice QCD, with a smallest pion mass of 354 MeV. This result is then extrapolated to the physical pion mass with the help of chiral perturbation theory. The pion mass of 354 MeV is still too large to allow for reliable extrapolations, but the feasibility has been demonstrated and more progress can be expected for the near future. In a lattice calculation of a very different kind, the nucleon-nucleon (*NN*) potential was studied. The central part of the potential shows a repulsive core plus attraction of intermediate range. This is a very promising result, but it must be noted that also in this investigation still rather large pion masses are being used. In any case, advanced lattice QCD calculations are under way and continuously improved. However, since these calculations are very time-consuming and expensive, they can only be used to check a few representative key-issues. For everyday nuclear structure physics, a more efficient approach is needed.

Chiral Effective Field Theory Approach to Nuclear Forces

The efficient approach is an effective field theory (EFT). For the development of an EFT, it is crucial to identify a separation of scales. In the hadron spectrum, a large gap between the masses of the pions and the masses of the vector mesons, like $\rho(770)$ and $\omega(782)$, can clearly be identified. Thus, it is natural to assume that the pion mass sets the "soft scale", $Q{\sim}m\pi$, and the rho mass the "hard scale", $\Lambda\chi{\sim}m\rho$, also known as the chiral-symmetry breaking scale. This is suggestive of considering an expansion in terms of the soft scale over the hard scale, $Q/\Lambda\chi$. Concerning the relevant degrees of freedom, we noticed already that, for the ground state and the low-energy excitation spectrum of an atomic nucleus as well as for conventional nuclear reactions, quarks and gluons are ineffective degrees of freedom, while nucleons and pions are the appropriate ones (this may include also low-lying nucleon resonances). To make sure that this EFT is not just another phenomenology, it must have a firm link with QCD. The link is established by having the EFT observe all relevant symmetries of the underlying theory. This requirement is based upon a "folk theorem" by Weinberg:

> "If one writes down the most general possible Lagrangian, including all terms consistent with assumed symmetry principles, and then calculates matrix elements with this Lagrangian to any given order of perturbation theory, the result will simply be the most general possible S-matrix consistent with analyticity, perturbative unitarity, cluster decomposition, and the assumed symmetry principles."

The EFT program consists of the following steps:

1. Identify the soft and hard scales, and the degrees of freedom appropriate for (low-energy) nuclear physics.
2. Identify the relevant symmetries of low-energy QCD and investigate if and how they are broken.
3. Construct the most general Lagrangian consistent with those symmetries and symmetry breakings.
4. Design an organizational scheme that can distinguish between more and less important contributions: a low-momentum expansion.
5. Guided by the expansion, calculate Feynman diagrams for the problem under consideration to the desired accuracy.

Note that we are presenting here the chiral EFT which scales with the ρ-mass and includes pion degrees of freedom. However, the ρ-mass is not the only possible (hard) scale. Depending on what energies we are interested in, other scales may be more appropriate. For example, if we wish to focus on a nuclear scenario at very low energy ($< m\pi$), then it is suggestive to choose the pion-mass, $m\pi$, as scale. Such EFT can, of course, not have pion degrees of freedom anymore and is, therefore, known as pionless EFT. It consists of contact terms, only. Thus, this pionless EFT may appear forbiddingly simplistic at first glance, but, as it turns out, it has some surprisingly intriguing features, particularly, with regard to renormalization and power counting.

Symmetries of Low-energy QCD and their Breakings

The QCD Lagrangian reads,

$$\mathcal{L}_{QCD} = \bar{q}(i\gamma^{\mu}D_{\mu} - M)q - \frac{1}{4}G_{\mu\nu,a}G_{a}^{\mu\nu}$$

with the gauge-covariant derivative,

$$D_\mu = \partial_\mu - ig\frac{\lambda_a}{2}A_{\mu,a}$$

and the gluon field strength tensor (note that for $SU(N)$ group indices, we use letters $a,b,c,...,i,$-$j,k,...$, and, in general, do not distinguish between subscripts and superscripts),

$$G_{\mu\nu,a} = \partial_\mu A_{\nu,a} - \partial_\nu A_{\mu,a} + gf_{abc}A_{\mu,b}A_{\nu,c}$$

In the above, q denotes the quark fields and M the quark mass matrix. Further, g is the strong coupling constant and $A_{\mu,a}$ are the gluon fields. The λ_a are the Gell-Mann matrices and the f_{abc} the structure constants of the $SU(3)_{\text{color}}$ Lie algebra (a, b, c = 1, ..., 8); summation over repeated indices is always implied. The gluon-gluon term in the last equation arises from the non-Abelian nature of the gauge theory and is the reason for the peculiar features of the color force.

The masses of the up (u), down (d), and strange (s) quarks are:

$$m_u = 2.3 \pm 0.7\,\text{MeV},$$

$$m_d = 4.8 \pm 0.7\,\text{MeV},$$

$$m_s = 95 \pm 5\,\text{MeV}.$$

These masses are small as compared to a typical hadronic scale, i.e., a scale of low-mass hadrons which are not Goldstone bosons, e.g. $m_\rho = 0.78\,\text{GeV} \approx 1\,\text{GeV}$.

It is therefore of interest to discuss the QCD Lagrangian in the limit of vanishing quark masses:

$$\mathcal{L}^0_{QCD} = \bar{q}i\gamma^\mu D_\mu q - \frac{1}{4}G_{\mu\nu,a}G^{\mu\nu}_a.$$

Defining right- and left-handed quark fields,

$$q_R = P_R q, \quad q_L = P_L q,$$

with,

$$P_R = \frac{1}{2}(1+\gamma_5), P_L = \frac{1}{2}(1-\gamma_5),$$

we can rewrite the Lagrangian as follows:

$$\mathcal{L}^0_{QCD} = \bar{q}_R i\gamma^\mu D_{\mu qR} + \bar{q}_L i\gamma^\mu D_{\mu qL} - \frac{1}{4}G_{\mu\nu,a}G^{\mu\nu}_a$$

This equation reveals that the right- and left-handed components of massless quarks do not *mix* in the QCD Lagrangian. For the two-flavor case, this is $SU(2)_R \times SU(2)_L$ symmetry, also known as chiral symmetry. However, this symmetry is broken in two ways: explicitly and spontaneously.

Explicit Symmetry Breaking

The mass term $-\bar{q}Mq$ in the QCD Lagrangian equation $\mathcal{L}_{QCD} = \bar{q}(i\gamma^{\mu}D_{\mu} - M)q - \frac{1}{4}G_{\mu\nu,a}G_a^{\mu\nu}$ breaks chiral symmetry explicitly. To better see this, let's rewrite M for the two-flavor case,

$$\begin{aligned}\mathcal{M} &= \begin{pmatrix} m_u & 0 \\ 0 & md \end{pmatrix} \\ &= \frac{1}{2}(m_u + m_d)\begin{pmatrix} 1 & 0 \\ 0 & 1 \end{pmatrix} + \frac{1}{2}(m_u - m_d)\begin{pmatrix} 1 & 0 \\ 0 & -1 \end{pmatrix} \\ &= \frac{1}{2}(m_u + m_d)I + \frac{1}{2}(m_u - m_d)\tau_3.\end{aligned}$$

The first term in the last equation in invariant under $SU(2)_V$ (isospin symmetry) and the second term vanishes for *mu=md*. Thus, isospin is an exact symmetry if *mu=md*. However, both terms in equation $\mathcal{M} = \begin{pmatrix} m_u & 0 \\ 0 & md \end{pmatrix}\ldots\ldots I + \frac{1}{2}(m_u - m_d)\tau_3$ break chiral symmetry. Since the up and down quark masses [equations: $m_u = 2.3 \pm 0.7\,\text{MeV}$ and $m_d = 4.8 \pm 0.7\,\text{MeV}$] are small as compared to the typical hadronic mass scale of ~1 GeV, the explicit chiral symmetry breaking due to non-vanishing quark masses is very small.

Spontaneous Symmetry Breaking

A (continuous) symmetry is said to be spontaneously broken if a symmetry of the Lagrangian is not realized in the ground state of the system. There is evidence that the (approximate) chiral symmetry of the QCD Lagrangian is spontaneously broken for dynamical reasons of nonperturbative origin which are not fully understood at this time. The most plausible evidence comes from the hadron spectrum.

From chiral symmetry, one naively expects the existence of degenerate hadron multiplets of opposite parity, i.e., for any hadron of positive parity one would expect a degenerate hadron state of negative parity and vice versa. However, these 'parity doublets' are not observed in nature. For example, take the ρ-meson which is a vector meson of negative parity ($J^P=1^-$) and mass 776 MeV. There does exist a 1^+ meson, the a_1, but it has a mass of 1230 MeV and, therefore, cannot be perceived as degenerate with the ρ. On the other hand, the ρ meson comes in three charge states (equivalent to three isospin states), the $\rho^{\pm}$ and the ρ^{0}, with masses that differ by at most a few MeV. Thus, in the hadron spectrum, $SU(2)_V$(isospin) symmetry is well observed, while axial symmetry is broken: $SU(2)_R \times SU(2)_L$ is broken down to $SU(2)_V$.

A spontaneously broken global symmetry implies the existence of (massless) Goldstone bosons. The Goldstone bosons are identified with the isospin triplet of the (pseudoscalar) pions, which explains why pions are so light. The pion masses are not exactly zero because the up and down quark masses are not exactly zero either (explicit symmetry breaking). Thus, pions are a truly remarkable species: they reflect spontaneous as well as explicit symmetry breaking.

Goldstone bosons interact weakly at low energy. They are degenerate with the vacuum and, therefore, interactions between them must vanish at zero momentum and in the chiral limit ($m\pi \rightarrow 0$).

Chiral Effective Lagrangians

The next step in the EFT program is to build the most general Lagrangian consistent with the (broken) symmetries. An elegant formalism for the construction of such Lagrangians was developed by who worked out the group-theoretical foundations of non-linear realizations of chiral symmetry. It is characteristic for these non-linear realizations that, whenever functions of the Goldstone bosons appear in the Lagrangian, they are always accompanied with at least one space-time derivative. The Lagrangians given are built upon the Callan et al. formalism.

The relevant degrees of freedom are pions (Goldstone bosons) and nucleons. Since the interactions of Goldstone bosons must vanish at zero momentum transfer and in the chiral limit ($m\pi \rightarrow 0$), the low-energy expansion of the Lagrangian is arranged in powers of derivatives and pion masses. The hard scale is the chiral-symmetry breaking scale, $\Lambda\chi \approx 1$ GeV. Thus, the expansion is in terms of powers of $Q/\Lambda\chi$ where Q is a (small) momentum or pion mass. This is the essence of chiral perturbation theory (ChPT).

The effective Lagrangian can formally be written as,

$$\mathcal{L}_{\text{eff}} = \mathcal{L}_{\pi\pi} + \mathcal{L}_{\pi} N + \mathcal{L}_{NN} + \ldots,$$

where Lππ deals with the dynamics among pions, LπN describes the interaction between pions and a nucleon, and LNN contains two-nucleon contact interactions which consist of four nucleon fields (four nucleon legs) and no meson fields. The ellipsis stands for terms that involve two nucleons plus pions and three or more nucleons with or without pions, relcvant for nuclear many-body forces, (last two terms of equation $\hat{\mathcal{L}}^{\Delta=1} = \bar{N}\left\{\frac{\vec{\nabla}^2}{2M_N} - \frac{ig_A}{4M_N f_\pi}\tau \cdot \left[\vec{\sigma} \cdot \left(\overleftarrow{\nabla}\partial_0 \pi - \partial_0 \pi \vec{\nabla}\right)\right] \ldots\ldots\right.$ $\left(\bar{N}\tau N\right) + \ldots,$).

The individual Lagrangians are organized as follows:

$$\mathcal{L}_{\pi\pi} = \mathcal{L}_{\pi\pi}^{(2)} + \mathcal{L}_{\pi\pi}^{(4)} + \ldots,$$
$$\mathcal{L}_{\pi N} = \mathcal{L}_{\pi N}^{(1)} + \mathcal{L}_{\pi N}^{(2)} + \mathcal{L}_{\pi N}^{(3)} + \ldots,$$

and

$$\mathcal{L}_{NN} = \mathcal{L}_{NN}^{(\,)} + \mathcal{L}_{NN}^{(\,)} + \mathcal{L}_{NN} + \ldots,$$

where the superscript refers to the number of derivatives or pion mass insertions (chiral dimension) and the ellipsis stands for terms of higher dimensions.

Above, we have organized the Lagrangians by the number of derivatives or pion-mass insertions. This is the standard way, appropriate particularly for considerations of π-π and π-N scattering. As it turns out, for interactions among nucleons, it is sometimes more useful to consider the so-called index of the interaction,

$$\Delta \equiv d + \frac{n}{2} - 2$$

where d is the number of derivatives or pion-mass insertions and n the number of nucleon field operators (nucleon legs). We will now write down the Lagrangian in terms of increasing values of the parameter Δ and we will do so using the so-called Heavy-Baryon (HB) formalism which we indicate by a "hat".

The lowest-index Lagrangian reads,

$$\begin{aligned}\hat{\mathcal{L}}^{\Delta=0} &= \frac{1}{2}\partial_\mu \pi \cdot \partial^\mu \pi - \frac{1}{2} m_\pi^2 \pi^2 \\ &+ \frac{1-4\alpha}{2f_\pi^2}(\pi \cdot \partial_\mu \pi)(\pi \cdot \partial^\mu \pi) - \frac{\alpha}{f_\pi^2}\pi^2 \partial_\mu \pi \cdot \partial^\mu \pi + \frac{8\alpha - 1}{8 f_\pi^2} m_\pi^2 \pi^4 \\ &+ \bar{N}\left[i\partial_0 - \frac{g_A}{2f_\pi^2}\tau \cdot (\vec{\sigma} \cdot \vec{\nabla})\pi - \frac{1}{4f_\pi^2}\tau \cdot (\pi \times \partial_0 \pi)\right] N \\ &+ \bar{N}\left\{ \frac{g_A(4\alpha - 1)}{4f_\pi^3}(\tau \cdot \pi)\left[\pi \cdot (\vec{\sigma} \cdot \vec{\nabla})\pi\right] + \frac{g_A \alpha}{2f_\pi^3}\pi^2\left[\tau \cdot (\vec{\sigma} \cdot \vec{\nabla})\pi\right]\right\} N \\ &- \frac{1}{2} Cs\, \bar{N}N\bar{N}N - \frac{1}{2} C_T (\bar{N}\vec{\sigma}N) \cdot (\bar{N}\vec{\sigma}N) + \ldots,\end{aligned}$$

and higher-index Lagrangians are,

$$\begin{aligned}\hat{\mathcal{L}}^{\Delta=1} &= \bar{N}\left\{ \frac{\vec{\nabla}^2}{2M_N} - \frac{i g_A}{4 M_N f_\pi}\tau \cdot \left[\vec{\sigma} \cdot (\vec{\nabla}\partial_0 \pi - \partial_0 \pi \vec{\nabla})\right] \right. \\ &\left. - \frac{i}{8 M_N f_\pi^2}\tau \cdot \left[\vec{\nabla} \cdot (\pi \times \vec{\nabla}\pi) - (\pi \times \vec{\nabla}\pi) \cdot \vec{\nabla}\right]\right\} N \\ &+ \bar{N}\left[4c_1 m_\pi^2 - \frac{2c_1}{f_\pi^2} m_\pi^2 \pi^2 \left(c_2 - \frac{g_A^2}{8M_N}\right)\frac{1}{f_\pi^2}(\partial_0 \pi \cdot \partial_0 \pi) \right. \\ &\left. + \frac{c_3}{f_\pi^2}(\partial_\mu \pi \cdot \partial^\mu \pi) - \left(c_4 + \frac{1}{4M_N}\right)\frac{1}{2f_\pi^2} \epsilon^{ijk} \epsilon^{abc} \sigma^i \tau^a (\partial^j \pi^b)(\partial^k \pi^c)\right] N \\ &- \frac{D}{4f_\pi} Cs(\bar{N}N)\bar{N}\left[\tau \cdot (\vec{\sigma} \cdot \vec{\nabla})\pi\right] N - \frac{1}{2} E(\bar{N}N)(\bar{N}\tau N) \cdot (\bar{N}\tau N) + \ldots,\end{aligned}$$

$$\hat{\mathcal{L}}^{\Delta=2} = \hat{\mathcal{L}}^{(4)}_{\delta\delta} + \hat{\mathcal{L}}^{(3)}_{\delta\delta} + \hat{\mathcal{L}}^{(2)}_{\delta\delta} + \ldots,$$

$$\hat{\mathcal{L}}^{\Delta=4} = \hat{\mathcal{L}}^{(4)}_{NN} + \ldots,$$

where the ellipses represent terms that are irrelevant for the derivation of nuclear forces up to fourth order. The Lagrangian $\hat{\mathcal{L}}_{\pi N}^{(3)}$ can be found in and the *NN* contact Lagrangians are given below. The pion fields are denoted by π and the heavy baryon nucleon field by N $\left(\bar{N} = N^{\dagger}\right)$. Furthermore, g_A, f_π, m_π, and M_N are the axial-vector coupling constant, pion decay constant, pion mass, and nucleon mass, respectively. Numerical values for these quantities will be given later. The *ci* are Low-Energy Constants (LECs) from the dimension two πN Lagrangian and α is a parameter that appears in the expansion of the pion fields are independent of α. The πNN coupling constant, $f_{\pi NN}$, used in the derivation of the one-pion-exchange potential, is related to the above quantities by,

$$\frac{f_{\pi NN}}{m_\pi} = \frac{g_A}{2f_\pi}$$

equation $V_{1\pi}(\vec{p}\,', \vec{p}) = -\frac{g_A^2}{4f_\pi^2}\tau_1 \cdot \tau_2 \frac{\vec{\sigma}_1 \cdot \vec{q} \cdot \vec{\sigma}_2 \cdot \vec{q}}{q^2 + m_\pi^2}$

The lowest order (or leading order) *NN* Lagrangian has no derivatives and reads,

$$\hat{\mathcal{L}}_{NN}^{(0)} = -\frac{1}{2}C_s\,\bar{N}N\bar{N}N - \frac{1}{2}C_T(\bar{N}\vec{\sigma}N)\cdot(\bar{N}\vec{\sigma}N),$$

where C_S and C_T are unknown constants which are determined by a fit to the *NN* data.

The second order *NN* Lagrangian can be stated as follows,

$$\begin{aligned}
\hat{\mathcal{L}}_{NN}^{(2)} = & -C_1'\left[\left(\bar{N}\vec{\nabla}N\right)^2 + \left(\overline{\vec{\nabla}N}N\right)^2\right] - C_2'\left(\bar{N}\vec{\nabla}N\right)\cdot\left(\overline{\vec{\nabla}N}N\right) - C_3'\,\bar{N}N\left[\bar{N}\vec{\nabla}^2N + \overline{\vec{\nabla}^2N}N\right] \\
& - iC_4'\left[\bar{N}\vec{\nabla}N\cdot(\overline{\vec{\nabla}N}\times\vec{\sigma}N) + (\overline{\vec{\nabla}N})N\cdot(\bar{N}\vec{\sigma}\times\vec{\nabla}N)\right] \\
& - iC_5'\,\bar{N}N(\overline{\vec{\nabla}N}\cdot\vec{\sigma}\times\vec{\nabla}N) + iC_6'(\bar{N}\vec{\sigma}N)\cdot(\overline{\vec{\nabla}N}\times\vec{\nabla}N) \\
& - (C_7'\,\delta_{ik}\delta_{jl} + C_8'\delta_{il}\delta_{kj} + C_9'\delta_{ij}\delta_{kl})\left[\bar{N}\sigma_k\partial_i N\bar{N}\sigma_l\partial_j N + \overline{\sigma_i N}_{\sigma k}N\overline{\partial_j N}_{\sigma 1}N\right] \\
& - (C_{10}'\,\delta_{ik}\delta_{jl} + C_{11}'\delta_{il}\delta_{kj} + C_{12}'\delta_{ij}\delta_{kl})\,\bar{N}\sigma_k\partial_i N\overline{\partial_j N}_{\sigma 1}N \\
& - \left(\frac{1}{2}C_{13}'\,\delta_{ik}\delta_{jl} + \delta_{il}\delta_{kj} + C_{14}'\delta_{ij}\delta_{kl}\right)\left[\partial_i N\sigma_k\partial_j N + \overline{\partial_j N}_{\sigma k}\partial_i N\right]\bar{N}_{\sigma l}N.
\end{aligned}$$

Similar to C_S and C_T of equation $\hat{\mathcal{L}}_{NN}^{(0)} = -\frac{1}{2}C_s\,\bar{N}N\bar{N}N - \frac{1}{2}C_T(\bar{N}\vec{\sigma}N)\cdot(\bar{N}\vec{\sigma}N),$, the C_i' of equation above are unknown constants which are fixed by a fit to the *NN* data. Obviously, these contact Lagrangians blow up quite a bit with increasing order, which is why we do not give $\hat{L}_{NN}^{(4)}$ explicitly here. The *NN* contact potentials that emerge from these Lagrangians are given.

Power Counting

Effective Lagrangians have infinitely many terms, and an unlimited number of Feynman graphs can be calculated from them. Therefore, we need a scheme that makes the theory manageable and calculable. This scheme which tells us how to distinguish between large (important) and small (unimportant) contributions is chiral perturbation theory.

In ChPT, graphs are analyzed in terms of powers of small external momenta over the large scale: $(Q/\Lambda_\chi)^\nu$, where Q is generic for a momentum (nucleon three-momentum or pion four-momentum) or a pion mass and $\Lambda_\chi \sim 1$ GeV is the chiral symmetry breaking scale (hadronic scale, hard scale). Determining the power ν has become known as power counting.

The nuclear potential is assembled from irreducible graphs. By definition, an irreducible graph is a diagram that cannot be separated into two by cutting only nucleon lines. Following the Feynman rules of covariant perturbation theory, a nucleon propagator is Q^{-1}, a pion propagator Q^{-2}, each derivative in any interaction is Q, and each four-momentum integration Q^4. This is also known as naive dimensional analysis. Applying then some topological identities, one obtains for the power of an irreducible diagram involving A nucleons,

$$\nu = -2 + 2A - 2C + 2L + \sum_i \Delta i$$

with,

$$\Delta_i \equiv d_i + \frac{n_i}{2} - 2,$$

where C denotes the number of separately connected pieces and L the number of loops in the diagram; di is the number of derivatives or pion-mass insertions and n_i the number of nucleon fields (nucleon legs) involved in vertex i; the sum runs over all vertices i contained in the diagram under consideration. Note that $\Delta_i \geq 0$ for all interactions allowed by chiral symmetry. Purely pionic interactions have at least two derivatives $(d_i \geq 2, n_i = 0)$; interactions of pions with a nucleon have at least one derivative $(d_i \geq 1, n_i = 2)$; and nucleon-nucleon contact terms $(n_i = 4)$ have $d_i \geq 0$. This demonstrates how chiral symmetry guarantees a low-energy expansion.

The power formula equation ($\nu = -2 + 2A - 2C + 2L + \sum_i \Delta i$) allows to predict the leading orders of connected multi-nucleon forces. Consider a m-nucleon irreducibly connected diagram (m-nucleon force) in an A-nucleon system ($m \leq A$). The number of separately connected pieces is $C = A - m + 1$. Inserting this into equation ($\nu = -2 + 2A - 2C + 2L + \sum_i \Delta i$) together with $L = 0$ and $\sum_i \Delta_i = 0$ yields $\nu = 2m - 4$. Thus, two-nucleon forces ($m = 2$) start at $\nu = 0$, three-nucleon forces ($m = 3$) at $\nu = 2$ (but they happen to cancel at that order), and four-nucleon forces at $\nu = 4$ (they don't cancel).

For later purposes, we note that for an irreducible *NN* diagram ($A = 2, C = 1$), the power formula collapses to the very simple expression,

$$\nu = 2L + \sum_i \Delta i.$$

The chief point of the ChPT expansion is that, at a given order ν, there exists only a finite number of graphs. This is what makes the theory calculable. The expression $(Q/\Lambda_\chi)^{\nu+1}$ provides a rough estimate of the relative size of the contributions left out and, thus, of the accuracy at order ν. In this sense, the theory can be calculated to any desired accuracy and has predictive power.

Hierarchy of Nuclear Forces

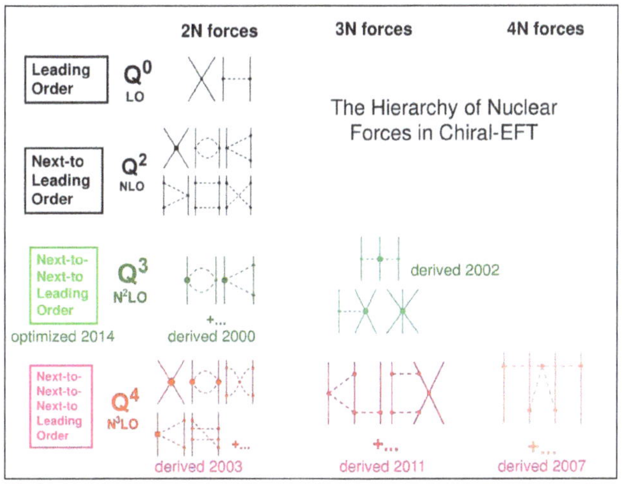

Hierarchy of nuclear forces in ChPT. Solid lines represent nucleons and dashed lines pions. Small dots, large solid dots, solid squares, and solid diamonds denote vertices of index Δi= 0, 1, 2, and 4, respectively.

Chiral perturbation theory and power counting imply that nuclear forces emerge as a hierarchy controlled by the power ν.

In lowest order, better known as leading order (LO, ν = 0), the *NN* potential is made up by two momentum-independent contact terms $(\sim Q^0)$, represented by the four-nucleon-leg graph with a small dot shown in the first row of figure. Furthermore, there is the static one-pion exchange (1PE), second diagram in the first row of the figure. This is, of course, a rather rough approximation to the two-nucleon force (2NF), but accounts already for some important features. The 1PE provides the tensor force, necessary to describe the deuteron, and it explains *NN* scattering in peripheral partial waves of very high orbital angular momentum. At this order, the two contacts, which contribute only in *S*-waves, provide the short- and intermediate-range interaction, which is somewhat crude.

In the next order, $\nu = 1$, all contributions vanish due to parity and time-reversal invariance.

Therefore, the next-to-leading order (NLO) is ν = 2. Two-pion exchange (2PE) occurs for the first time ("leading 2PE") and, thus, the creation of a more sophisticated description of the intermediate-range interaction is starting here. Since the loop involved in each 2PE-diagram implies already $\nu = 2$ [equation $(\nu = 2L + \sum_i \Delta i.)$], the vertices must have $\Delta_i = 0$. Therefore, at this order, only the lowest order πNN and $\pi\pi NN$ vertices are allowed which is why the leading 2PE is rather weak. Furthermore, there are seven new contact terms of $O(Q^2)$, shown in the figure by the four-nucleon-leg graph with a solid square, which contribute in *S* and *P* waves. The operator structure of these contacts include a spin-orbit term besides central, spin-spin, and tensor terms. Thus, essentially all spin-isospin structures necessary to describe the two-nucleon force phenomenologically have been generated. The main deficiency at this stage of development is an insufficient intermediate-range attraction.

This problem is finally fixed at order three (ν = 3), next-to-next-to-leading order (NNLO). The 2PE involves now the two-derivative $\pi\pi NN$ seagull vertices (proportional to the *ci* LECs) denoted by a large solid dot in figure. These vertices represent correlated 2PE as well as intermediate Δ(1232)-isobar contributions. It is well-known from the meson phenomenology of nuclear forces that these two contributions are crucial for a realistic and quantitative 2PE model. Consequently, the 2PE now assumes a realistic size and describes the intermediate-range attraction of the nuclear force about right. Moreover, first relativistic corrections come into play at this order. There are no new contacts, because contacts appear only at even orders.

The reason why we talk of a hierarchy of nuclear forces is that two- and many-nucleon forces are created on an equal footing and emerge in increasing number as we go to higher and higher orders. At NNLO, the first set of nonvanishing three-nucleon forces (3NFs) occur, cf. column '3N Force' of figure. In fact, at the previous order, NLO, irreducible 3N graphs appear already, however, it can be shown that these diagrams all cancel. Since nonvanishing 3NF contributions happen first at order $(Q/\Lambda_\chi)^3$, they are very weak as compared to the 2NF which starts at $(Q/\Lambda_\chi)^0$.

More 2PE is produced at ν = 4, next-to-next-to-next-to-leading order (N³LO), of which we show only a few symbolic diagrams in figure. Two-loop 2PE graphs show up for the first time and so does three-pion exchange (3PE) which necessarily involves two loops. The 3PE is negligible at this order. Most importantly, 15 new contact terms $\sim Q^4$ arise and are represented by the four-nucleon-leg graph with a solid diamond. They include a quadratic spin-orbit term and contribute up to *D*-waves.

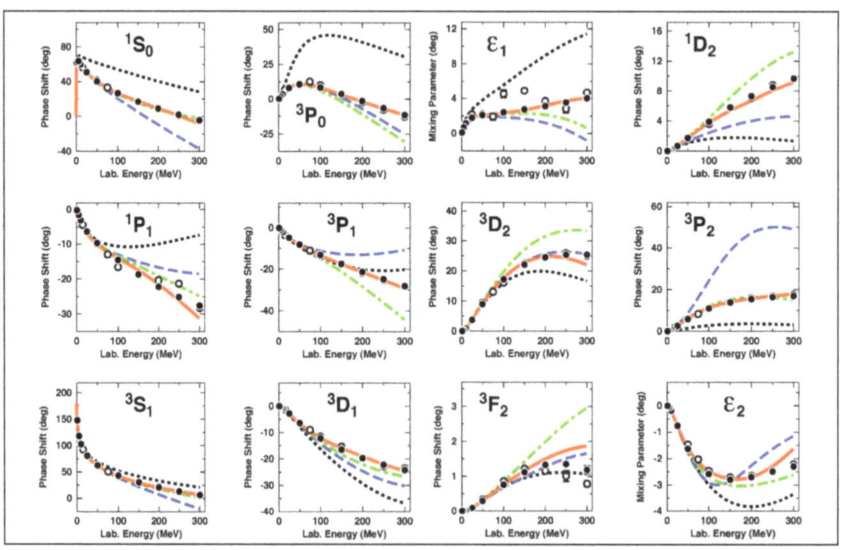

Phase shifts of *np* scattering as calculated from *NN* potentials at different orders of ChPT. The black dotted line is LO, the blue dashed is NLO, the green dash-dotted NNLO, and the red solid N³LO. Partial waves with total angular momentum $J \leq 2$ are displayed. The solid dots and open circles are the results from phase shift analysis.

Mainly due to the larger number of contact terms, a quantitative description of the two-nucleon interaction up to about 300 MeV lab energy is possible, at N³LO. This is demonstrated in figure, where we show the order by order improvement of the *NN* phase shift predictions from LO to

N^3LO. Table quantifies this order by order improvement by providing the χ^2/datum for the fit of *NN* data from NLO to N^3LO.

Table: χ^2/datum for the reproduction of the np data below 290 MeV by chiral NN potentials at NLO, NNLO, and N^3LO. T_{lab} denotes the kinetic energy of the incident nucleon in the laboratory system.

T_{lab} bin (MeV)	number of np data	N^3LO	NNLO	NLO
0--100	1058	1.05	1.7	4.5
100--190	501	1.08	22	100
190--290	843	1.15	47	180
0--290	2402	1.10	20	86

Moreover, there are more 3NF contributions at N^3LO, and four-nucleon forces (4NFs) start at this order. Since the leading 4NFs come into existence one order higher than the leading 3NFs, 4NFs are weaker than 3NFs. Thus, ChPT provides a straightforward explanation for the empirically known fact that $2\,NF \gg 3\,NF \gg 4\,NF$.

Comparison with Conventional Meson Theory

We have now two approaches at hand that can both describe the *NN* interaction quantitatively, which is a non-trivial result. It is then natural to ask, in which way the two approaches differ. There is a clear and revealing answer.

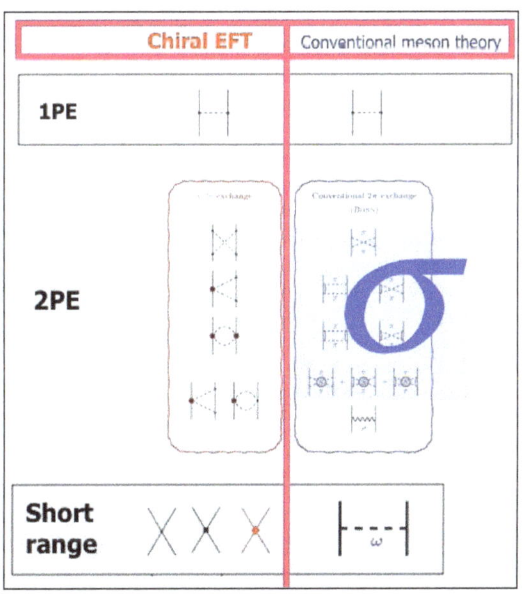

Comparing the chiral EFT approach to the *NN* interaction (left) with conventional meson theory (right).

In chiral EFT, the nuclear potential is expanded in terms of increasing powers of small momenta, $(Q/\Lambda_\chi)^\nu$. In meson theory, the expansion is in terms of Yukawas, $Y(m_\alpha r)$, of increasing masses m_α, corresponding to decreasing ranges $1/m_\alpha$.

Since both approaches describe the same complicated object quantitatively, they should be equivalent to a large extent. This is demonstrated in Figure. First, there is a 1PE in both cases, which is trivial. The 2PE may look diagrammatically quite different, but the figure shows the correspondence between the contributions. The main difference is that, in chiral EFT, the 2PE is build up order by order, while in conventional meson theory it comes as one set. Finally, the short-range contributions appear graphically very different with heavy boson exchange (like ω-exchange) in meson theory and contacts in chiral EFT. However, since $Q \ll m_\omega \approx \Lambda_\chi$, the propagator of a heavy-meson can be expanded into a power series generating contacts of increasing order, as demonstrated in figure.

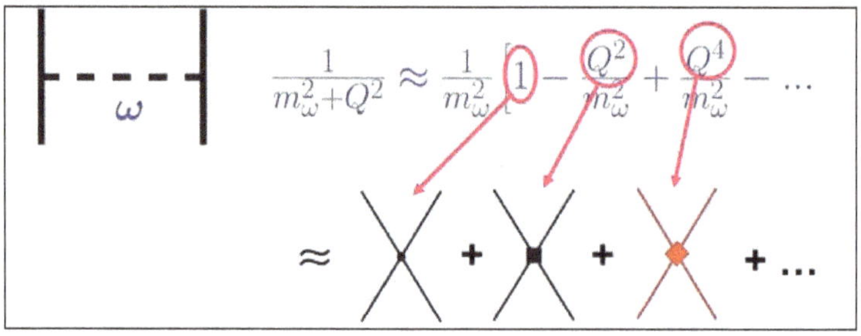

Relationship between heavy boson exchange and the contact terms of chiral EFT.

Although the two approaches can be regarded as equivalent, there are arguments why chiral EFT may be perceived as superior. Chiral EFT:

- Is more closely connected to QCD via chiral symmetry;
- Comes with an organizational scheme (power counting) that allows to estimate the accuracy of the predictions (at a given order);
- Generates two- and many-body forces on an equal footing.

Two-nucleon Forces: Doing the Math

Here we will fill in the mathematical details we left out when presenting the overview over the chiral hierarchy. Up to N^{3}LO, the various irreducible diagrams, which were shown in figure and define the chiral *NN* potential order by order, are given by:

$$\begin{aligned}
V_{LO} &= V_{ct}^{(0)} + V_{1\pi}^{(0)} \\
V_{NLO} &= V_{LO} + V_{ct}^{(2)} + V_{1\pi}^{(2)} + V_{2\pi}^{(2)} \\
V_{NNLO} &= V_{NLO} + V_{1\pi}^{(3)} + V_{2\pi}^{(3)} \\
V_{N^3LO} &= V_{NNLO} + V_{ct}^{(4)} + V_{1\pi}^{(4)} + V_{2\pi}^{(4)} + V_{3\pi}^{(4)}
\end{aligned}$$

where the superscript denotes the order v of the low-momentum expansion. Contact potentials carry the subscript "ct" and pion-exchange potentials can be identified by an obvious subscript.

The charge-independent 1PE potential reads:

$$V_{1\pi}(\vec{p}\,', \vec{p}) = -\frac{g_A^2}{4f_\pi^2}\,\tau_1 \cdot \tau_2\,\frac{\vec{\sigma}_1 \cdot \vec{q} \cdot \vec{\sigma}_2 \cdot \vec{q}}{q^2 + m_\pi^2},$$

Numerical vales for the constants are $f_\pi = 92.4$ MeV, $g_A = 1.29$; and $m_\pi = 138$ MeV for the average pion mass (figure and equation ($\frac{f_{\pi NN}}{m_\pi} = \frac{g_A}{2f_\pi}$)). Since higher order corrections contribute only to mass and coupling constant renormalizations and since, on shell, there are no relativistic corrections, the on-shell 1PE has the same form as in equaton ($V_{1\pi}(\vec{p}\,', \vec{p}) = -\frac{g_A^2}{4f_\pi^2}\tau_1 \cdot \tau_2 \frac{\vec{\sigma}_1 \cdot \vec{q} \cdot \vec{\sigma}_2 \cdot \vec{q}}{q^2 + m_\pi^2}$) at all orders.

The two zero-order contact terms at LO are:

$$V_{ct}^{(0)}(\vec{p}\,', \vec{p}) = C_s + C_T \vec{\sigma}_1 \cdot \vec{\sigma}_2 \cdot$$

To state the mathematical expressions for 2PE contributions, we use the following scheme:

$$\begin{aligned} V_{2\pi}^{(v)}(\vec{p}\,', \vec{p}) = {} & V_C^{(v)} + \tau_1 \cdot \tau_2 W_C^{(v)} \\ & + \left[V_S^{(v)} + \tau_1 \cdot \tau_2 W_S^{(v)}\right] \vec{\sigma}_1 \cdot \vec{\sigma}_2 \\ & + \left[V_{LS}^{(v)} + \tau_1 \cdot \tau_2 W_{LS}^{(v)}\right] \left(-i\vec{S} \cdot (\vec{q} \times \vec{k})\right) \\ & + \left[V_T^{(v)} + \tau_1 \cdot \tau_2 W_T^{(v)}\right] \vec{\sigma}_1 \cdot \vec{q} \cdot \vec{\sigma}_2 \cdot \vec{q} \\ & + \left[V_{\sigma L}^{(v)} + \tau_1 \cdot \tau_2 W_{\sigma L}^{(v)}\right] \vec{\sigma}_1 \cdot (\vec{q} \times \vec{k}) \vec{\sigma}_2 \cdot (\vec{q} \times \vec{k}) \end{aligned}$$

where $\vec{p}\,'$ and $\vec{p}$ denote the final and initial nucleon momenta in the CMS, respectively; moreover,

$\vec{q} \equiv \vec{p}\,' - \vec{p}$ is the momentum transfer,

$\vec{k} \equiv \frac{1}{2}(\vec{p}\,' + \vec{p})$ the average momentum,

$\vec{S} \equiv \frac{1}{2}(\vec{\sigma}_1 + \vec{\sigma}_2)$ the total spin,

and $\vec{\sigma}_{1,2}$ and $\tau_{1,2}$ are the spin and isospin operators, respectively, of nucleon 1 and 2. For on-energy-shell scattering, V_i and $W_i\,(i = C, S, LS, T, \sigma L)$ can be expressed as functions of *q* and *k*, only (with $q \equiv |\vec{q}|$ and $k \equiv |\vec{k}|$).

Using the above scheme, the contribution from the five NLO 2PE diagrams can be stated in an amazingly compact form, namely,

$$W_C^{(2)} = -\frac{L(q)}{384\pi^2 f_\pi^4}\left[4m_\pi^2(5g_A^4 - 4g_A^2 - 1) + q^2(23g_A^4 - 10g_A^2 - 1) + \frac{48g_A^4 m_\pi^4}{w^2}\right]$$

$$V_T^{(2)} = -\frac{1}{q^2}V_s^{(2)} = -\frac{3g_A^4 L(g)}{64\pi^2 f_\pi^4},$$

where,

$$L(q) \equiv \frac{w}{q} \ln \frac{w+q}{2m_\pi},$$

and

$$w \equiv \sqrt{4m_\pi^2 + q^2}\cdot$$

Note that all 2π loops involve a four-dimensional integral, which is divergent. Thus regularization is required. All 2π contributions are obtained by applying dimensional regularization (DR).

The seven NLO contact terms are:

$$\begin{aligned} V_{\text{ct}}^{(2)}(\vec{p}\,',\vec{p}) = & \; C_1\, q^2 + C_2\, k^2 \\ & + \left(C_3\, q^2 + C_4\, k^2\right)\vec{\sigma}_1\cdot\vec{\sigma}_2 \\ & + C_5\left(-i\vec{S}\cdot(\vec{q}\times\vec{k})\right) \\ & + C_6\left(\,\vec{\sigma}_1\cdot\vec{q}\,\right)\left(\,\vec{\sigma}_2\cdot\vec{q}\right) \\ & + C_7\left(\,\vec{\sigma}_1\cdot\vec{k}\,\right)\left(\,\vec{\sigma}_2\cdot\vec{k}\right). \end{aligned}$$

The coefficients C_i used here in the contact potential are, of course, related to the coefficients C_i' that occur in the Lagrangian $\hat{\mathcal{L}}_{NN}^{(2)}$, equation:

$$\begin{aligned} \hat{\mathcal{L}}_{NN}^{(2)} = & -C_1'\left[\left(\bar{N}\vec{\nabla}N\right)^2 + \left(\overline{\vec{\nabla}N}N\right)^2\right] - C_2'\left(\bar{N}\vec{\nabla}N\right)\cdot\left(\overline{\vec{\nabla}N}N\right) - C_3'\,\bar{N}N\left[\bar{N}\vec{\nabla}^2N + \overline{\vec{\nabla}^2N}N\right] \\ & - iC_4'\left[\bar{N}\vec{\nabla}N\cdot(\overline{\vec{\nabla}N}\times\vec{\sigma}N) + (\overline{\vec{\nabla}N})N\cdot(\bar{N}\vec{\sigma}\times\vec{\nabla}N)\right] \\ & - iC_5'\,\bar{N}N(\overline{\vec{\nabla}N}\cdot\vec{\sigma}\times\vec{\nabla}N) + iC_6'(\bar{N}\vec{\sigma}N)\cdot(\overline{\vec{\nabla}N}\times\vec{\nabla}N) \\ & - (C_7'\,\delta_{ik}\delta_{jl} + C_8'\delta_{il}\delta_{kj} + C_9'\delta_{ij}\delta_{kl})\left[\bar{N}\sigma_k\partial_i N\bar{N}\,\sigma_l\,\partial_j N + \overline{\sigma_i N}_{\sigma k}N\overline{\partial_j N}_{\sigma 1}N\right] \\ & - (C_{10}'\,\delta_{ik}\delta_{jl} + C_{11}'\delta_{il}\delta_{kj} + C_{12}'\delta_{ij}\delta_{kl})\,\bar{N}\sigma_k\partial_i N\overline{\partial_j N}_{\sigma 1}N \\ & - \left(\frac{1}{2}C_{13}'\,\delta_{ik}\delta_{jl} + \delta_{il}\delta_{kj} + C_{14}'\delta_{ij}\delta_{kl}\right)\left[\partial_i N\sigma_k\partial_j N + \overline{\partial_j N}_{\sigma k}\partial_i N\right]\bar{N}_{\sigma l}N. \end{aligned}$$

But, the exact relationship is unimportant.

The NNLO 2PE is represented by the following expressions:

$$V_C^{(3)} = \frac{3g_A^2}{16\pi f_\pi^4}\left\{\frac{g_A^2 m_\pi^5}{16M_N w^2} - \left[2m_\pi^2(2c_1 - c_3) - q^2\left(c_3 + \frac{3g_A^2}{16M_N}\right)\right]\tilde{w}^2 A(q)\right\},$$

$$W_C^{(3)} = \frac{g_A^2}{128\pi M_N f_\pi^4}\left\{3g_A^2 m_\pi^5 w^{-2} - \left[4m_\pi^2 + 2q^2 - g_A^2(4m_\pi^2 + 3q^2)\right]\tilde{w}^2 A(q)\right\},$$

$$V_T^{(3)} = \frac{1}{q^2}V_S^{(3)} = \frac{9g_A^4 \tilde{w}^2 A(q)}{512\pi M_N f_\pi^4},$$

$$W_T^{(3)} = -\frac{1}{q^2}W_S^{(3)} = -\frac{g_A^2 A(q)}{32\pi f_\pi^4}\left[\left(c_4 + \frac{1}{4M_N}\right)w^2 - \frac{g_A^2}{8M_N}(10m_\pi^2 + 3q^2)\right],$$

$$V_{LS}^{(3)} = \frac{3g_A^4 \tilde{w}^2 A(q)}{32\pi M_N f_\pi^4},$$

$$W_{LS}^{(3)} = \frac{g_A^2(1-g_A^2)}{32\pi M_N f_\pi^4} w^2 A(q),$$

with,

$$A(q) \equiv \frac{1}{2q}\arctan\frac{q}{2m_\pi}$$

and

$$\tilde{w} \equiv \sqrt{2m_\pi^2 + q^2}\,.$$

This contribution to the 2PE is the crucial one, because it provides an intermediate-range attraction of proper strength. The iso-scalar central potential, $V_C^{(3)}$, is strong and attractive due to the LEC c_3, which is negative and of large magnitude. Via resonance saturation, c_3 is associated with $\pi-\pi$ correlations ('σ meson') and virtual Δ-isobar excitations, which create the most crucial contributions to 2PE in the frame work of conventional meson theory. First relativistic $1/M_N$ corrections come also into play at this order; they are included in the above potential expressions.

The contacts at N³LO are:

$$\begin{aligned} V_{ct}^{(4)}(\vec{p}\,',\vec{p}) = {} & D_1 q^4 + D_2 k^4 + D_3 q^2 k^2 + D_4(\vec{q}\times\vec{k})^2 \\ & + \left(D_5 q^4 + D_6 k^4 + D_7 q^2 k^2 + D_8(\vec{q}\times\vec{k})^2\right)\vec{\sigma}_1\cdot\vec{\sigma}_2 \\ & + \left(D_9 q^2 + D_{10}k^2\right)\left(-i\vec{S}\cdot(\vec{q}\times\vec{k})\right) \\ & + \left(D_{11}q^2 + D_{12}k^2\right)(\vec{\sigma}_1\cdot\vec{q})(\vec{\sigma}_2\cdot\vec{q}) \\ & + \left(D_{13}q^2 + D_{14}k^2\right)(\vec{\sigma}_1\cdot\vec{k})(\vec{\sigma}_2\cdot\vec{k}) \\ & + D_{15}\left(\vec{\sigma}_1\cdot(\vec{q}\times\vec{k})\vec{\sigma}_2\cdot(\vec{q}\times\vec{k})\right). \end{aligned}$$

The 2PE potential at N³LO, $V(4)2\pi$, is very involved, which is why will not give its expressions here. It can be found in the N³LO 3PE contributions, $V_{2\pi}^{(4)}$, are negligible.

The two-nucleon system is characterized by large scattering lengths and a shallow bound states (the deuteron), which cannot be calculated by perturbation theory. Therefore, the *NN* potential must be inserted into a Schrödinger or Lippmann-Schwinger (LS) equation to obtain the *NN* amplitude,

$$\hat{T}(\vec{p}\,',\vec{p}) = \hat{V}(\vec{p}\,',\vec{p}) + \int d^3p''\hat{V}(\vec{p}\,',\vec{p}\,'')\frac{M_N}{p^2 - p''^2 + i\in}\hat{T}(\vec{p}\,'',\vec{p}),$$

where the definitions,

$$\hat{V}(\vec{p}\,',\vec{p}) \equiv \frac{1}{(2\pi)^3}\sqrt{\frac{M_N}{E_{p'}}}\,V(\vec{p}\,',\vec{p})\sqrt{\frac{M_N}{E_p}}$$

and

$$\hat{T}(\vec{p}\,',\vec{p}) \equiv \frac{1}{(2\pi)^3}\sqrt{\frac{M_N}{E_{p'}}}\,T(\vec{p}\,',\vec{p})\sqrt{\frac{M_N}{E_p}},$$

are used, with $E_p \equiv \sqrt{M_N^2 + p^2}$. Iteration of $\hat{V}$ in the LS equation, equation $\hat{T}(\vec{p}\,',\vec{p}) = \hat{V}(\vec{p}\,',\vec{p}) + \int d^3p''\hat{V}(\vec{p}\,',\vec{p}\,'')\frac{M_N}{p^2 - p''^2 + i\in}\hat{T}(\vec{p}\,'',\vec{p})$, requires cutting $\hat{V}$ off for high momenta to avoid infinities. This is consistent with the fact that ChPT is a low-momentum expansion which is valid only for momenta $Q < \Lambda_\chi \approx 1$ GeV. Therefore, the potential $\hat{V}$ is multiplied with a regulator function $f(p',p)$,

$$\hat{V}(\vec{p}\,',\vec{p}) \mapsto \hat{V}(\vec{p}\,',\vec{p})\; f(p',p)$$

with, for example,

$$f(p',p) = \exp\left[-(p'/\Lambda)^{2n} - (p/\Lambda)^{2n}\right].$$

Typical choices for the cutoff parameter Λ that appears in the regulator are,

$$\Lambda \approx \sim 0.5\text{ GeV} < \Lambda_\chi \approx \sim 1\text{ GeV}.$$

It is pretty obvious that results for the $\hat{T}$-matrix may then depend sensitively on the regulator and its cutoff parameter, which is undesirable. The removal of such regulator dependence is known as renormalization. Note that renormalizability is crucial for the validity of an EFT. The quantitative chiral *NN* potentials currently in use apply two regularization schemes: In the derivation of the potential, dimensional regularization is used (where the cutoff is taken to infinity), while in the LS equation, $\hat{T}(\vec{p}\,',\vec{p}) = \hat{V}(\vec{p}\,',\vec{p}) + \int d^3p''\hat{V}(\vec{p}\,',\vec{p}\,'')\frac{M_N}{p^2 - p''^2 + i\in}\hat{T}(\vec{p}\,'',\vec{p})$, the regulator equation ($f(p',p) = \exp\left[-(p'/\Lambda)^{2n} - (p/\Lambda)^{2n}\right]$) is applied with a finite cutoff. This scheme has produced useful *NN* potentials, but the way regularization and renormalization is handled is controversial. In spite of almost two decades of research by a large variety of theoretical physicists, there is still no consensus in the community on how to conduct the renormalization of chiral EFT based nuclear forces in a satisfactory way. In this context, the pionless EFT has turned out to be enlightening, since it allows for a more transparent renormalization procedure (because it consists of contacts only and does not include pion loops).

A related unresolved issue is the proper counting of the powers of the low-energy expansion. Notice that the power given in equation ($\nu = -2+2A-2C+2L+\sum_i \Delta i$) is the one of the perturbatively calculated potential. However, the quantity of physical relevance is the $\hat{T}$-matrix, which is obtained by iterating the potential in the LS equation, equation $\hat{T}(\vec{p}\,',\vec{p}) = \hat{V}(\vec{p}\,',\vec{p}) + \int d^3p''\hat{V}(\vec{p}\,',\vec{p}\,'')\frac{M_N}{p^2 - p''^2 + i\in}\hat{T}(\vec{p}\,'',\vec{p})$, which may change the power. Also here the pionless theory has been helpful since, due to its simplicity, it allows for analytic solutions of the LS equation revealing the power explicitly.

Three-nucleon Forces

In microscopic calculations of nuclear structure and reactions, the 2NF makes, of course, the largest contribution. However, from *ab-initio* studies it is well-known that certain few-nucleon reactions and nuclear structure issues require 3NFs for their precise microscopic explanation. In short, we need 3NFs. As noted before, an important advantage of the EFT approach to nuclear forces is that it creates two- and many-nucleon forces on an equal footing.

For a 3NF, we have A=3 and C=1 and, thus, Eq. ($\nu = -2+2A-2C+2L+\sum_i \Delta i$) implies:

$$\nu = 2 + 2L + \sum_i \Delta_i$$

We will use this equation to analyze 3NF contributions order by order.

The lowest possible power is obviously ν=2 (NLO), which is obtained for no loops (L=0) and only leading vertices ($\sum_i \Delta i = 0$). As it turns out, this contribution vanishes.

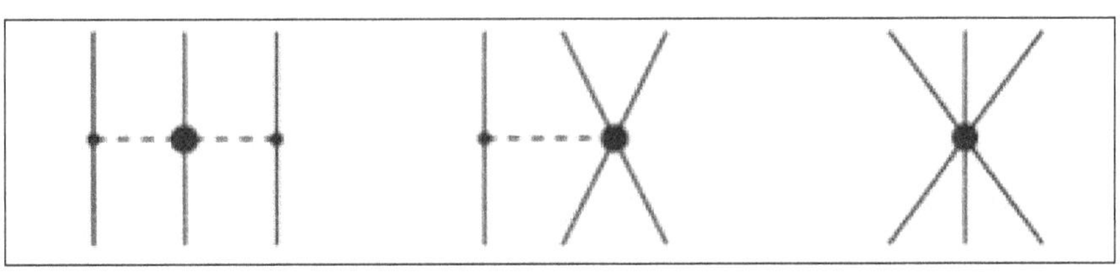

3NF at NNLO (ν=3). From left to right: 2PE, 1PE-contact and contact diagrams.

The first non-vanishing 3NF appears at NNLO (ν=3). The power ν=3 is obtained when there are no loops ($L=0$) and $\sum_i \Delta_i = 1$, i.e., $\Delta_i = 1$ for one vertex while $\Delta_i = 0$ for all other vertices. There are three topologies which fulfill this condition, known as the 2PE, 1PE, and contact graphs.

The 2PE 3N-potential is derived to be,

$$V_{2PE}^{3NF} = \left(\frac{g_A}{2f_\pi}\right)^2 \frac{1}{2} \sum_{i\neq j\neq k} \frac{(\vec{\sigma}_i \cdot \vec{q}_i)(\vec{\sigma}_j \cdot \vec{q}_j)}{(q_i^2 + m_\pi^2)(q_j^2 + m_\pi^2)} F_{ijk}^{ab} \tau_i^a \tau_j^b$$

with $\vec{q}_i \equiv \vec{p}_i' - \vec{p}_i$, where $\vec{p}_i$ and $\vec{p}_i'$ are the initial and final momenta of nucleon i, respectively and,

$$F_{ijk}^{ab} = \delta^{ab}\left[-\frac{4c_1 m_\pi^2}{f_\pi^2} + \frac{2c_3}{f_\pi^2}\vec{q}_i \cdot \vec{q}_j\right] + \frac{c_4}{f_\pi^2}\sum_c \in^{abc} \tau_k^c \vec{\sigma}_k \cdot \left[\vec{q}_i \times \vec{q}_j\right].$$

There are great similarities between this force and derivations of 2PE 3NFs from conventional meson theory.

The other two 3NF contributions are easily derived by taking the last two terms of the Δ=1 Lagrangian, equation ($\widehat{\mathcal{L}}^{\Delta=0} = ...\left(\bar{N}\tau N\right)+...,$), into account. The 1PE contribution is,

$$V_{1\text{PE}}^{3\text{NF}} = -D\frac{g_A}{8f_\pi^2}\sum_{i\neq j\neq k}\frac{\vec{\sigma}_j \cdot \vec{q}_j}{q_j^2 + m_\pi^2}(\tau_i \cdot \tau_j)(\vec{\sigma}_i \cdot \vec{q}_j)$$

and the 3N contact potential reads,

$$V_{\text{ct}}^{3\text{NF}} = E\frac{1}{2}\sum_{j\neq k}\tau_j \cdot \tau_k$$

These 3NF terms involve the two new parameters D and E, which do not appear in the 2N problem. There are many ways to pin these two parameters down. Using the triton binding energy and the nd doublet scattering length and $2a_{\text{nd}}$ is one possibility. One may also choose the binding energies of 3H and 4He or an optimal over-all fit of the properties of light nuclei. Once D and E are fixed, the results for other 3N, 4N, etc. observables are predictions.

The 3NF at NNLO has been applied in calculations of few-nucleon reactions, structure of light- and medium-mass nuclei, and nuclear and neutron matter with a good deal of success. Yet, the problem with the underprediction of the analyzing power of nucleon-deuteron and p-^{3}He scattering, which has become known as the 'A_y puzzle', is not resolved by this 3NF. Furthermore, the spectra of light nuclei leave room for improvement. Therefore, 3NFs of higher orders are needed for at least two reasons: to hopefully resolve outstanding problems in microscopic structure and reactions and for consistency with the 2NF (recall that a precise 2NF is of order N^{3}LO).

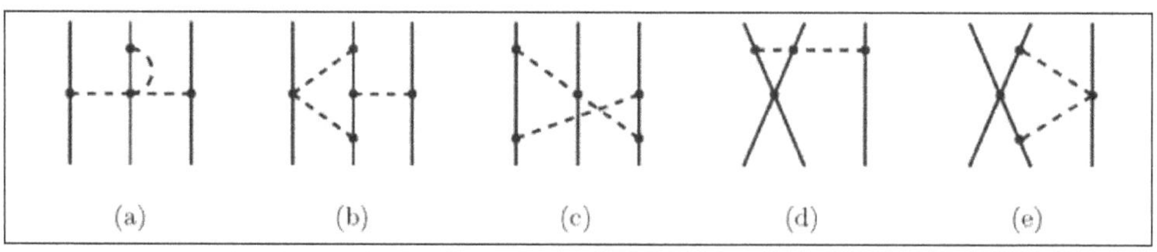

3NF at N3LO (ν=4). We show one representative diagram for each of five topologies, which are: (a) 2PE, (b) 2PE-1PE, (c) ring, (d) 1PE-contact, and (e) 2PE-contact.

The next order is N3LO (ν=4), where we are faced with a very large number of loop diagrams. For those loops, L is one and, therefore, all Δ_i have to be zero to ensure $\nu = 4$. Thus, these one-loop 3NF diagrams can include only leading order vertices, the parameters of which are fixed from πN and NN analysis. There are five loop topologies. In figure we show one sample diagram for each topology. Note, however, that each topology consists of many diagrams such that the total number of diagrams is between 50 and 100, depending on how the diagrams are represented. Preliminary applications of the 3N potentials derived from these diagrams indicate that the N³LO 3NF is fairly weak and does not solve the A_y puzzle.

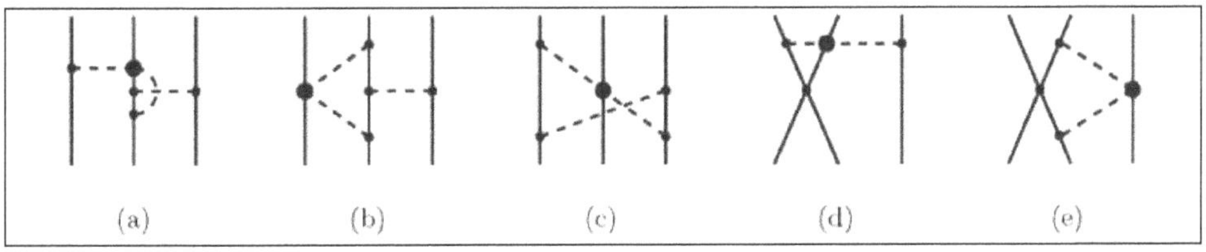

3NF loop contributions at N⁴LO (ν=5). We show one representative diagram for each of five topologies, which are: (a) 2PE, (b) 2PE-1PE, (c) ring, (d) 1PE-contact, and (e) 2PE-contact.

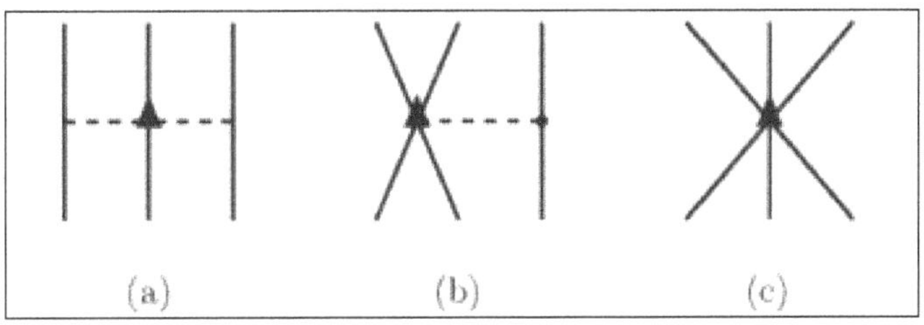

3NF 'tree' graphs at N⁴LO (ν=5) denoted by: (a) 2PE, (b) 1PE-contact, and (c) contact. Solid triangles represent vertices of index Δi=3.

Since we are dealing with a perturbation theory, it is natural to turn to the next order when looking for further improvements. Thus, we proceed to order N⁴LO (ν = 5). The loop contributions that occur at this order are obtained by replacing in the N³LO loops one vertex by a Δi=1 vertex (with LEC *ci*), which is why these loops may be more sizable than the N³LO loops. Again, there are five loop topologies, each of which consists of many diagrams. In addition, we have three 'tree' topologies which include a new set of 3N contact interactions [graph (c)]. Contact terms are typically simple (as compared to loop diagrams) and their coefficients are unconstrained (except for naturalness). The N⁴LO 3NF terms include all possible spin-isospin-momentum structures that a 3NF can have. Thus, there is hope that the 3NF at N⁴LO may provide the missing pieces in the 3NF puzzle. However, a problem is how to deal with the explosion of 3NF contributions that emerge at N³LO and N⁴LO.

Four-nucleon Forces

For connected (C=1) A=4 diagrams, equation ($\nu = -2 + 2A - 2C + 2L + \sum_i \Delta i$) yields:

$$\nu = 4 + 2L + \sum_i \Delta_i$$

Therefore, a connected 4NF appears for the first time at $\nu = 4$ (N^{3}LO), with no loops and only leading vertices. This 4NF includes no new parameters and does not vanish. Some graphs in figure appear to be reducible (iterative). Note, however, that these are Feynman diagrams, which are best analyzed in terms of time-ordered perturbation theory. The various time-orderings include also some irreducible topologies (which are, by definition, 4NFs). Or, in other words, the Feynman diagram minus the reducible part of it yields the (irreducible) contribution to the 4NF.

Assuming a good rate of convergence, a contribution of order $(Q/\Lambda_\chi)^4$ is expected to be rather small. Thus, ChPT predicts 4NF to be essentially insignificant, consistent with experience. Still, nothing is fully proven in physics unless we have performed explicit calculations. The leading

4NF has been applied in a calculation of the ⁴He binding energy, where it was found to contribute about 0.1 MeV. This is small as compared to the full 4He binding energy of 28.3 MeV.

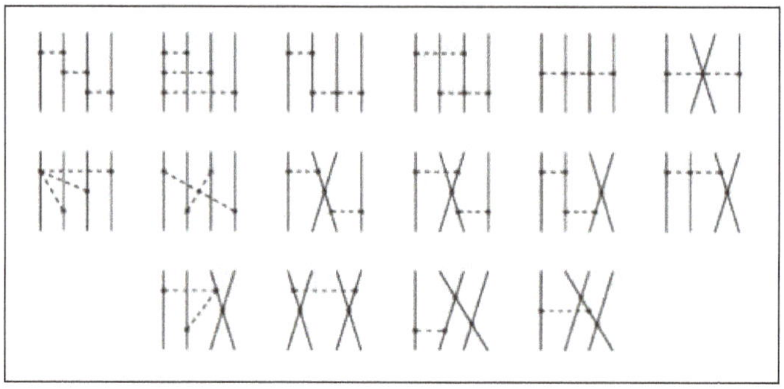

Leading 4NF at N3LO.

Δ-isobar Degrees of Freedom

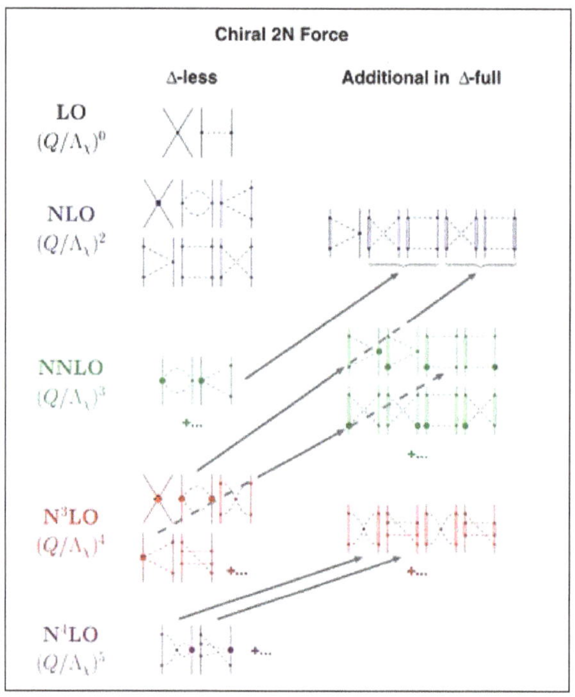

Chiral 2NF without and with Δ-isobar degrees of freedom. Arrows indicate the shift of strength when explicit Δ's are added to the theory. Note that the Δ-full theory consists of the diagrams involving Δ's *plus* the Δ-less ones. Double lines represent Δ-isobars.

The lowest excited state of the nucleon is the Δ(1232) resonance or isobar (a $\pi - N$ P-wave resonance with both spin and isospin 3/2) with an excitation energy of $\Delta M = M\Delta - MN = 293\,MeV$. Because of its strong coupling to the $\pi - N$ system and low excitation energy, it is an important ingredient for models of pion-nucleon scattering in the Δ-region and pion production from the two-nucleon system at intermediate energies, where the particle production proceeds prevailingly through the formation of Δ isobars. At low energies, the more sophisticated conventional models for the 2π-exchange contribution to the *NN* interaction include the virtual excitation of Δ's, which in these models accounts for about 50% of the intermediate-range attraction of the nuclear force as demonstrated by the Bonn potential.

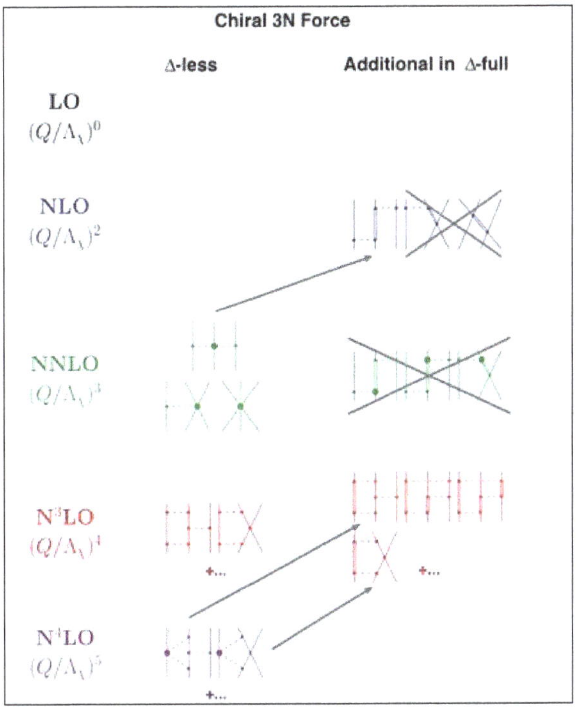

Chiral 3NF without and with Δ-isobar degrees of freedom.

Because of its relatively small excitation energy, it is not clear from the outset if, in an EFT, the Δ should be taken into account explicitly or integrated out as a heavy degree of freedom. If it is included, then $\Delta M \sim m_\pi$ is considered as another small expansion parameter, besides the pion mass and small external momenta. This scheme has become known as the small scale expansion (SSE). Note, however, that this extension is of phenomenological character, since ΔM does not vanish in the chiral limit.

In the chiral EFT discussed so far (also known as the Δ-less theory), the effects due to Δ isobars are taken into account implicitly. Note that the dimension-two LECs, c_i, have unnaturally large values. The reason for this is that the Δ-isobar (and some meson resonances) contribute considerably to the c_i-a mechanism that has become known as resonance saturation. Therefore, the explicit inclusion of the Δ [the so-called Δ-full theory takes strength out of these LECs and moves this strength to a lower order. As a consequence, the convergence of the expansion improves, which is another motivation for introducing explicit Δ-degrees of freedom. In the Δ-less theory, the subleading 2PE and 3PE contributions to the 2NF are larger than the leading ones. The promotion of large contributions by one order in the Δ-full theory fixes this problem.

In analogy to the $NN\pi$ coupling, which is proportional to $g_A / 2f_\pi$ and includes one derivative, the $N\Delta\pi$ coupling is proportional to $h_A / 2f_\pi$ and includes one derivative. The LECs of the πN Lagrangian are usually extracted in the analysis of π-N scattering data and clearly come out differently in the Δ-full theory as compared to the Δ-less one. While in the Δ-less theory the magnitude of the LECs c_3 and c_4 is about 3-5 GeV^{-1}, they turn out to be around 1 GeV^{-1} in the Δ-full theory.

In the 2NF, the virtual excitation of Δ-isobars requires at least one loop and, thus, the contribution occurs first at v=2 (NLO). The consistency between the Δ-full and Δ-less theories has been verified by showing that the contributions due to intermediate Δ-excitations, expanded in powers of $1/\Delta M$, can be absorbed into a redefinition of the LECs of the Δ-less theory. The corresponding shift of the LECs c_3,c_4 is given by,

$$c_3 = -2\,c_4 = \frac{h_A^2}{9\Delta M}.$$

Using $h_A = 3g_A/\sqrt{2}$ (large *Nc* value), almost all of c_3 and an appreciable part of c_4 is explained by the Δ resonance.

Several studies have confirmed that a large amount of the intermediate-range attraction of the 2NF is shifted from NNLO to NLO with the explicit introduction of the Δ-isobar. However, it is also found that the NNLO 2PE potential of the Δ-less theory provides a very good approximation to the NNLO potential in the Δ-full theory.

The Δ isobar also changes the 3NF scenario. The leading 2PE 3NF is promoted to NLO. In the Δ-full theory, this term has the same mathematical form as the corresponding term in the Δ-less theory, equations,

$$V_{2PE}^{3NF} = \left(\frac{g_A}{2f_\pi}\right)^2 \frac{1}{2}\sum_{i\neq j\neq k} \frac{(\vec{\sigma}_i\cdot\vec{q}_i)(\vec{\sigma}_j\cdot\vec{q}_j)}{(q_i^2+m_\pi^2)(q_j^2+m_\pi^2)} F_{ijk}^{ab}\,\tau_i^a\tau_j^b \text{ and}$$

$$F_{ijk}^{ab} = \delta^{ab}\left[-\frac{4c_1m_\pi^2}{f_\pi^2} + \frac{2c_3}{f_\pi^2}\vec{q}_i\cdot\vec{q}_j\right] + \frac{c_4}{f_\pi^2}\sum_c \in^{abc} \tau_k^c\vec{\sigma}_k\cdot\left[\vec{q}_i\times\vec{q}_j\right],$$

provided one chooses c_1=0 and c_3, c_4 according to equation ~($c_3 = -2\,c_4 = \frac{h_A^2}{9\Delta M}$). Note that the other two NLO 3NF terms involving Δ's vanish as a consequence of the antisymmetrisation of the 3N states. The Δ contributions to the 3NF at NNLO vanish at this order, because the subleading $N\Delta\pi$ vertex contains a time-derivative, which demotes the contributions by one order. However, substantial 3NF contributions are expected at N³LO from one-loop diagrams with one, two, or three intermediate Δ-excitations, which correspond to diagrams of order N⁴LO, N⁵LO, and N⁶LO, respectively, in the Δ-less theory.

The inclusion of explicit Δ degrees of freedom does certainly improve the convergence of the chiral expansion by shifting sizable contributions from NNLO to NLO. On the other hand, at NNLO the results for the Δ-full and Δ-less theory are essentially the same. Note that the Δ-full theory consists of the diagrams involving Δ's plus all diagrams of the Δ-less theory. Thus, the Δ-full theory is much more involved. Moreover, in the Δ-full theory, $1/M_N$ 2NF corrections appear at NNLO (not shown in figure), which were found to be uncomfortably large. Thus, it appears that up to NNLO, the Δ-less theory is more manageable.

The situation could, however, change at N³LO where potentially large contributions enter the picture. It may be more efficient to calculate these terms in the Δ-full theory, because in the Δ-less theory they are spread out over N³LO, N⁴LO and, in part, N⁵LO. These higher order contributions are a crucial test for the convergence of the chiral expansion of nuclear forces and represent a challenging topic for future research.

Baryon-baryon Interactions

All baryons interact strongly with each other. Therefore, besides interactions between nucleons, there are many more strong baryon-baryon interactions. Traditionally, one focus has been the forces between nucleons and hyperons (strange baryons) and hyperons and hyperons. Furthermore, the interaction between a baryon and an anti-baryon has drawn considerable interest, for which the nucleon-antinucleon interaction is the most studied example. As in the case of the nucleon-nucleon interaction, the approaches that have been tried to explain the baryon-(anti)baryon interactions include: phenomenology, meson theory, quark models, lattice QCD, and effective field theory.

Nuclear Stability

Nuclear Stability is a concept that helps to identify the stability of an isotope. To identify the stability of an isotope it is needed to find the ratio of neutrons to protons. To determine the stability of an isotope you can use the ratio neutron/proton (N/Z). Also to help understand this concept there is a chart of the nuclides, known as a Segre chart. This chart shows a plot of the known nuclides as a function of their atomic and neutron numbers. It can be observed from the chart that there are more neutrons than protons in nuclides with Z greater than about 20 (Calcium). These extra neutrons are necessary for stability of the heavier nuclei. The excess neutrons act somewhat like nuclear glue.

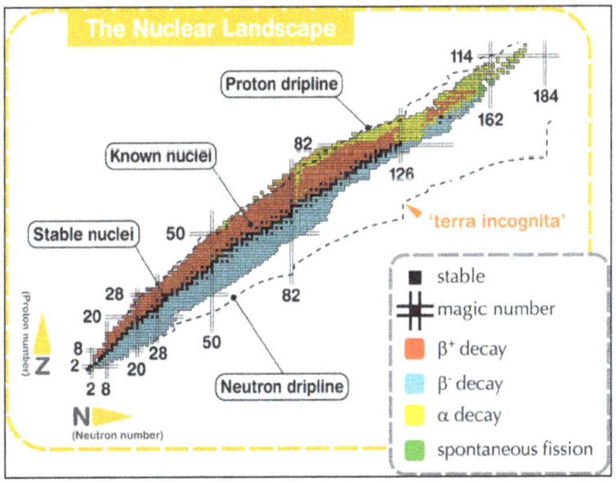

Atomic nuclei consist of protons and neutrons, which attract each other through the nuclear force, while protons repel each other via the electric force due to their positive charge. These two forces compete, leading to various stability of nuclei. There are only certain combinations of neutrons and protons, which forms stable nuclei.

Neutrons stabilize the nucleus, because they attract each other and protons, which helps offset the electrical repulsion between protons. As a result, as the number of protons increases, an increasing ratio of neutrons to protons is needed to form a stable nucleus. If there are too many or too few neutrons for a given number of protons, the resulting nucleus is not stable and it undergoes radioactive decay. Unstable isotopes decay through various radioactive decay pathways, most commonly alpha decay, beta decay, or electron capture. Many other rare types of decay, such as spontaneous fission or neutron emission are known. It should be noted that all of these decay pathways may be accompanied by the subsequent emission of gamma radiation. Pure alpha or beta decays are very rare.

Positive Beta Decay

Nuclei, such as ^{15}O, which are lacking in neutrons (consist of 8 protons and 7 neutrons) undergo positron decay (positive beta decay). In this process, one of the protons in the nucleus is transformed into a neutron, positron and neutrino.The positron and the neutrino are emitted. The number of protons is thus reduced from 8 to 7 (number of neutrons is increased from 7 to 8), so that the resulting nucleus is an isotope of nitrogen, ^{15}N, which is stable.

Negative Beta Decay

On the other hand nuclei, such as ^{19}O, which have excess of neutrons, decay by negative beta decay, emitting a negative electron and an antineutrino. In this process, one of the neutrons in the nucleus is transformed into a proton. The number of protons is thus increased from 8 to 9 (number of neutrons is reduced from 11 to 10), so that the resulting nucleus is an isotope of fluor, ^{19}F, which is stable. It should be noted that in both positive or negative beta decays the atomic mass number remains the same.

Nuclear Stability: Periodic Table

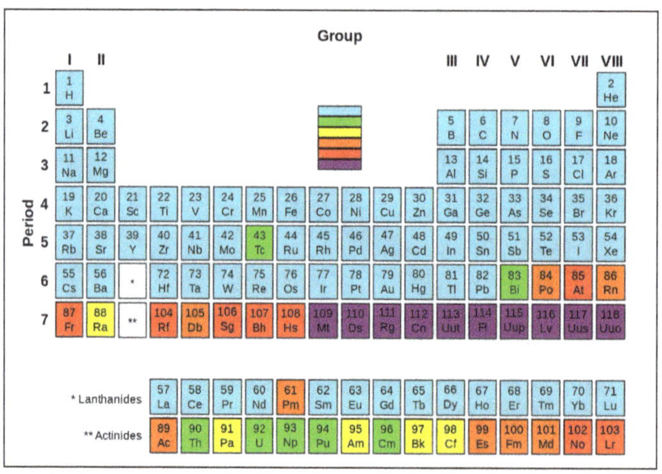

Periodic table with elements colored according to the half-life of their most stable isotope.

Of the first 82 elements in the periodic table, 80 have isotopes considered to be stable. Technetium, promethium and all the elements with an atomic number over 82 are unstable and decompose through radioactive decay. No undiscovered heavy elements (with atomic number over 110) are expected to be stable, therefore lead is considered the heaviest stable element. For each of the 80 stable elements, the number of the stable isotopes is given. For example, tin has 10 such stable isotopes.

There are 80 elements with at least one stable isotope, but 114 to 118 chemical elements are known. All elements to element 98 are found in nature, and the remainder of the discovered elements are artificially produced, with isotopes all known to be highly radioactive with relatively short half-lives.

Bismuth, thorium, uranium and plutonium are primordial nuclides because they have half-lives long enough to still be found on the Earth, while all the others are produced either by radioactive decay or are synthesized in laboratories and nuclear reactors. Primordial nuclides are nuclides found on the Earth that have existed in their current form since before Earth was formed. Primordial nuclides are residues from the Big Bang, from cosmogenic sources, and from ancient supernova explosions which occurred before the formation of the solar system. Only 288 such nuclides are known.

Connection between Nuclear Stability and Radioactive Decay

The nuclei of radioisotopes are unstable. In an attempt to reach a more stable arrangement of its neutrons and protons, the unstable nucleus will spontaneously decay to form a different nucleus. If the number of neutrons changes in the process (number of protons remains), a different isotopes is formed and an element remains (e.g. neutron emission). If the number of protons changes (different atomic number) in the process, then an atom of a different element is formed. This decomposition of the nucleus is referred to as radioactive decay. During radioactive decay an unstable nucleus spontaneosly and randomly decomposes to form a different nucleus (or a different energy state – gamma decay), giving off radiation in the form of atomic partices or high energy rays. This decay occurs at a constant, predictable rate that is referred to as half-life. A stable nucleus will not undergo this kind of decay and is thus non-radioactive.

Magic Numbers

The Octet Rule was formulated from the observation that atoms with eight valence electrons were especially stable (and common). A similar situation applies to nuclei regarding the number of neutron and proton numbers that generate stable (non-radioactive) isotopes. These "magic numbers" are natural occurrences in isotopes that are particularly stable. Table list of numbers of protons and neutrons; isotopes that have these numbers occurring in either the proton or neutron are stable. In some cases there the isotopes can consist of magic numbers for both protons and neutrons; these would be called double magic numbers. The double numbers only occur for isotopes that are heavier, because the repulsion of the forces between the protons. The magic numbers are:

- Proton: 2, 8, 20, 28, 50, 82, 114.
- Neutron: 2, 8, 20, 28, 50, 82, 126, 184.

Also, there is the concept that isotopes consisting a combination of even-even, even-odd, odd-even, and odd-odd are all stable. There are more nuclides that have a combination of even-even than odd-odd. Just like there exist violations to the octet rule, many isotopes with no magic numbers of nucleons are stable.

Table: Distribution of stable and unstable isotopes based on neutron and proton numbers.

Proton number (Z)	Neutron Number	Number of stable Isotopes
Even	Even	163
Even	Odd	53
Odd	Even	50
Odd	Odd	4

Although rare, four stable odd-odd nuclides exist: ${}^{2}_{1}H, {}^{6}_{3}Li, {}^{10}_{5}B, {}^{14}_{7}N$.

Nuclear Spin

The concept of spin is regularly addressed in subatomic particle physics. However, to most people spin seems like an abstract concept. This is due to the fact there is no macroscopic equivalent of what spin is. Electrons are subatomic particles which have spin intrinsic to them. The nucleus is not much different. Spin is just another form of angular momentum. The nucleus consists of protons and neutrons and neutrons and protons are comprised of subatomic particles known as quarks and gluons. The neutron has 2 quarks with a -e/3 charge and one quark with a +2e/3 charge resulting in a total charge of 0. The proton however, has 2 quarks with +2e/3 charge and only one quark with a -e/3 charge giving it a net positive charge. Both protons and neutrons are spin=1/2.

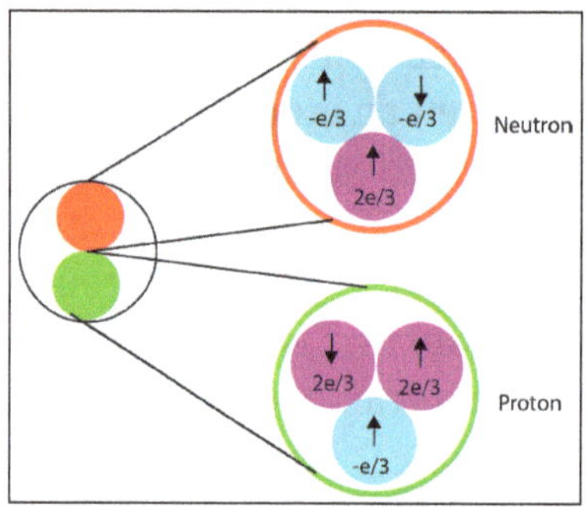

The atomic nucleus (black) of ^{2}H. The proton (green) and neutron (red) are composed of quarks (purple and teal) which have a charge and spin (arrow).

For any system consisting of n multiple parts, each with an angular momentum the total angular momentum can be described by J where,

$$J = |J_1 + J_2 + \ldots + J_n|, |J_1 + J_2 + \ldots + J_n| - 1, \ldots |J_1 - J_2 - \ldots - J_n|$$

$$\alpha = mx^2$$

Here are some examples using the isotopes of hydrogen:

- ^{1}H = 1 proton so J=1/2,
- ^{2}H= 1 proton and 1 neutron so J=1 or 0.

For larger nuclei, it is not immediately evident what the spin should be as there are a multitude of possible values. For the remainder of the discussion we will attribute the spin of the nucleus, I, to be an intrinsic value. There are some rules that the nuclei do follow with respect to nuclear spin. They are summarized in the table below.

Table: General rules for determination of nuclear spin quantum numbers.

Mass Number	Number of Protons	Number of Neutrons	Spin (I)	Example
Even	Even	Even	0	^{16}O(NMR.3)
	Odd	Odd	Integer (1,2,...)	^{2}H(NMR.4)
Odd	Even	Odd	Half-Integer (1/2, 3/2,...)	^{13}C(NMR.5)
	Odd	Even	Half-Integer (1/2, 3/2,...)	^{15}N(NMR.6)

Nuclear Spin Angular Momentum and Quantum Numbers

Spin is a type of angular momentum. Nuclear spin angular momentum was first reported by Pauli. Analogous to the angular momentum commonly encountered in electron, the angular momentum is a vector which can be described by a magnitude L and a direction, m. The magnitude is given by,

$$L = \hbar\sqrt{I(I+1)}$$

The projection of the vector on the z axis (arbitrarily chosen), takes on discretized values according to m, where,

$$m = -I, -I+1, -I+2, \ldots +I$$

The angular momentum along the z-axis is now,

$$I_z = m\hbar$$

Pictorially, this is represented in the figure below for three values of II.

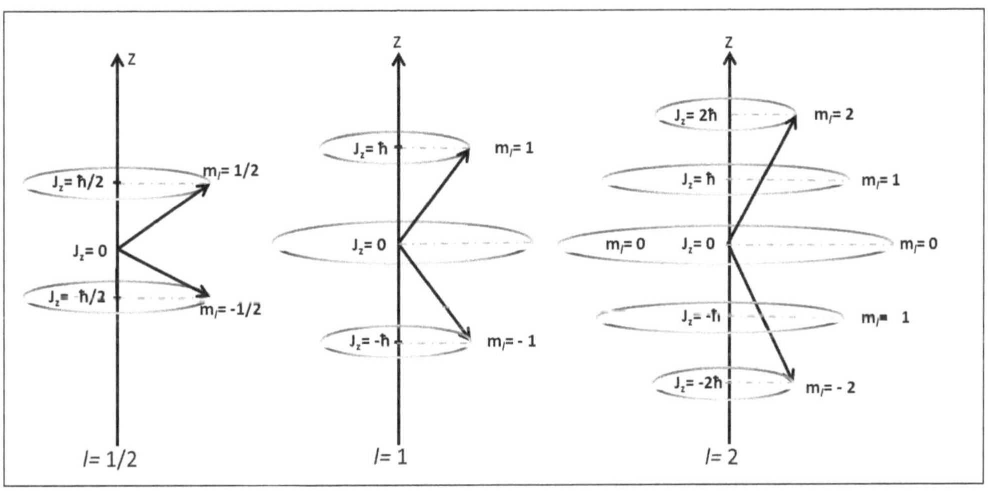

The quantized angular momentum values for a I=1/2 (left) I=1(middle) and I=2 (right). The magnitude is denoted by the arrow while the projection along the z-axis is denoted by the circle.

Table: The quantum numbers of the nucleus are denoted below.

Interaction	Symbol	Quantum Numbers
Nuclear Spin Angular Momentum	I	0<I<9/2 by 1/2
Spin Angular Momentum Magnitude	L	$L = \hbar\sqrt{I(I+1)}$
Spin Angular Momentum Direction	m	$m = -I, -I+1, -I+2, \ldots +I$

Magnetic Moment of a Nucleus

We have now established that the nucleus has spin which can be denoted using specific quantum numbers. As with all charged particles, if the nucleus is moved in a loop it will generate a magnetic field. The magnetic moment μμ is related to the angular momentum of the nucleus by,

$$\mu = \gamma I$$

where γγ is the gyromagnetic ratio, a proportionality constant unique to each nucleus. The table below shows some of the gyromagnetic ratios for some commonly studies nuclei.

Nuclei	ã$10^7 T^{-1} s^{-1}$	v/ MHz	Natural Abundance
^1H	26.75	400	99.985
^2H	4.11	61.4	0.015
^{13}C	6.73	100.6	1.108
^{15}N	-2.71	40.5	0.37
^{19}F	25.18	376.5	100.0
^{31}P	10.84	162.1	100.0

The net or bulk magnetization of the sample is given by M and is the sum of each individual magnetic vector,

$$M = \sum \mu$$

since these magnetic moments are vectors and are randomly aligned, the bulk magnetization arising from the nucleus is zero. There may be unpaired electrons which give rise to paramagnetic, anti ferromagnetic, or ferromagnetic properties. However, if an external magnetic field is applied, the nuclei will align either with or against the field and result in a non-zero bulk magnetization.

Nuclear Energy Levels in a Magnetic Field

We now understand why the nucleus has a magnetic moment associated with it. Now we are getting to the crux of NMR, the use of an external magnetic field. Initially, the nucleus is in the nuclear ground state which is degenerate. The degeneracy of the ground state is 2l+1. The application of a magnetic field splits the degenerate 2l+1 nuclear energy levels. The energy of a particular level is:

$$E=-\mu \cdot Bo,$$

where B_o is the external magnetic field. Along the z-direction, which we assume the magnetic field is applied,

$$E= - \mu B_o$$

by substitution,

$$E = - m\hbar\gamma B_o$$

The magnitude of the splitting therefore depends on the size of the magnetic field. In most labs this magnetic field is somewhere between 1 and 21T. Those spins which align with the magnetic field are lower in energy, while those that align against the field are higher in energy.

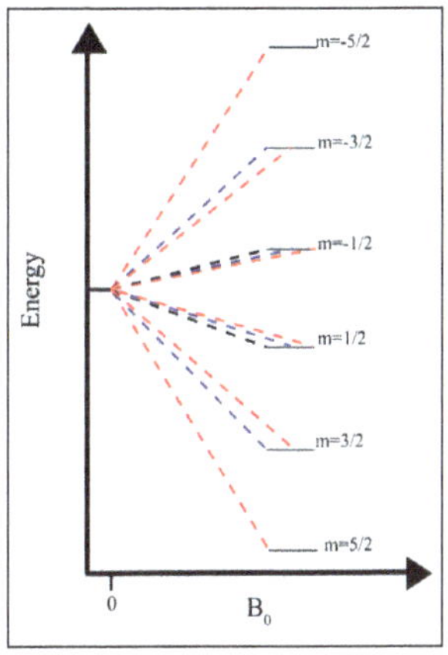

Splitting of the energy levels for a I=1/2 (black dashed lines), I= 3/2 (blue dashed lines), and I=5/2 (red dashed lines). Note how the energy level is dependent on the applied magnetic field. The offset of the red and blue lines is for illustrative purposes only.

Energy Level Spin Distribution

In the absence of a magnetic field the magnetic dipoles are oriented randomly and there is no net magnetization (vector sum of μ is zero). Application of an external magnetic field, as was shown above, creates distinct energy levels based on the spin angular momentum of the nucleus. Each energy level is populated by the spins which have the same angular momentum. To illustrate this, consider a I=1/2 system. There are two energy levels, +1/2 and -1/2, which are populated by spins that have aligned against or with the external magnetic field, respectively.

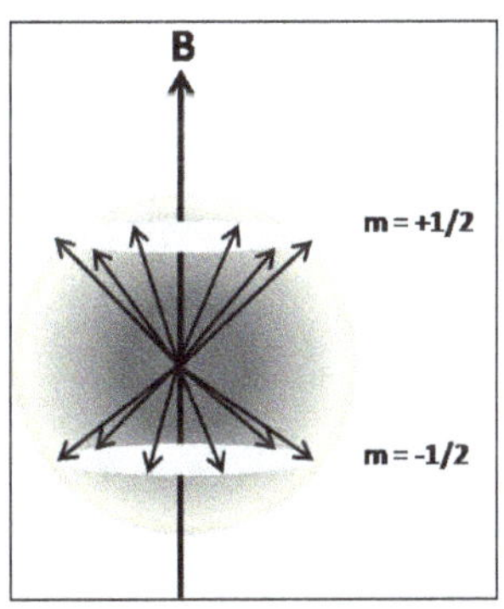

Spins configurations according to applied magnetic field.

The energy separation between these states is relatively small and the energy from thermal collisions is sufficient to place many nuclei into higher energy spin states. The number of nuclei in each spin state can be described by the Boltzmann distribution. The Boltzmann equation expresses the relationship between temperature and the related energy as shown below.

$$\frac{N_{\text{upper}}}{N_{\text{lower}}} = e^{\frac{-\Delta E}{kT}} = e^{\frac{-hv}{kT}}$$

Where N_{upper} and N_{lower} represent the population of nuclei in upper and lower energy states, E is the energy difference between the spin states, k is the Boltzmann constant (1.3805 x 10-23 J/Kelvin) and T is the temperature in K. At room temperature, the number of spins in the lower energy level, N_{lower}, slightly outnumbers the number in the upper level, N_{upper}.

Selection Rules

The selection rule in NMR is,

$$\Delta m = \pm 1$$

For a nucleus with I=1/2 there is only one allowed transition. For nuclei with I>1/2, there are multiple transitions which can take place. Consider the case of I=3/2. The following transitions can take place,

$$-\frac{3}{2} \leftrightarrow -\frac{1}{2}$$

$$-\frac{1}{2} \leftrightarrow \frac{1}{2}$$

$$\frac{1}{2} \leftrightarrow \frac{3}{2}$$

which is illustrated below.

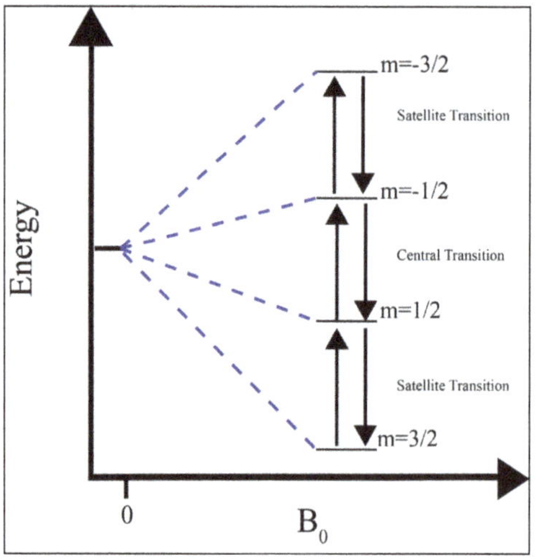

The transition from,

$$-\frac{3}{2} \leftrightarrow -\frac{1}{2}$$

$$\frac{1}{2} \leftrightarrow \frac{3}{2}$$

are known as satellite transitions, while the,

$$-\frac{1}{2} \leftrightarrow \frac{1}{2}$$

transition is known as the central transition. The central transition is primarily observed in an NMR experiment.

References

- Nuclear-physics, entry: newworldencyclopedia.org, Retrieved 29 June, 2019
- Atomic-nucleus, ar-bo: chemistryexplained.com, Retrieved 30 April, 2019
- Proton, entry: newworldencyclopedia.org, Retrieved 11 January, 2019
- Neutron, entry: newworldencyclopedia.org, Retrieved 14 July, 2019
- What-is-nuclear-size, nuclear-and-radiation-chemistry, chemistry: thebigger.com, Retrieved 18 April, 2019
- Nuclear-radius, radioactivity, a-level-physics-as-a2: physicsnet.co.uk, Retrieved 7 May, 2019
- Density-nuclear-matter, what-is-density-physics, thermodynamic-properties, thermodynamics, nuclear-engineering: nuclear-power.net, Retrieved 21 July, 2019
- Nuclear-binding-energy-and-mass-defect, introchem: lumenlearning.com, Retrieved 13 July, 2019
- Nuclear-stability, atomic-nuclear-physics, reactor-physics, nuclear-power: nuclear-power.net, Retrieved 24 March, 2019

2

Nuclear Models

A nuclear model is a theoretical description which is used to predict the properties of nuclei. Some of the major nuclear models are nuclear shell model, liquid-drop model and collective model. This chapter closely examines these nuclear models to provide an extensive understanding of the subject.

Nuclear model refers to any of several theoretical descriptions of the structure and function of atomic nuclei (the positively charged, dense cores of atoms). Each of the models is based on a plausible analogy that correlates a large amount of information and enables predictions of the properties of nuclei.

Nuclear models can be classified into two main groups. In those of the first group, called independent-particle models, the main assumption is that little or no interaction occurs between the individual particles that constitute nuclei; each proton and neutron moves in its own orbit and behaves as if the other nuclear particles were passive participants. The shell nuclear model and its variations fall into this group.

In a second group, called strong-interaction, or statistical models, the main assumption is that the protons and neutrons are mutually coupled to each other and behave cooperatively in a way that reflects the short-ranged strong nuclear force between them. The liquid-drop model and compound-nucleus model are examples of this group.

Other nuclear models incorporate aspects of both groups, such as the collective model, which is a combination of the shell model and the liquid-drop model.

Nuclear Shell Model

In nuclear physics, the nuclear shell model is a theoretical model to describe the atomic nucleus. The nuclear shell model was proposed by Dmitry Ivanenko in 1932 and further developed independently by several physicists such as Maria Goeppert-Mayer, Eugene Paul Wigner and J. Hans D. Jensen in 1949. It must be noted this model is based on the Pauli exclusion principle to describe the structure of the nucleus in terms of energy levels.

The nuclear shell model is partly analogous to the atomic shell model which describes the arrangement of electrons in an atom, in that a filled shell results in greater stability. Nucleons are added to shells which increase with energy that orbit around a central potential. In the atomic shell model the central potential around which the electrons orbit is generated by the nucleus. Nucleons are added to shells which increase with energy that orbit around a central potential.

In comparison to atomic shell model, the atomic nucleus governed by two different forces. The residual strong force, also known as the nuclear force, acts to hold neutrons and protons together in nuclei. In nuclei, this force acts against the enormous repulsive electromagnetic force of the protons. The term residual is associated with the fact, it is the residuum of the fundamental strong interaction between the quarks that make up the protons and neutrons. The strong interaction is very complicated interaction, because it significantly varies with distance. At distances comparable to the diameter of a proton, the strong force is approximately 100 times as strong as electromagnetic force. At smaller distances, however, the strong force between quarks becomes weaker, and the quarks begin to behave like independent particles. In particle physics, this effect is known as asymptotic freedom.

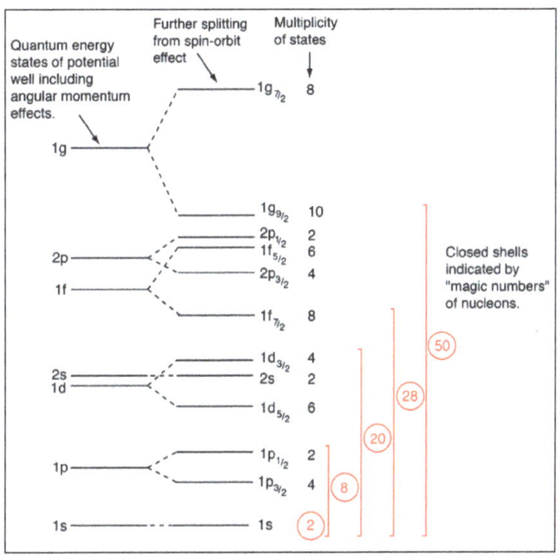

With the enormous strong force acting between individual nucleons and with so many nucleons to collide with, how can nucleons orbit a central potential without interacting? This problem is explained by the Pauli exclusion principle, which states that two fermions cannot occupy the same quantum state. In other words, the interaction will not occur, if the higher energy shells are fully occupied and the energy imparted to the nucleon during the collision is insufficient to promote the nucleon to an unfilled orbit. As a result, the nucleons orbit independent of one another.

The nuclear shell model was able to describe many phenomena like the magic numbers, the ground state spin and parity etc.

Simple Shell Model

The basic assumption of the shell model is that the effects of internuclear interactions can be represented by a single-particle potential. One might think that with very high density and strong forces, the nucleons would be colliding all the time and therefore cannot maintain a single-particle orbit. But, because of Pauli exclusion the nucleons are restricted to only a limited number of allowed orbits. A typical shell-model potential is,

$$V(r) = -\frac{V_o}{1+\exp[(r-R)/a]}$$

where typical values for the parameters are $V_o \sim 57$ Mev, $R \sim 1.25A^{1/3}$ F, a ~ 0.65 F. In addition one can consider corrections to the well depth arising from: (i) symmetry energy from an unequal number of neutrons and protons, with a neutron being able to interact with a proton in more ways than n-n or p-p (therefore n-p force is stronger than n-n and p-p), and (ii) Coulomb repulsion. For a given spherically symmetric potential V(r), one can examine the bound-state energy levels that can be calculated from radial wave equation for a particular orbital angular momentum ℓ,

$$-\frac{\hbar}{2m}\frac{d^2u_\ell}{dr^2}+\left[\frac{\ell(\ell+1)\hbar^2}{2mr^2}+V(r)\right]u_\ell(r)=Eu_\ell(r)$$

Figure shows the energy levels of the nucleons for an infinite spherical well and a (harmonic oscillator potential, $V(r) = m\omega^2 r^2/2$. While no simple formulas can be given for the former, for the latter one has the expression,

$$E_v = \hbar\omega(v+3/2) = \hbar\omega(n_x+n_y+n_z+3/2)$$

Where v = 0, 1, 2, ..., and n_x, n_y, n_z = 0, 1, 2, ... are quantum numbers. One should notice the degeneracy in the oscillator energy levels. The quantum number v can be divided into radial quantum number n (1, 2, ...) and orbital quantum numbers ℓ1 (0, 1, ...) as shown in below figure. One can see from these results that a central force potential is able to account for the first three magic numbers, 2, 8, 20, but not the remaining four, 28, 50, 82, 126. This situation does not change when more rounded potential forms are used. The implication is that something very fundamental about the single-particle interaction picture is missing in the description.

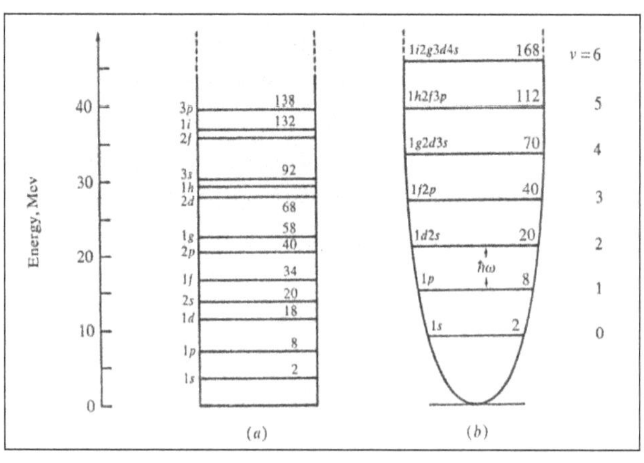

Figure above shows energy levels of nucleons in: (a) infinite spherical well (range R = 8F) and (b) a parabolic potential well. In the spectroscopic notation (n, ℓ), n refers to the number of times the orbital angular momentum state ℓ has appeared. Also shown at certain levels are the cumulative number of nucleons that can be put into all the levels up to the indicated level.

Shell Model with Spin-orbit Coupling

An essential missing piece is an attractive interaction between the orbital angular momentum and the intrinsic spin angular momentum of the nucleon. To take into account this interaction we add a term to the Hamiltonian H,

$$H = \frac{p^2}{2m} + V(r) + V_{so}(r)\underline{s} \cdot \underline{L}$$

where V_{so} is another central potential (known to be attractive). This modification means that the interaction is no longer spherically symmetric; the Hamiltonian now depends on the relative orientation of the spin and orbital angular momenta. It is beyond the scope of this class to go into the bound-state calculations for this Hamiltonian. In order to understand the meaning of the results of such calculations (eigenvalues and eigenfunctions) we need to digress somewhat to discuss the addition of two angular momentum operators.

The presence of the spin-orbit coupling term in ($H = \frac{p^2}{2m} + V(r) + V_{so}(r)\underline{s} \cdot \underline{L}$) means that we will have a different set of eigenfunctions and eigenvalues for the new description. What are these new quantities relative to the eigenfunctions and eigenvalues we had for the problem without the spin-orbit coupling interaction? We first observe that in labeling the energy levels in figure we had already taken into account the fact that the nucleon has an orbital angular momentum (it is in a state with a specified ℓ), and that it has an intrinsic spin of ½ (in unit of $\hbar$). For this reason the number of nucleons that we can put into each level has been counted correctly. For example, in the 1s ground state one can put two nucleons, for zero orbital angular momentum and two spin orientations (up and down). For a state of given ℓ, the number of nucleons that can go into that state is $2(2\ell + 1)$. This comes about because the eigenfunctions we are using to describe the system is a representation that diagonalizes the orbital angular momentum operator L^2, its z-component, L_z, the intrinsic spin angular momentum operator S?, and its z-component S_z. Let us use the following notation to label these eigenfunctions (or representation),

$$|\ell, m_\ell, s, m_s\rangle \equiv Y_\ell^{m_\ell}\ \chi_s^{m_s}$$

where is the spherical harmonic we encountered in Lec4, and we know it is the eigenfunction of the orbital angular momentum operator L^2 (it is also the eigenfunction of L_z). The function $\chi_s^{m_s}$ is the spin eigenfunction with the expected properties,

$$S^2\ \chi_s^{m_s} = s(s+1)\hbar^2 \chi_s^{m_s},\ \ s = 1/2$$

$$S_z\ \chi_s^{m_s} = m_s \hbar \chi_s^{m_s}, \qquad -s \le m_s \le s$$

The properties of $\chi_s^{m_s}$ with respect to operations by S^2 and S_z completely mirror the properties of $Y_\ell^{m_\ell}$ with respect to L^2 and L_z. Going back to our representation ($|\ell, m_\ell, s, m_s\rangle \equiv Y_\ell^{m_\ell}\ \chi_s^{m_s}$) we will see that the eigenfunction is a "ket" with indices which are the good quantum numbers for the problem, namely, the orbital angular momentum and its projection (sometimes called the magnetic quantum number m, but here we use a subscript to denote that it goes with the orbital angular momentum), the spin (which has the fixed value of ½) and its projection (which can be +1/2 or -1/2).

The representation given in ($|\ell, m_\ell, s, m_s\rangle \equiv Y_\ell^{m_\ell}\, \chi_s^{m_s}$) is no longer a good representation when the spin-orbit coupling term is added to the Hamiltonian. It turns out that the good representation is just a linear combination of the old representation. It is sufficient for our purpose to just know this, without going into the details of how to construct the linear combination. To understand the properties of the new representation we now discuss angular momentum addition.

The two angular momenta we want to add are obviously the orbital angular momentum operator $\underline{L}$ and the intrinsic spin angular momentum operator $\underline{S}$, since they are the only angular momentum operators in our problem. Why do we want to add them? The reason lies in ($H = \frac{p^2}{2m} + V(r) + V_{so}(r)\underline{s} \cdot \underline{L}$). Notice that if we define the total angular momentum as:

$$\underline{j} = \underline{S} + \underline{L}$$

we can then write,

$$\underline{S} \cdot \underline{L} = \left(j^2 - S^2 - L^2\right)/2$$

So the problem of diagonalizing ($H = \frac{p^2}{2m} + V(r) + V_{so}(r)\underline{s} \cdot \underline{L}$) is the same as diagonalizing j^2, S^2 and L^2. This is then the basis for choosing our new representation. In analogy to ($|\ell, m_\ell, s, m_s\rangle \equiv Y_\ell^{m_\ell}\, \chi_s^{m_s}$) we will denote the new eigenfunctions by $|jm_j\, \ell s\rangle$, which has the properties:

$$j^2 \left|jm_j\, \ell s\right\rangle = j(j+1)\hbar^2 \left|jm_j\, \ell s\right\rangle, \quad |\ell - s| \le j \le \ell + s$$

$$j_z \left|jm_j\, \ell s\right\rangle = m_j \hbar \left|jm_j\, \ell s\right\rangle, \quad -j \le m_j \le j$$

$$L^2 \left|jm_j\, \ell s\right\rangle = \ell(\ell+1)\hbar^2 \left|jm_j\, \ell s\right\rangle, \quad \ell = 0, 1, 2, \ldots$$

$$S^2 \left|jm_j\, \ell s\right\rangle = s(s+1)\hbar^2 \left|jm_j\, \ell s\right\rangle, \quad s = \tfrac{1}{2}$$

In ($j^2 \left|jm_j\, \ell s\right\rangle = j(j+1)\hbar^2 \left|jm_j\, \ell s\right\rangle, \quad |\ell - s| \le j \le \ell + s$) we indicate the values that j can take for given ℓ and s (=1/2), the lower (upper) limit corresponds to when $\underline{S}$ and $\underline{L}$ are antiparallel (parallel) as shown in the sketch.

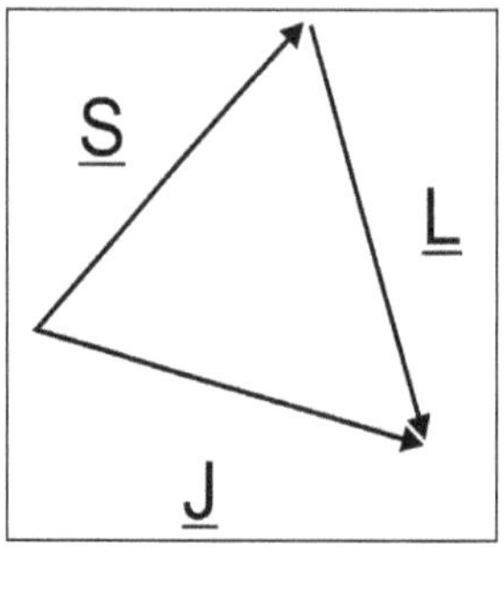

Returning now to the energy levels of the nucleons in the shell model with spin-orbit coupling we can understand the conventional spectroscopic notation where the value of j is shown as a subscript.

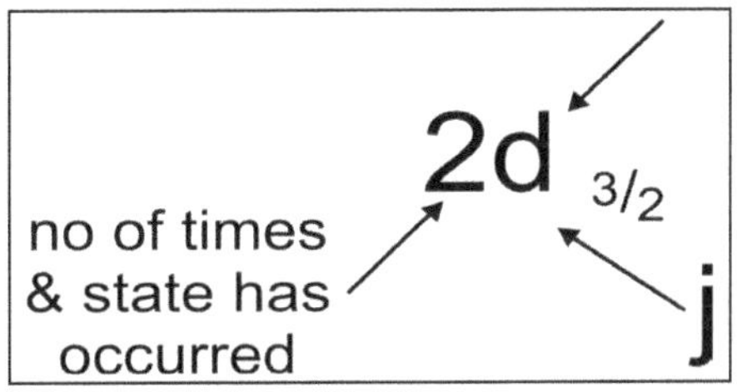

This is then the notation in which the shell-model energy levels are displayed in figure.

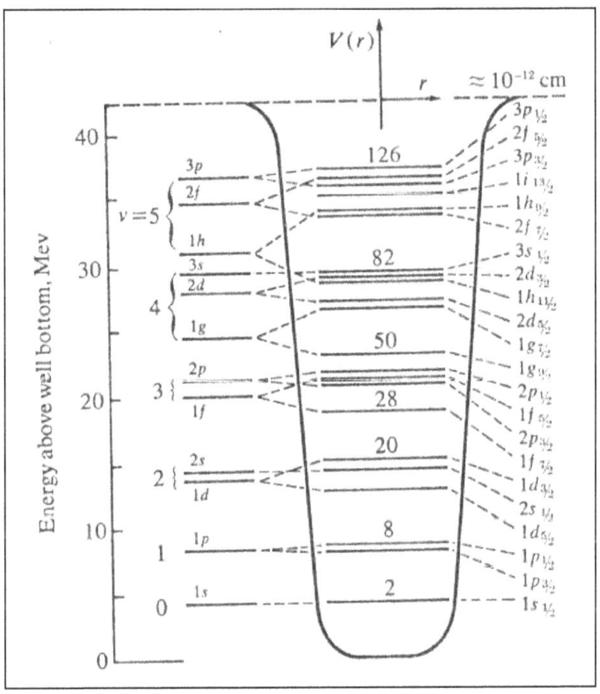

Energy levels of nucleons in a smoothly varying potential well with a strong spin-orbit coupling term.

For a given (n, ℓ, j) level, the nucleon occupation number is 2j+1. It would appear that having 2j+1 identical nucleons occupying the same level would violate the Pauli exclusion principle. But this is not the case since each nucleon would have a distinct value of m_j (this is why there are 2j+1 values of mj for a given j).

We see in figure the shell model with spin-orbit coupling gives a set of energy levels having breaks at the seven magic numbers. This is considered a major triumph of the model, for which Mayer and Jensen were awarded the Noble prize in physics. For our purpose we will use the results of the shell model to predict the ground-state spin and parity of nuclei.

1. The shell model is most useful when applied to closed-shell or near closed-shell nuclei.
2. Away from closed-shell nuclei collective models taking into account the rotation and vibration of the nucleus are more appropriate.

3. Simple versions of the shell model do not take into account pairing forces, the effects of which are to make two like-nucleons combine to give zero orbital angula momentum.

4. Shell model does not treat distortion effects (deformed nuclei) due to the attraction between one or more outer nucleons and the closed-shell core. When the nuclear core is not spherical, it can exhibit "rotational" spectrum.

Prediction of Ground-state Spin and Parity

There are three general rules for using the shell model to predict the total angular momentum (spin) and parity of a nucleus in the ground state. These do not always work, especially away from the major shell breaks.

1. Angular momentum of odd-A nuclei is determined by the angular momentum of the last nucleon in the species (neutron or proton) that is odd.
2. Even-even nuclei have zero ground-state spin, because the net angular momentum associated with even N and even Z is zero, and even parity.
3. In odd-odd nuclei the last neutron couples to the last proton with their intrinsic spins in parallel orientation.

To illustrate how these rules work, we consider an example for each case. Consider the odd-A nuclide Be^9 which has 4 protons and 5 neutrons. Since the last nucleon is the fifth neutron, we see in figure that this nucleon goes into the state $1p_{3/2}\left(\ell=1, j=3/2\right)$. Thus we would predict the spin and parity of this nuclide to be $3/2^-$. For an even-even nuclide we can take A^{36}, with 18 protons and neutrons, or Ca^{40}, with 20 protons and neutrons. For both cases we would predict spin and parity of 0^+. For an odd-odd nuclide we take Cl^{38}, which has 17 protons and 21 neutrons. In figure we see that the 17th proton goes into the state $1d_{3/2}\left(\ell=2, j=3/2\right)$, while the 21st neutron goes into the state $1f_{7/2}\left(\ell=3, j=7/2\right)$. From the ℓ and j values we know that for the last proton the orbital and spin angular momenta are pointing in opposite direction (because j is equal to ℓ -1/2). For the last neutron the two momenta are point in the same direction (j = ℓ +1/2). Now the rule tells us that the two spin momenta are parallel, therefore the orbital angular momentum of proton is pointing in the opposite direction from the orbital angular momentum of the neutron, with the latter in the same direction as the two spins. Adding up the four angular momenta, we have +3+1/2+1/2-2 = 2. Thus the total angular momentum (nuclear spin) is 2. What about the parity? The parity of the nuclide is the product of the two parities, one for the last proton and the other for the last neutron. Recall that the parity of a state is determined by the orbital angular momentum quantum number $\ell, \pi=\left(-1\right)^{\ell}$. So with proton in a state with ℓ = 2, its parity is even, while the neutron in a state with ℓ = 3 has odd parity. The parity of the nucleus is therefore odd. Our prediction for Cl^{38} is then 2^-.

Potential Wells for Neutrons and Protons

The qualitative features of the potential wells for neutrons and protons. If we exclude the Coulomb interaction for the moment, then the well for a proton is known to be deeper than that for a neutron. The reason is that in a given nucleus usually there are more neutrons than protons, especially for the heavy nuclei, and the n-p interactions can occur in more ways than either the n-n or p-p interactions on account of the Puali exclusion principle. The difference in well depth ΔV_s is called the symmetry s energy; it is approximately given by,

$$\Delta V_s = \pm 27 \frac{(N-Z)}{A} \text{Mev}$$

where the (+) and (-) signs are for protons and neutrons respectively. If we now consider the Coulomb repulsion between protons, its effect is to raise the potential for a proton. In other words, the Coulomb effect is a positive contribution to the nuclear potential which is larger at the center than at the surface.

Combining the symmetry and the Coulomb effects we have a sketch of the potential for a neutron and a proton as indicated in figure. One can also estimate the potential for a neutron and a proton.

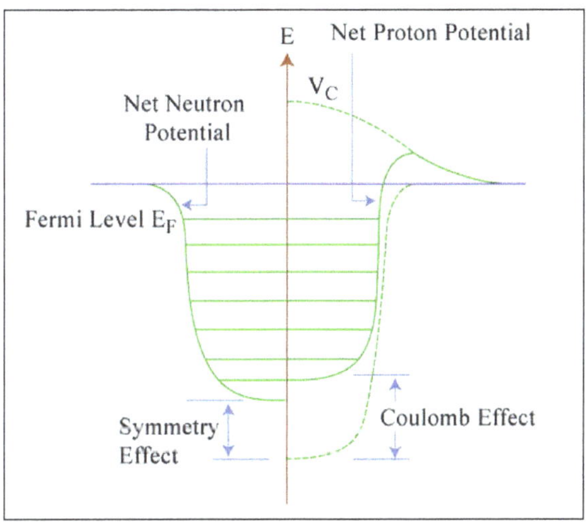

Schematic showing the effects of symmetry and Coulomb interactions on the potential for a neutron and a proton.

Well depth in each case using the Fermi Gas model. One assumes the nucleons of a fixed kind behave like a fully degeneragte gas of fermions (degeneracy here means that the states are filled continuously starting from the lowest energy state and there are no unoccupied states below the occupied ones), so that the number of states occupied is equal to the number of nucleons in the particular nucleus. This calculation is carried out separately for neutrons and protons. The highest energy state that is occupied is called the Fermi level, and the magnitude of the difference between this state and the ground state is called the Fermi energy E_F. It turns out that E_F is proportional to $n^{2/3}$, where n is the number of nucleons of a given kind, therefore E_F (neutron) > E_F (proton). The sum of EF and the separation energy of the last nucleon provides an estimate of the well depth. (The separation energy for a neutron or proton is about 8 Mev for many nuclei.) Based on these considerations one obtains the results shown in figure.

We have so far considered only a spherically symmetric nuclear potential well. (We know there is in addition a cen 2 trifugal contribution of the form $\ell(\ell+1)\hbar^2/2mr^2$ and a spin-orbit contribution. As a result of the former the well becomes narrower and shallower for the higher orbital angular momentum states. Since the spin-orbit coupling is attractive, its effect depends on whether $\underline{S}$ is parallel or anti-parallel to $\underline{L}$. The effects are illustrated in figures. Notice that for $\ell = 0$ both are absent.

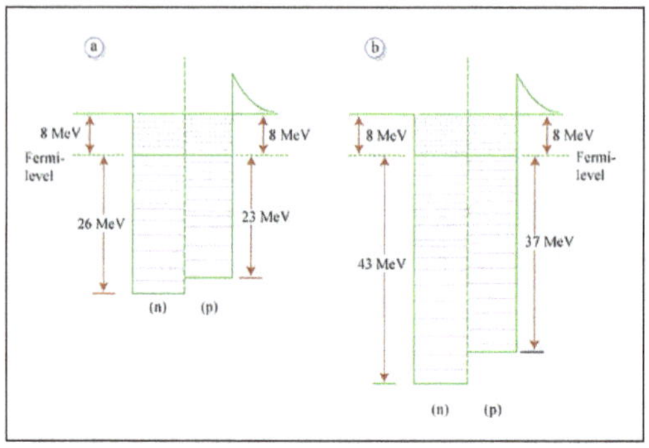

Nuclear potential wells for neutrons and protons according to the Fermi-gas model, assuming the mean binding energy per nucleon to be 8 Mev, the mean relative nucleon admixture to be $N/A \sim 1/1.8m \;\; Z/A \sim 1/2.2$, and a range of 1.4 F (a) and 1.1 F (b).

In addition to the bound states in the nuclear potential well there exist also virtual states (levels) which are not positive energy states in which the wave function is large within the potential well. This can happen if the deBroglie wavelength is such that approximately standing waves are formed within the well. (Correspondingly, the reflection coefficient at the edge of the potential is large.) A virtual level is therefore not a bound state; on the other hand, there is a non-negligible probability that inside the nucleus a nucleon can be found in such a state.

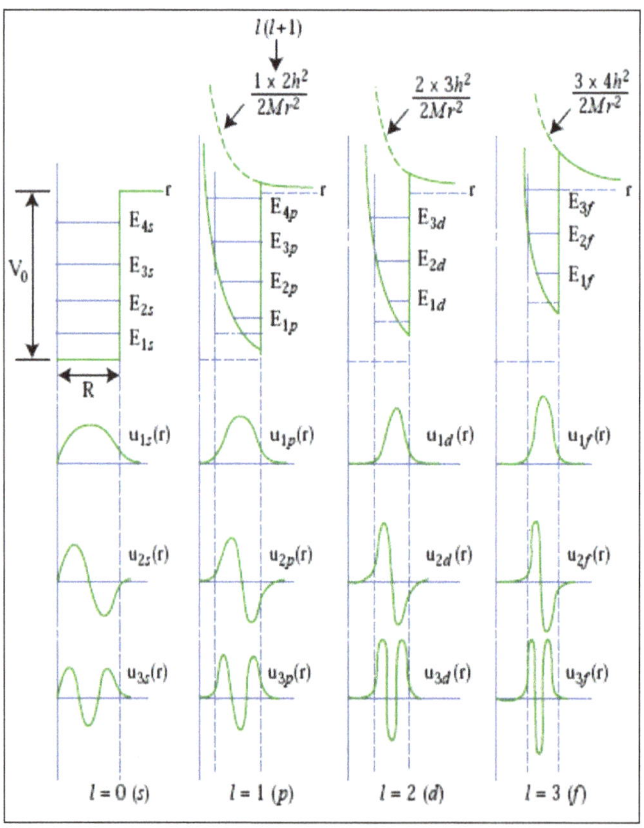

Energy levels and wave functions for a square well for l = 0, 1, 2, and 3.

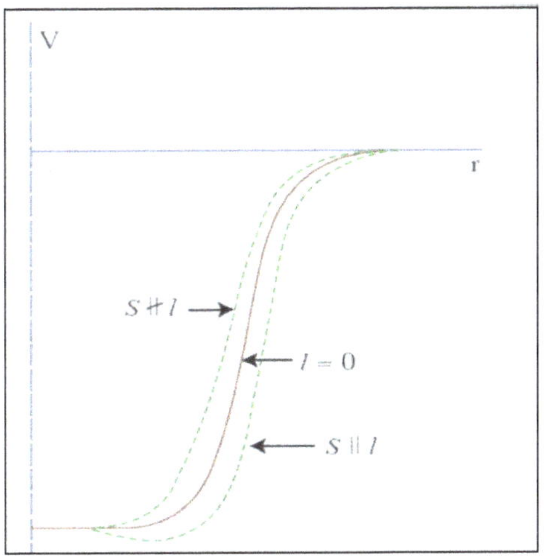

The effect of spin-orbit interaction on the shell-model potential.

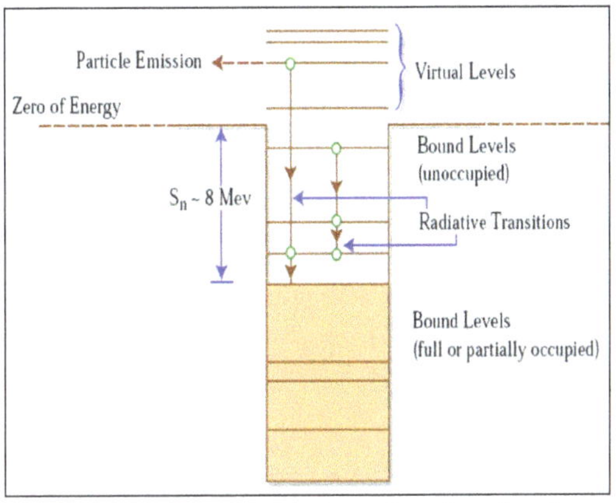

Schematic representation of nuclear levels.

Deformed Shell Model

The properties of a deformed nucleus can be understood from the shell model viewpoint if a deformed mean field is used. This idea was first pursued by Nilsson, who used a deformed harmonic oscillator as a mean field along with a spin-orbit coupling and a l^2 term. The single particle Hamiltonian (also known as Nilsson Hamiltonian) assuming axial symmetry of the potential around z axis can be written as,

$$h_{nilsson} = \frac{\hbar^2}{2m}\Delta + \frac{1}{2}m\omega_{\perp}^2(x^2+y^2) + \frac{1}{2}m\omega_z^2 z^2 + C1.s + D\left(1^2 - \langle 1 \rangle_N^2\right)$$

The Hamiltonian can also be written in cylindrical coordinates by introducing a deformation-dependent oscillator length $b(\delta) = \left(\hbar / m\omega_o(\delta)\right)^{1/2}$ and dimensionless coordinates $r' = r/b$:

$$h_{nilsson} = \hbar\omega_o(\delta)\left(-\frac{1}{2}\Delta' + \frac{1}{2}r'^2 - \beta r'^2 Y_{20}\right) - k\hbar\bar{\omega}_o\left(21.s + \mu\left(1^2 - \langle 1 \rangle_N^2\right)\right)$$

The constants C and D are given in the form,

$$C = -2k\hbar\overline{\omega}_o, \quad D = -k\mu\hbar\overline{\omega}_o$$

and β is given by,

$$\beta = \frac{1}{3}\sqrt{\frac{16\pi}{5}}\delta + .. = 1.057\,\delta$$

The anisotropy of the potential is achieved by using different values for $\omega_\perp$ and ω_Z which are written in terms of deformation parameter δ as,

$$\omega_\perp^2 = \omega_0^2(\delta)\left(1 + \frac{2}{3}\delta\right)$$

$$\omega_z^2 = \omega_0^2(\delta)\left(1 - \frac{4}{3}\delta\right)$$

where ω_0^2 is given by,

$$\omega_o(\delta) = \overline{\omega}_o\left(1 + \frac{2}{3}\delta^2\right)$$

using the volume conservation condition,

$$\omega_\perp^2\omega_z = \overline{\omega}_o^3 = \text{constant}$$

The consequence of using a deformed potential is that the magnetic substates of total angular momentum are no longer degenerate. For the potential having a prolate shape defined by a positive value of δ, the orbitals having smaller projection K of total angular momentum on the symmetry axis are lowered more in energy. Whereas, the opposite is true for the potential having an oblate shape defined by a negative value of δ. The reason lie in the fact that the deformed potential can be thought as being produced by a deformed core having the same shape. Therefore, the orbitals having probability distribution much closer to the density distribution of the core feel more attractive force and hence arc lowered more in energy. This is depicted in figure for a prolate shape potential.

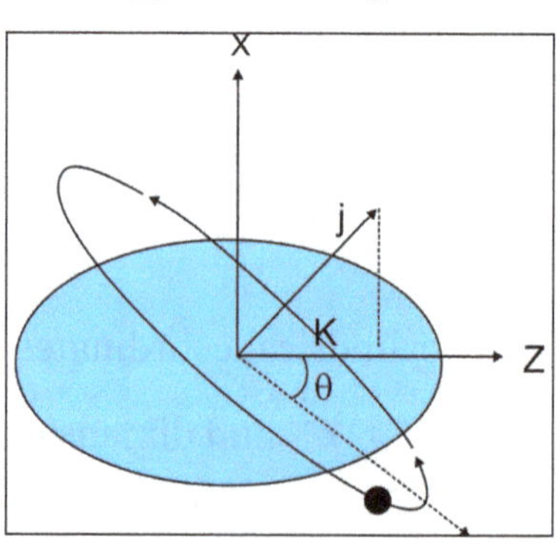

A schematic diagram of a Nilsson orbital in a deformed potential. The angle θ between the classical orbit and the symmetry axis is approximately given by $\sin^{-1}$ (K/j).

Since the total angular momentum j is not a conserved quantity in Nilsson model, it can no longer be used to label the Nilsson wave functions. However, the projection of the total angular momentum K on the symmetry axis and parity 1r are still conserved and arc used to label the wave functions. The Nilsson wave function are also labeled with asymptotic quantum numbers N denoting the principal quantum number of the shell from where the Nilsson wave function have originated at zero deformation, n_z the number of phonons along the z axis and Λ the projection of orbital angular momcntmn on the symmetry axis. The later two become good quantum numbers only at large deformation where the spin-orbit coupling can be neglected. A typical Nilsson orbit is written as:

$$K^{\pi}[Nn_z\ \Lambda]$$

Liquid-drop Model

One of the first models which could describe very well the behavior of the nuclear binding energies and therefore of nuclear masses was the mass formula of von Weizsaecker (also called the semi-empirical mass formula – SEMF), that was published in 1935 by German physicist Carl Friedrich von Weizsäcker. This theory is based on the liquid drop model proposed by George Gamow.

According to this model, the atomic nucleus behaves like the molecules in a drop of liquid. But in this nuclear scale, the fluid is made of nucleons (protons and neutrons), which are held together by the strong nuclear force. The liquid drop model of the nucleus takes into account the fact that the nuclear forces on the nucleons on the surface are different from those on nucleons in the interior of the nucleus. The interior nucleons are completely surrounded by other attracting nucleons. Here is the analogy with the forces that form a drop of liquid.

In the ground state the nucleus is spherical. If the sufficient kinetic or binding energy is added, this spherical nucleus may be distorted into a dumbbell shape and then may be splitted into two fragments. Since these fragments are a more stable configuration, the splitting of such heavy nuclei must be accompanied by energy release. This model does not explain all the properties of the atomic nucleus, but does explain the predicted nuclear binding energies.

The nuclear binding energy as a function of the mass number A and the number ofprotons Z based on the liquid drop model can be written as:

$$E_b\ (MeV) = a_V A - a_s A^{\frac{2}{3}} - a_c \frac{Z^2}{A^{\frac{1}{3}}} - a_A \frac{(A-2Z)^2}{A} \pm \delta\ (A, Z)$$

$$\delta(A,Z) = \begin{cases} +\delta_0 \ \mathit{for}\ Z, N\ \mathit{even} \\ 0 \\ -\delta_0 \ \mathit{for}\ Z, N\ \mathit{odd} \end{cases}$$

This formula is called the Weizsaecker Formula (or the semi-empirical mass formula). The physical meaning of this equation can be discussed term by term.

Volume Term

Volume term – $a_V.A$: The first two terms describe a spherical liquid drop of an incompressible fluid with a contribution from the volume scaling with A and from the surface, scaling with $A^{2/3}$. The first positive term $a_V.A$ is known as the volume term and it is caused by the attracting strong forces between the nucleons. The strong force has a very limited range and a given nucleon may only interact with its direct neighbours. Therefore this term is proportional to A, instead of A^2. The coefficient a_V is usually about ~ 16 MeV.

Surface Term

Surface term – $a_{sf}.A^{2/3}$: The surface term is also based on the strong force. It is, in fact, a correction to the volume term. The point is that particles at the surface of the nucleus are not completely surrounded by other particles. In the volume term, it is suggested that each nucleon interacts with a constant number of nucleons, independent of A. This assumption is very nearly true for nucleons deep within the nucleus, but causes an overestimation of the binding energy on the surface. By analogy with a liquid drop this effect is indicated as the surface tension effect. If the volume of the nucleus is proportional to A, then the geometrical radius should be proportional to $A^{1/3}$ and therefore the surface term must be proportional to the surface area i.e. proportional to $A^{2/3}$.

Coulomb Term

Coulomb term – $a_C.Z^2.A^{-1/3}$: This term describes the Coulomb repulsion between the uniformly distributed protons and is proportional to the number of proton pairs Z^2/R, whereby R is proportional to $A^{1/3}$. This effect lowers the binding energy because of the repulsion between charges of equal sign.

Asymmetry Term

Asymmetry term – $a_A.(A-2Z)^2/A$: This term cannot be described as 'classically' as the first three. This effect is not based on any of the fundamental forces, this effect is based only on the Pauli exclusion principle (no two fermions can occupy exactly the same quantum state in an atom). The heavier nuclei contain more neutrons than protons. These extra neutrons are necessary for stability of the heavier nuclei. They provide (via the attractive forces between the neutrons and protons) some compensation for the repulsion between the protons. On the other hand, if there are significantly more neutrons than protons in a nucleus, some of the neutrons will be higher in energy level in the nucleus. This is the basis for a correction factor, the so-called symmetry term.

Pairing Term

Pairing term – $\delta(A,Z)$: The last term is the pairing term $\delta(A,Z)$. This term captures the effect of spin-coupling. Nuclei with an even number of protons and an even number of neutrons are (due to Pauli exclusion principle) very stable thanks to the occurrence of 'paired spin'. On the other hand, nuclei with an odd number of protons and neutrons are mostly unstable.

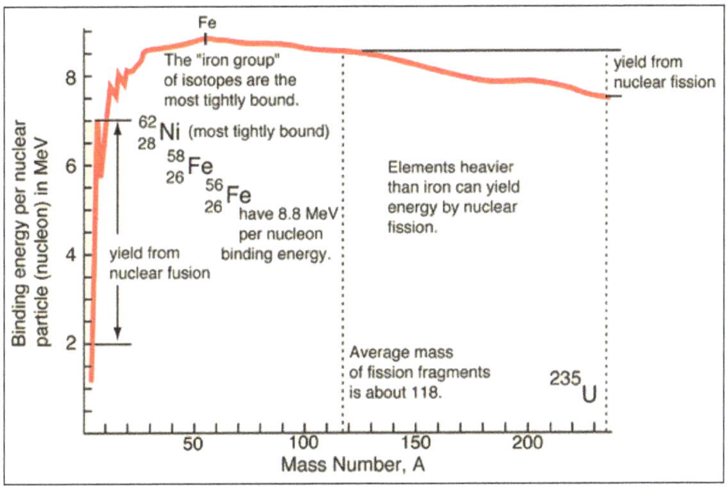

Nuclear binding energy curve.

With the aid of the Weizsaecker formula the binding energy can be calculated very well for nearly all isotopes. This formula provides a good fit for heavier nuclei. For light nuclei, especially for ^{4}He, it provides a poor fit. The main reason is the formula does not consider the internal shell structure of the nucleus.

In order to calculate the binding energy, the coefficients a_V, a_S, a_C, a_A and a_P must be known. The coefficients have units of megaelectronvolts (MeV) and are calculated by fitting to experimentally measured masses of nuclei. They usually vary depending on the fitting methodology. The coefficients in the equation are following:

$$E_b\,(MeV) = 15.76A - 17.81A^{2/3} - 0.711\frac{Z^2}{A^{1/3}} - 23.7\frac{(N-Z)^2}{A} \pm 34A^{-3/4}$$

Using the Weizsaecker formula, also the mass of an atomic nucleus can be derived and is given by:

$$m = Z.m_p + N.m_n - E_b / c^2$$

where m_p and m_n are the rest mass of a proton and a neutron, respectively, and E_b is the nuclear binding energy of the nucleus.From the nuclear binding energy curve and from the table it can be seen that, in the case of splitting a ^{235}U nucleus into two parts, the binding energy of the fragments ($A \approx 120$) together is larger than that of the original ^{235}U nucleus.

According to the Weizsaecker formula, the total energy released for such reaction will be approximately $235\,x\,(8.5 - 7.6) \approx 200$ MeV.

Table of binding energies fo some nuclides. Calculated according to the semi-empirical mass formula.

	$^{40}_{20}Ca$	$^{107}_{47}Ag$	$^{238}_{92}U$
Volume term	630	1686	3751
Surface term	-208	-401	-684

Coulomb term	-83	-331	-971
symmetry term	0	-37	-290
pairing term	+2	0	+0.6
calculated E_B	341	917	1806
measured E_B	342	915	1802
measured E_B/A	8.6	8.6	7.6

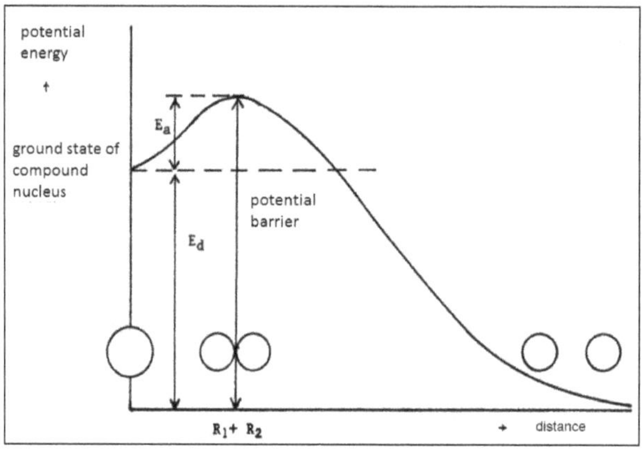

The minimum excitation energy required for fission to occur is known as the critical energy (E_{crit}) or threshold energy.

Target Nucleus	Critical Energy E_{crit}	Binding Energy of Last Neutron BE_n	$B_{en} - E_{crit}$
${}^{232}_{90}Th$	7.5Mev	5.4 Mev	- 2.1 Mev
${}^{238}_{92}U$	7.0 Mev	5.5 Mev	- 1.5 Mev
${}^{235}_{92}U$	6.5 Mev	6.8 Mev	+ 0.3 Mev
${}^{233}_{92}U$	6.0 Mev	7.0 Mev	+ 1.0 Mev
${}^{239}_{94}Pu$	5.0 Mev	6.6 Mev	+ 1.6 Mev

This table shows critical energies compared to binding energies of the last neutron of a number of nuclei.

Collective Model

Collective model, also called unified model, description of atomic nuclei that incorporates aspects of both the shell nuclear model and the liquid-drop model to explain certain magnetic and electric properties that neither of the two separately can explain.

The Shell Model has its shortcomings. This is particularly true for heavier nuclei. We have already seen that the Shell Model does not predict magnetic dipole moments or the spectra of excited states very well.

One further failing of the Shell Model are the predictions of electric quadrupole moments, which in the Shell Model are predicted to be very small. However, heavier nuclei with A in the range 150 - 190 and for A > 220, these electric quadrupole moments are found to be rather large.

The failure of the Shell Model to correctly predict electric quadrupole moments arises from the assumption that the nucleons move in a spherically symmetric potential.

The Collective Model generalises the result of the Shell Model by considering the effect of a non-spherically symmetric potential, which leads to substantial deformations for large nuclei and consequently large electric quadrupole moments.

One of the most striking consequences of the Collective Model is the explanation of low-lying excited states of heavy nuclei. These are of two types:

Rotational States: A nucleus whose nucleon density distributions are spherically symmetric (zero quadrupole moment) cannot have rotational excitations (this is analogous to the application of the principle of equipartition of energy to monatomic molecules for which there are no degrees of freedom associated with rotation).

On the other hand a nucleus with a non-zero quadrupole moment can have excited levels due to rotational perpendicular to the axis of symmetry.

For an even-even nucleus whose ground state has zero spin, these states have energies:

$$E_{rot} = \frac{I(I+1)\hbar^2}{2\mathcal{I}},$$

where $\mathcal{I}$ is the moment of inertia of the nucleus about an axis through the centre perpendicular to the axis of symmetry.

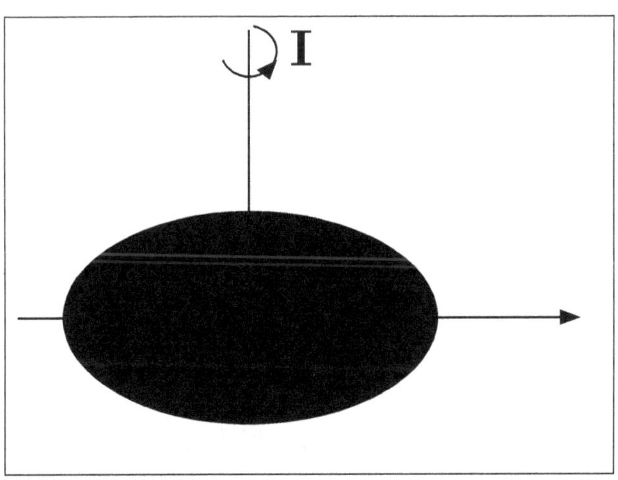

It turns out that the rotational energy levels of an even-even nucleus can only take even values of I. For example the nuclide ${}^{170}_{72}$Hf (hafnium) has a series of rotational states with excitation energies:

$$\mathrm{E(KeV)} : \quad 100, \quad 321, \quad 641$$

These are almost exactly in the ratio $2\times3:4\times5:6\times7$, meaning that these are states with rotational spin equal to 2, 4, 6 respectively. The relation is not exact because the moment of inertia changes as the spin increases.

We can extract the moment of inertia for each of these rotational states from equation $E_{rot}=\frac{I(I+1)\hbar^2}{2\mathcal{I}}$. We could express this in SI units, but more conveniently nuclear moments of inertia are quoted in MeV/c² fm², with the help of the relation,

$$\hbar c = 197.3 \text{ MeV fm}.$$

Therefore the moment inertia of the I = 2 state, whose excitation energy is 0.1 MeV, is given - inseting I = 2 into equation $E_{rot}=\frac{I(I+1)\hbar^2}{2\mathcal{I}}$ by,

$$\mathcal{I}=2\times3\times\frac{\hbar^2c^2}{2c^2\,\mathrm{E_{rot}}}=\frac{6}{2}\frac{197.3^2}{0.1}=1.17\times10^6\ \mathrm{MeV/\,c^2fm^2}$$

For odd-a nuclides for which the spin of the ground state I_0 is non-zero, the rotational levels have excitation levels of,

$$\mathrm{E_{rot}}=\frac{1}{2\mathcal{I}}\left(I\left(I+1\right)-I_0(I_0+1\right)\hbar^2,$$

where I can take the values I_0+1, I_0+2 etc. For example the first two rotational excitation energies of $^{143}_{60}$Nd (neodymium), whose ground state has spin $\frac{7}{2}$, have energies 128 KeV and 290 KeV. They correspond to rotational levels with nuclear spin $\frac{9}{2}$ and $\frac{11}{2}$ respectively. The ratio of these two excitation energies (2.27) is almost exactly equal to,

$$\frac{\frac{11}{2}\times\frac{13}{2}-\frac{7}{2}\times\frac{9}{2}}{\frac{9}{2}\times\frac{11}{2}-\frac{7}{2}\times\frac{9}{2}}=2.22$$

Shape oscillations: These are modes of vibration in which the deformation of the nucleus oscillates - the electric quadrupole moment oscillates about its mean value. It could be that this mean value is very small, in which case the nucleus is oscillating between an oblate and a prolate spheroidal shape. It is also possible to have shape oscillations with different shapes.

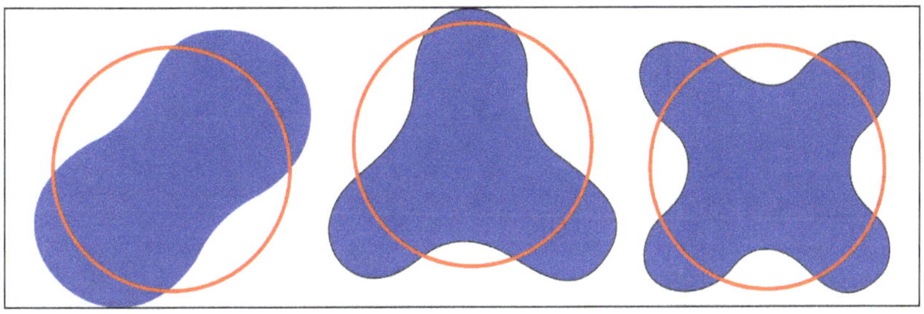

The small oscillations about the equilibrium shape perform simple harmonic motion. The energy levels of such modes are equally spaced. Thus an observed sequence of equally spaced energy levels within the spectrum of a nuclide is interpreted as a manifestation of such shape oscillations.

References

- Nuclear-model, science: britannica.com, Retrieved 13 June, 2019
- Nuclear-shell-model, atomic-nucleus, atom-properties-of-atoms, reactor-physics, nuclear-power: nuclear-power.net, Retrieved 21 March , 2019
- Liquid-drop-model, fission, nuclear-power: nuclear-power.net, Retrieved 12 April, 2019
- Shell, course, teaching: personal.soton.ac.uk, Retrieved 23 June, 2019

3

Nuclear Reactions

The process wherein two nuclei collide and produce one or more different nuclides is termed as a nuclear reaction. They are classified into a few major types such as direct reactions and compound nucleus reactions. The topics elaborated in this chapter will help in gaining a better perspective about these types of nuclear reactions.

A nuclear reaction is considered to be the process in which two nuclear particles (two nuclei or a nucleus and a nucleon) interact to produce two or more nuclear particles or γ-rays (gamma rays). Thus, a nuclear reaction must cause a transformation of at least one nuclide to another. Sometimes if a nucleus interacts with another nucleus or particle without changing the nature of any nuclide, the process is referred to a nuclear scattering, rather than a nuclear reaction. Perhaps the most notable nuclear reactions are the nuclear fusion reactions of light elements that power the energy production of stars and the Sun. Natural nuclear reactions occur also in the interaction between cosmic rays and matter.

The most notable man-controlled nuclear reaction is the fission reaction which occurs in nuclear reactors. Nuclear reactors are devices to initiate and control a nuclear chain reaction, but there are not only manmade devices. The world's first nuclear reactor operated about two billion years ago. The natural nuclear reactor formed at Oklo in Gabon, Africa, when a uranium-rich mineral deposit became flooded with groundwater that acted as a neutron moderator, and a nuclear chain reaction started. These fission reactions were sustained for hundreds of thousands of years, until a chain reaction could no longer be supported. This was confirmed by existence of isotopes of the fission-product gas xenon and by different ratio of U-235/U-238 (enrichment of natural uranium).

Notation of Nuclear Reactions

Standard nuclear notation shows the chemical symbol, the mass number and the atomic number of the isotope.

If the initial nuclei are denoted by *a* and *b*, and the product nuclei are denoted by *c* and *d*, the reaction can be represented by the equation:

$$a + b \rightarrow c + d$$

$${}^{10}_{5}B + {}^{1}_{0}n \rightarrow {}^{7}_{3}Li + {}^{4}_{2}He + 2.8\ Mev$$

This equation describes neutron capture in the boron, which is diluted in the coolant. Boric acid is used in nuclear power plants as a long-term compensator of nuclear fuel reactivity.

Instead of using the full equations in the style above, in many situations a compact notation is used to describe nuclear reactions. This style of the form a(b,c)d is equivalent to a + b producing c + d.

Light particles are often abbreviated in this shorthand, typically p means proton, n means neutron, d means deuteron, α means an alpha particle or helium-4, β means beta particle or electron, γ means gamma photon, etc. The reaction above would be written as 10B(n,α)7Li.

Basic Classification of Nuclear Reactions

In order to understand the nature of neutron nuclear reactions, the classification according to the time scale of of these reactions has to be introduced. Interaction time is critical for defining the reaction mechanism.

There are two extreme scenarios for nuclear reactions (not only neutron reactions):

- A projectile and a target nucleus are within the range of nuclear forces for the very short time allowing for an interaction of a single nucleon only. These type of reactions are called the direct reactions.
- A projectile and a target nucleus are within the range of nuclear forces for the time allowing for a large number of interactions between nucleons. These type of reactions are called the compound nucleus reactions.

In fact, there is always some non-direct (multiple internuclear interaction) component in all reactions, but the direct reactions have this component limited.

Basic Characteristics of Direct Reactions

- The direct reactions are fast and involve a single-nucleon interaction.
- The interaction time must be very short (~10^{-22} s).
- The direct reactions require incident particle energy larger than ~5 MeV/Ap. (Ap is the atomic mass number of a projectile).
- Incident particles interact on the surface of a target nucleus rather than in the volume of a target nucleus.
- Products of the direct reactions are not distributed isotropically in angle, but they are forward focused.
- Direct reactions are of importance in measurements of nuclear structure.

Basic Characteristics of Compound Nucleus Reactions

- The compound nucleus is a relatively long-lived intermediate state of particle-target composite system.
- The compound nucleus reactions involve many nucleon-nucleon interactions.
- The large number of collisions between the nucleons leads to a thermal equilibrium inside the compound nucleus.
- The time scale of compound nucleus reactions is of the order of 10^{-18} s – 10^{-16} s.
- The compound nucleus reactions is usually created if the projectile has low energy.
- Incident particles interact in the volume of a target nucleus.

- Products of the compound nucleus reactions are distributed near isotropically in angle (the nucleus loses memory of how it was created – the *Bohr's hypothesis of independence*).
- The mode of decay of compound nucleus do not depend on the way the compound nucleus is formed.
- Resonances in the cross-section are typical for the compound nucleus reaction.

Types of Nuclear Reactions

Although the number of possible nuclear reactions is enormous, nuclear reactions can be sorted by types. Most of nuclear reactions are accompanied by gamma emission. Some examples are:

- Elastic scattering: Occurs, when no energy is transferred between the target nucleus and the incident particle.

  $$208\,\mathrm{Pb}\left(\mathrm{n},\mathrm{n}\right)208\,\mathrm{Pb}$$

- Inelastic scattering: Occurs, when energy is transferred. The difference of kinetic energies is saved in excited nuclide.

  $$40\,\mathrm{Ca}\left(\alpha,\alpha'\right)40\,\mathrm{mCa}$$

- Capture reactions: Both charged and neutral particles can be captured by nuclei. This is accompanied by the emission of γ-rays. Neutron capture reaction produces radioactive nuclides (induced radioactivity).

  $$238\,\mathrm{U}\left(\mathrm{n},\gamma\right)239\,\mathrm{U}$$

- Transfer Reactions: The absorption of a particle accompanied by the emission of one or more particles is called the transfer reaction.

  $$4\,\mathrm{He}\left(\alpha,\mathrm{p}\right)7\,\mathrm{Li}$$

- Fission reactions: Nuclear fission is a nuclear reaction in which the nucleus of an atom splits into smaller parts (lighter nuclei). The fission process often produces free neutrons and photons (in the form of gamma rays), and releases a large amount of energy.

  $$235\,\mathrm{U}\left(\mathrm{n},3\,\mathrm{n}\right)\text{ fission products}$$

- Fusion reactions: Occur when, two or more atomic nuclei collide at a very high speed and join to form a new type of atomic nucleus.The fusion reaction of deuterium and tritium is particularly interesting because of its potential of providing energy for the future.

  $$3\,\mathrm{T}\left(\mathrm{d},\mathrm{n}\right)4\,\mathrm{He}$$

- Spallation reactions: Occur, when a nucleus is hit by a particle with sufficient energy and momentum to knock out several small fragments or, smash it into many fragments.

- Nuclear decay (Radioactive decay): Occurs when an unstable atom loses energy by emitting ionizing radiation. Radioactive decay is a random process at the level of single atoms, in that, according to quantum theory, it is impossible to predict when a particular atom will decay. There are many types of radioactive decay:
    - Alpha radioactivity: Alpha particles consist of two protons and two neutrons bound together into a particle identical to a helium nucleus. Because of its very large mass (more than 7000 times the mass of the beta particle) and its charge, it heavy ionizes material and has a very short range.
    - Beta radioactivity: Beta particles are high-energy, high-speed electrons or positrons emitted by certain types of radioactive nuclei such as potassium-40. The beta particles have greater range of penetration than alpha particles, but still much less than gamma rays. The beta particles emitted are a form of ionizing radiation also known as beta rays. The production of beta particles is termed beta decay.
    - Gamma radioactivity: Gamma rays are electromagnetic radiation of an very high frequency and are therefore high energy photons. They are produced by the decay of nuclei as they transition from a high energy state to a lower state known as gamma decay. Most of nuclear reactions are accompanied by gamma emission.
    - Neutron emission: Neutron emission is a type of radioactive decay of nuclei containing excess neutrons (especially fission products), in which a neutron is simply ejected from the nucleus. This type of radiation plays key role in nuclear reactor control, because these neutrons are delayed neutrons.

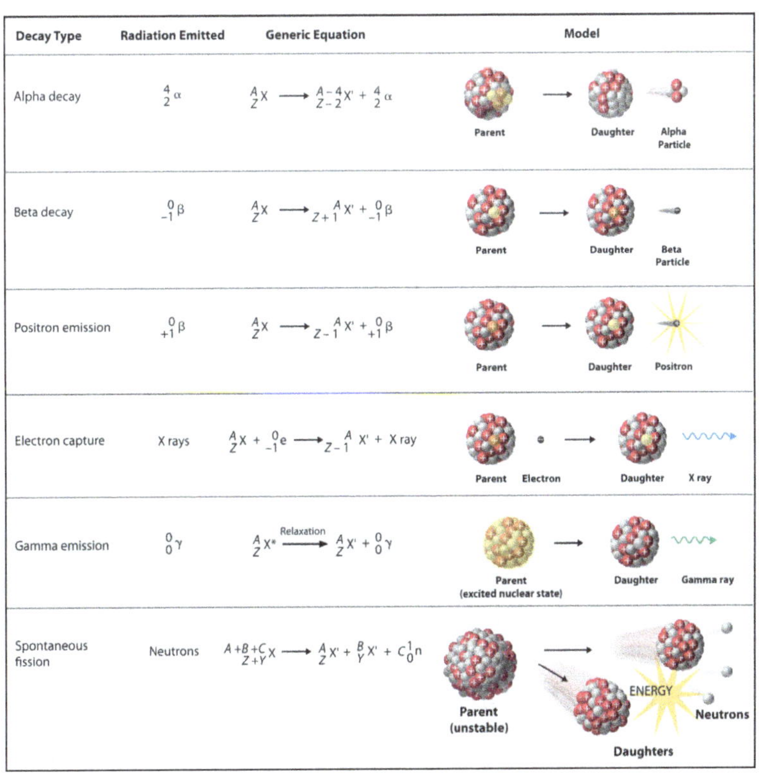

Radioactive decays.

Conservation Laws in Nuclear Reactions

In analyzing nuclear reactions, we apply the many conservation laws. Nuclear reactions are subject to classical conservation laws for charge, momentum, angular momentum, and energy(including rest energies). Additional conservation laws, not anticipated by classical physics, are:

- Law of Conservation of Lepton Number.
- Law of Conservation of Baryon Number.
- Law of Conservation of Electric Charge.

Law of Conservation of Lepton Number

Q/e	$L_e = -1$	$L_\mu = -1$	$L_\tau = -1$
0	$\bar{\nu}_e$	$\bar{\nu}_\mu$	$\bar{\nu}_\tau$
$+1$	e^+	μ^+	τ^+
Q/e	$L_e = 1$	$L_\mu = -1$	$L_\tau = -1$
0	ν_e	ν_μ	ν_τ
-1	e^-	μ^-	τ^-

In particle physics, the lepton number is used to denote which particles are leptons and which particles are not. Each lepton has a lepton number of 1 and each antilepton has a lepton number of -1. Other non-leptonic particles have a lepton number of 0. The lepton number is a conserved quantum number in all particle reactions. A slight asymmetry in the laws of physics allowed leptons to be created in the Big Bang.

The conservation of lepton number means that whenever a lepton of a certain generation is created or destroyed in a reaction, a corresponding antilepton from the same generation must be created or destroyed. It must be added, there is a separate requirement for each of the three generations of leptons, the electron, muon and tau and their associated neutrinos.

Consider the decay of the neutron. The reaction involves only first generation leptons: electrons and neutrinos:

$$n \rightarrow p + e \ + \bar{v}$$

$$\text{Lepton Number: } 0 \rightarrow 0 + 1 \ - 1$$

Since the lepton number must be equal to zero on both sides and it was found that the reaction is a three-particle decay (the electrons emitted in beta decay have a continuous rather than a discrete spectrum), the third particle must be an electron antineutrino.

Law of Conservation of Baryon Number

In particle physics, the baryon number is used to denote which particles are baryons and which particles are not. Each baryon has a baryon number of 1 and each antibaryonhas a baryon number of -1. Other non-baryonic particles have a baryon number of 0. Since there are exotic hadrons like pentaquarks and tetraquarks, there is a general definition of baryon number as:

$$B = \frac{1}{3}(n_q - n_{\bar{q}})$$

where n_q is the number of quarks, and $n_{\bar{q}}$ is the number of antiquarks.

The baryon number is a conserved quantum number in all particle reactions.

The law of conservation of baryon number states that:

The sum of the baryon number of all incoming particles is the same as the sum of the baryon numbers of all particles resulting from the reaction.

For example, the following reaction has never been observed:

$$\begin{array}{rl} & p + n \rightarrow p + p + \bar{p} \\ \text{Baryon Number:} & 1 + 1 \rightarrow 1 + 1 - 1 \end{array}$$

even if the incoming proton has sufficient energy and charge, energy, and so on, are conserved. This reaction does not conserve baryon number since the left side has B =+2, and the right has B =+1.

On the other hand, the following reaction (proton-antiproton pair production) does conserve B and does occur if the incoming proton has sufficient energy (the threshold energy = 5.6 GeV):

$$\begin{array}{rl} & p + p \rightarrow p + p + \bar{p} + p \\ \text{Baryon Number:} & 1 + 1 \rightarrow 1 + 1 - 1 + 1 \end{array}$$

As indicated, B = +2 on both sides of this equation.

From these and other reactions, the conservation of baryon number has been established as a basic principle of physics.

This principle provides basis for the stability of the proton. Since the proton is the lightest particle among all baryons, the hypothetical products of its decay would have to be non-baryons. Thus, the decay would violate the conservation of baryon number. It must be added some theories have suggested that protons are in fact unstable with very long half-life (~10^{30} years) and that they decay into leptons. There is currently no experimental evidence that proton decay occurs.

Law of Conservation of Electric Charge

The law of conservation of electric charge can be demonstrated also on positron-electron pair production. Since a gamma ray is electrically neutral and sum of the electric charges of electron and positron is also zero, the electric charge in this reaction is also conserved.

$$\gamma \rightarrow e^- + e^+$$

It must be added, in order for electron-positron pair production to occur, the electromagnetic energy of the photon must be above a threshold energy, which is equivalent to the rest mass of two electrons. The threshold energy (the total rest mass of produced particles) for electron-positron pair production is equal to 1.02 MeV (2 x 0.511 MeV) because the rest mass of a single electron is equivalent to 0.511 MeV of energy. If the original photon's energy is greater than 1.02 MeV, any energy above 1.02 MeV is according to the conservation law split between the kinetic energy of motion of the two particles. The presence of an electric field of a heavy atom such as lead or uranium is essential in order to satisfy conservation of momentum and energy. In order to satisfy both conservation of momentum and energy, the atomic nucleus must receive some momentum. Therefore a photon pair production in free space cannot occur.

Certain of these laws are obeyed under all circumstances, others are not. We have accepted conservation of energy and momentum. In all the examples given we assume that the number of protons and the number of neutrons is separately conserved. We shall find circumstances and conditions in which this rule is not true. Where we are considering non-relativistic nuclear reactions, it is essentially true. However, where we are considering relativistic nuclear energies or those involving the weak interactions, we shall find that these principles must be extended.

Some conservation principles have arisen from theoretical considerations, others are just empirical relationships. Notwithstanding, any reaction not expressly forbidden by the conservation laws will generally occur, if perhaps at a slow rate. This expectation is based on quantum mechanics. Unless the barrier between the initial and final states is infinitely high, there is always a non-zero probability that a system will make the transition between them.

For purposes of analyzing non-relativistic reactions, it is sufficient to note four of the fundamental laws governing these reactions:

1. Conservation of nucleons: The total number of nucleons before and after a reaction are the same.
2. Conservation of charge: The sum of the charges on all the particles before and after a reaction are the same.
3. Conservation of momentum: The total momentum of the interacting particles before and after a reaction are the same.
4. Conservation of energy: Energy, including rest mass energy, is conserved in nuclear reactions.

Elastic Nuclear Collision

A neutron (n) of mass 1.01 u traveling with a speed of 3.60×10^4 m/s interacts with a carbon (C) nucleus (m_C = 12.00 u) initially at rest in an elastic head-on collision.

What are the velocities of the neutron and carbon nucleus after the collision?

Solution: This is an elastic head-on collision of two objects with unequal masses. We have to use the conservation laws of momentum and of kinetic energy, and apply them to our system of two particles.

Conservation of momentum:

Total P before = total P after,

$$m_n v_n + 0 = m_n v_n^{'} + m_c v_c^{'}$$

Conservation of kinetic energy:

Total K before = total K after,

$$\frac{1}{2} m_n v_n^2 + 0 = \frac{1}{2} m_n v_n^{'2} + \frac{1}{2} m_c v_c^{'2}$$

This equation stated that the relative speed of the two objects after the collision has the same magnitude (but opposite direction) as before the collision, no matter what the masses are:

$$v_A + v_A^{'} = v_B^{'} + v_B$$

In our case:

$$v_n - 0 = v_c^{'} - v_n^{'}$$

Thus,

$$v_n^{'} = v_c^{'} - v_n$$

Substituting this into our momentum equation, we get:

$$m_n v_n = m_n v_c^{'} - m_n v_n + m_c v_c^{'}$$

Solving this equation for both velocities, we obtain:

$$v_c^{'} = \frac{2 m_n v_n}{m_n + m_c} = \frac{2 \, x \, 1.01}{13.01} \, x \, 3.6 \, x \, 10^4 = 0.56 \, x \, 10^4 \; m/s$$

$$v_c^{'} = v_c^{'} - v_n^{'} = 0.56 \, x \, 10^4 - 3.6 \, x \, 10^4 = -3.04 \, x \, 10^4 \; m/s$$

The minus sign for v' tells us that the neutron scatters back of the carbon nucleus, because the carbon nucleus is significantly heavier. On the other hand its speed is less than its initial speed. This process is known as the neutron moderation and it significantly depends on the mass of moderator nuclei.

Energetics of Nuclear Reactions: Q-value

In nuclear and particle physics the energetics of nuclear reactions is determined by the Q-value of that reaction. The Q-value of the reaction is defined as the difference between the sum of the masses of the initial reactants and the sum of the masses of the final products, in energy units (usually in MeV).

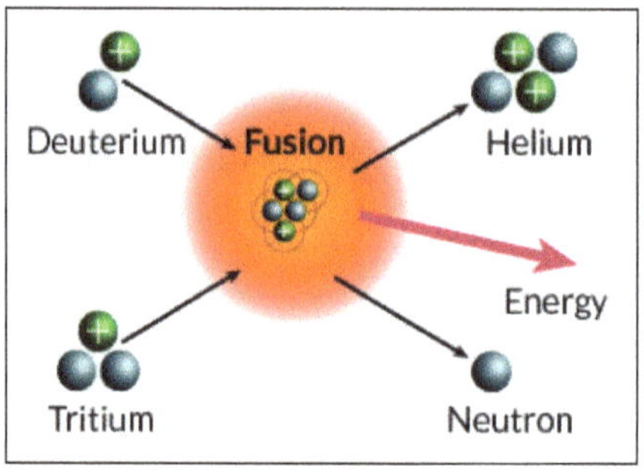

Nuclear fusion

Consider a typical reaction, in which the projectile a and the target A gives place to two products, B and b. This can also be expressed in the notation that we used so far, a + A → B + b, or even in a more compact notation, A(a,b)B.

The Q-value of this reaction is given by:

$$Q=[m_a + m_A - (m_b + m_B)]c^2$$

which is the same as the excess kinetic energy of the final products:

$$\begin{aligned} Q &= T_{final} - T_{initial} \\ &= T_b + T_B - (T_a + T_A) \end{aligned}$$

For reactions in which there is an increase in the kinetic energy of the products Q is positive. The positive Q reactions are said to be exothermic (or exergic). There is a net release of energy, since the kinetic energy of the final state is greater than the kinetic energy of the initial state.

For reactions in which there is a decrease in the kinetic energy of the products Q is negative. The negative Q reactions are said to be endothermic (or endoergic) and they require a net energy input.

The energy released in a nuclear reaction can appear mainly in one of three ways:

- Kinetic energy of the products.
- Emission of gamma rays: Gamma rays are emitted by unstable nuclei in their transition from a high energy state to a lower state known as gamma decay.
- Metastable state: Some energy may remain in the nucleus, as a metastable energy level.

A small amount of energy may also emerge in the form of X-rays. Generally, products of nuclear reactions may have different atomic numbers, and thus the configuration of their electron shells is different in comparison with reactants. As the electrons rearrange themselves and drop to lower energy levels, internal transition X-rays (X-rays with precisely defined emission lines) may be emitted.

Exothermic Reactions

DT Fusion

The DT fusion reaction of deuterium and tritium is particularly interesting because of its potential of providing energy for the future. Calculate the reaction Q-value.

3T (d, n) 4He

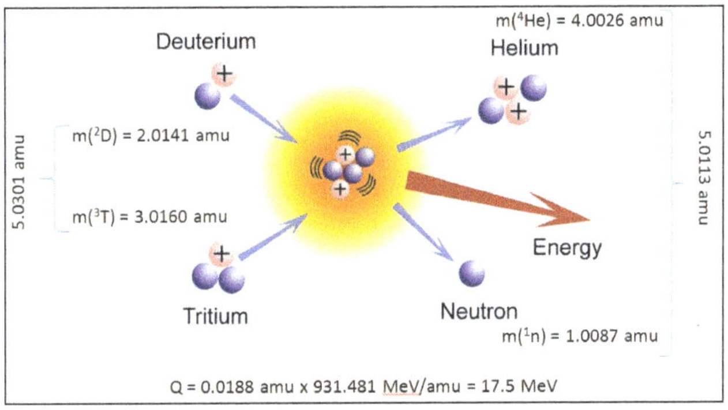

Q-value of DT fusion reaction.

The atom masses of the reactants and products are:

m(^{3}T) = 3.0160 amu

m(^{2}D) = 2.0141 amu

m(^{1}n) = 1.0087 amu

m(^{4}He) = 4.0026 amu

Using the mass-energy equivalence, we get the Q-value of this reaction as:

Q = {(3.0160+2.0141) [amu] – (1.0087+4.0026) [amu]} x 931.481 [MeV/amu]

= 0.0188 x 931.481 = 17.5 MeV

Tritium in Reactors

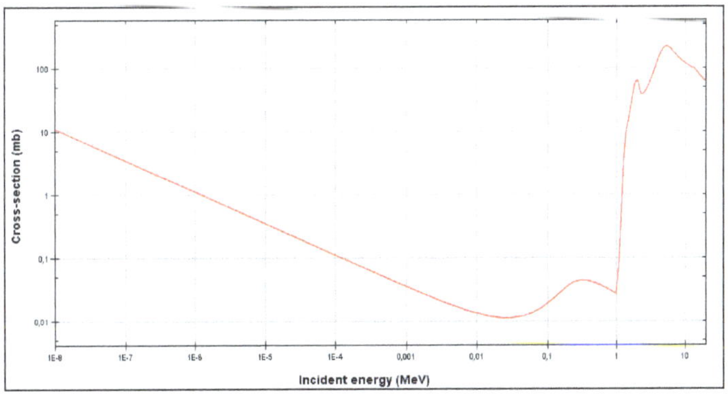

Cross-section of 10B(n,2 alpha)T reaction.

Tritium is a byproduct in nuclear reactors. Most of the tritium produced in nuclear power plants stems from the boric acid, which is commonly used as a chemical shim to compensate an excess of initial reactivity. Main reaction, in which the tritium is generated from boron is below:

10B(n,2* alpha)T

This reaction of a neutron with an isotope ^{10}B is the main way, how radioactive tritium in primary circuit of all PWRs is generated. Note that, this reaction is a threshold reaction due to its cross-section.

Calculate the reaction Q-value. The atom masses of the reactants and products are:

$m(^{10}B) = 10.01294$ amu

$m(^{1}n) = 1.00866$ amu

$m(^{3}T) = 3.01604$ amu

$m(^{4}He) = 4.0026$ amu

Using the mass-energy equivalence, we get the Q-value of this reaction as,

$$Q = \{(10.0129+1.00866)\ [amu] - (3.01604+2 \times 4.0026)\ [amu]\} \times 931.481\ [MeV/amu]$$

$$= 0.00036 \times 931.481 = 0.335\ MeV$$

Endothermic Reactions

Photoneutrons

In nuclear reactors the gamma radiation plays a significant role also in reactor kinetics and in a subcriticality control. Especially in nuclear reactors with D_2O moderator (CANDU reactors) or with Be reflectors (some experimental reactors). Neutrons can be produced also in (γ, n) reactions and therefore they are usually referred to as photoneutrons.

A high energy photon (gamma ray) can under certain conditions eject a neutron from a nucleus. It occurs when its energy exceeds the binding energy of the neutron in the nucleus. Most nuclei have binding energies in excess of 6 MeV, which is above the energy of most gamma rays from fission. On the other hand there are few nuclei with sufficiently low binding energy to be of practical interest. These are: ^{2}D, ^{9}Be, ^{6}Li, ^{7}Li and ^{13}C. As can be seen from the table the lowest threshold have ^{9}Be with 1.666 MeV and ^{2}D with 2.226 MeV.

Nuclides with low photodisintegration threshold energies.

Nuclide	Threshold(MeV)	Reaction
^{2}D	2.225	$^{2}H(\gamma, n)^{1}H$
^{6}Li	3.697	$^{6}Li\ (\gamma, n+p)^{4}He$
^{6}Li	5.67	$^{6}Li\ (\gamma, n)^{5}Li$
^{7}Li	7.251	$^{7}Li\ (\gamma, n)^{6}Li$
^{9}Be	1.667	$^{9}Be(\gamma, n)^{8}Be$
^{13}C	4.9	$^{13}C(\gamma, n)^{12}C$

In case of deuterium, neutrons can be produced by the interaction of gamma rays (with a minimum energy of 2.22 MeV) with deuterium:

$$^{2}_{1}D + \gamma \rightarrow {}^{1}_{1}H + n$$

The reaction Q-value is calculated below.

The atom masses of the reactant and products are,

m(^{2}D) = 2.01363 amu

m(^{1}n) = 1.00866 amu

m(^{1}H) = 1.00728 amu

Using the mass-energy equivalence, we get the Q-value of this reaction as,

Q = {2.01363 [amu] – (1.00866+1.00728) [amu]} x 931.481 [MeV/amu]

= -0.00231 x 931.481 = -2.15 MeV.

(α, n) Reaction

Calculate the reaction Q-value of the following reaction:

7Li (α, n) 10B

The atom masses of the reactants and products are:

m(^{4}He) = 4.0026 amu

m(^{7}Li) = 7.0160 amu

m(^{1}n) = 1.0087 amu

m(^{10}B) = 10.01294 amu

Using the mass-energy equivalence, we get the Q-value of this reaction as:

Q = {(7.0160+4.0026) [amu] – (1.0087+10.01294) [amu]} x 931.481 [MeV/amu]

= 0.00304 x 931.481 = -2.83 MeV.

Reaction Cross Section

Consider the situation where a beam of projectile nuclei of intensity ϕ_0 particles/sec is incident upon a thin foil of target nuclei with the result that the beam is attenuated by reactions in the foil such that the transmitted intensity is ϕ particles/sec.

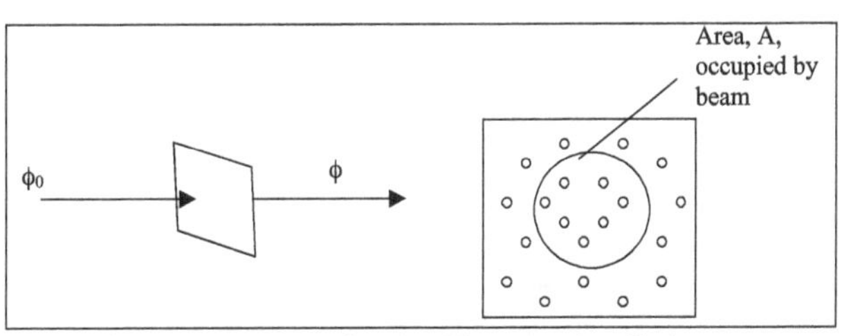

The attenuation of an incident projectile beam.

We can ask what fraction of the incident particles disappear from the beam, i.e, react in passing through the foil. Let us assume the beam intersects an area A on the foil. We can then assert that the fraction of beam particles that is blocked (reacts) is the fraction of the area A that is covered by target nuclei. If the foil contains N atoms/cm², then the area a that is covered by nuclei is N(atoms/cm²) x A (cm²) x (the effective area subtended by one atom) (cm²/atom). This latter term, the effective area subtended by one atom, is called the cross section, σ, for the reaction under study. Then the fraction of the area A that is blocked is a/A or σ (atoms/cm²) σ (cm²/atom). If we say the number of projectile nuclei absorbed per unit time is $\Delta\phi$, then we have:

$$\Delta\phi = -\phi \mathrm{N}\sigma$$

As an aside, we note the units of N are atoms/cm² or thickness, $\Delta x(cm)$ x density n(atoms/cm³). Expressing the above equation as a differential equation we have,

$$-d\phi = \phi N \sigma$$

Thus, upon rearranging we have,

$$d\phi / \phi = -N\sigma = -\mathrm{n}\sigma \mathrm{dx}$$

$$\int_{\phi_{\text{initial}}}^{\phi_{trans}} \frac{d\phi}{\phi} = -n\sigma \int_0^x dx$$

$$\ln \frac{\phi_{trans}}{\phi_{\text{initial}}} = -n\sigma x$$

$$\phi_{trans} = \phi_{\text{initial}} \, e^{-n\sigma x}$$

Thus we see exponential absorption of the incident projectile beam (and have thus derived a form of the Lambert-Beers Law). The number of reactions that are occurring is the difference between the initial and transmitted flux, i.e.

$$\phi_{\text{initial}} - \phi_{trans} = \phi_{\text{initial}}(1 - \exp(-n\sigma x))$$

The foregoing discussion focusses on the attenuation of the incident beam and thus refers to all reactions. In many cases, we are interested in only one of several reactions that may be taking place. We can refer to the cross section for that particular reaction. In addition, we may be interested not only in a specific product, but a particular product moving in a particular direction relative to the direction of the projectile beam.

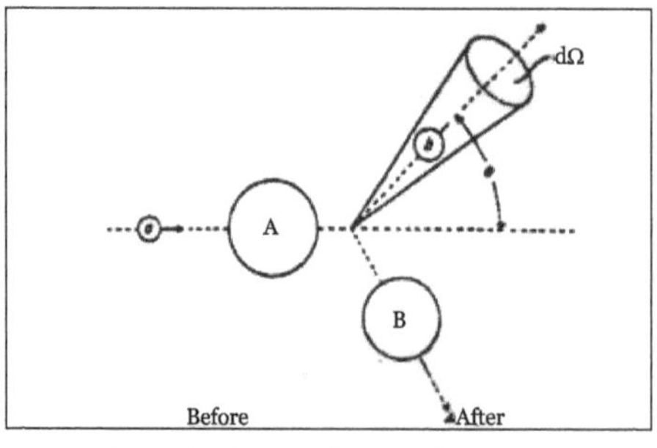

A typical experimental setup.

In this case, we can speak of a differential cross section, or the cross section per unit solid angle, $d\sigma / d\Omega$. For a thin target, we have,

$$dN / d\Omega = \phi n(d\sigma / d\Omega)dx$$

where $dN / d\Omega$ is the number of particles detected moving in a particular direction per unit solid angle. The total cross section, σ, is given as,

$$\sigma = \int_0^{2\pi}\int_0^{\pi} \frac{d\sigma}{d\Omega}(\theta)\sin\theta d\theta d\phi$$

The description given is appropriate for work at accelerators, where one has a beam of particles that is smaller than the target. In this case, the beam intensity, ϕ is given in particles/sec and the target density, N, is given in atoms/cm². In a nuclear reactor, we immerse a small target in a sea of neutrons. In this case, the neutron flux, ϕ, represents the number of neutrons passing through the target per cm² per sec and N is the total number of atoms in the target. Otherwise the arithmetic is the same. For charged particles from an accelerator, the beam intensity is usually measured as a current. Thus for a beam of protons with a current of 1 μampere, we have:

$$\phi = (1\ \mu ampere)(10^{-6}\ Coul/\sec/\ \mu ampere)\left(\frac{1}{1.602 \times 10^{-19}\ Coul\ /\ proton}\right) = 6.24 \times 10^{12}\ protons/\sec$$

For a beam of some other ion with charge q, one simply divides by the charge on the ion to get the projectile beam intensity. Thus, for a beam of 4 mamperes of Ar^{+17} ions, we have:

$$\phi = (4x\,10^{-6}\ Coul/\sec)\left(\frac{1}{17 \times 1.602 \times 10^{-19}\ Coul\ /\ Ar}\right) = 1.47 \times 10^{12}\ Ar/\sec$$

To put the intensities of beams of differing charges on a common footing, it is common to quote charge particle beam intensities in units of particle microamperes or particle nanoamperes where 1 particle microampere = $6.24\ x\ 10^{12}$ ions/ sec.

It is easy to calculate the number of product nuclei produced during an irradiation, N. If we assume the product nuclei are stable, then the number of nuclei produced is the (rate of production) x (length of the irradiation, t). For a thick target irradiation, we have:

$$\mathrm{N} = \phi(1 - \exp(-n\sigma\Delta x))t$$

For a thin target if we expand $(1 - \exp(-n\sigma\Delta x))$, we have:

$$\mathrm{N} = \phi n\sigma\Delta\,\mathrm{xt}$$

But, what if the products are radioactive? Then some of them will decay during the irradiation. In this case, we can setup the familiar differential equations,

$$\frac{dN}{dt} = (rate\ of\ production) - (rate\ of\ decay)$$

$$\frac{dN}{dt} = n\sigma\Delta x\phi - \lambda N$$

$$\frac{dN}{n\sigma\Delta x\phi - \lambda N} = dt$$

Multiplying by λ and rearranging,

$$\frac{d(\lambda N)}{\lambda N - n\,\sigma\Delta x\phi} = -\lambda dt$$

Integrating, we have:

$$\text{In}\,(\lambda N - n\,\sigma\Delta x\phi)\Big|_0^N = -\lambda t\Big|_0^t$$

$$\frac{\lambda N - n\,\sigma\Delta x\phi}{-n\,\sigma\Delta x\phi} = e^{-\lambda t}$$

$$A = \Delta N = n \mapsto \hat{=} xO\,(1 - e^{-\Delta t})$$

where A is the disintegration rate of product nuclei at the end of the irradiation. The number of product nuclei, N, present at the end of the irradiation is A/λ:

$$N = \frac{n\,\sigma\Delta x\phi}{\lambda}\left(1 - e^{-\lambda t}\right)$$

This relationship is shown in figure.

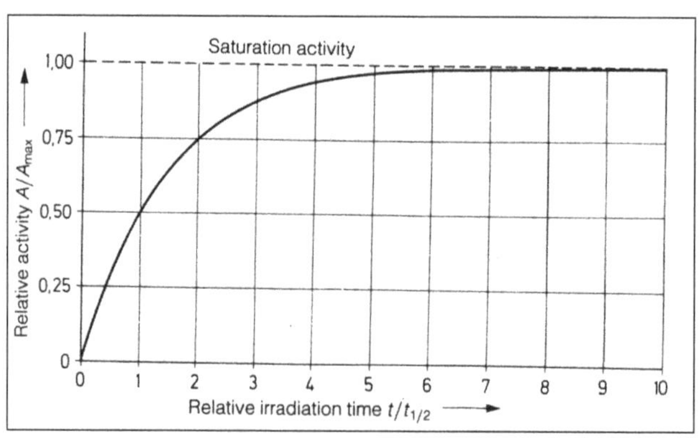

Variation of product activity during an irradiation.

Note that in the limit of infinitely long irradiation, $e^{-\lambda t} \to 0$, and thus the activity present is $n\,\sigma\Delta x\phi$ which is termed the saturation activity. Note also that for very short times (compared to the half-life of the product nuclei), $e^{-\lambda t} \to 1 - \lambda t + \ldots$ thus the activity increases linearly with time. In general, we note that we achieve ½ the saturation activity after an irradiation of one half-life, 3/4 of the saturation activity after irradiating two half-lives, 7/8 of the saturation activity after irradiating 3 half- lives, etc. Thus it does not pay to make the irradiation longer than 1-2 half-lives. (This effect can be used to tune the length of the irradiation to maximize the yield of the product of interest relative to the other reaction products).

Sample Problem: Calculate the activity of ^{254}No ($t_{1/2}$ = 55 s) produced in a one-minute irradiation of ^{208}Pb by ^{48}Ca. Assume the ^{208}Pb target thickness is 0.5 mg/cm², the ^{48}Ca beam current is 0.5 particle microamperes and the ^{208}Pb(^{48}Ca, 2n) reaction cross section is 3.0 μb.

Solution:

$$A = N\sigma\phi(1-e^{-\lambda t_i})$$

$$\mathrm{N} = (0.5 \text{ x } 10^{-3} \text{ g/ cm}^2)\,(6.02 \text{ x} 10^{23} \text{ atoms/ g-at-wt}) / 208 \text{ g/ g-at-wt} - 1.44 \text{ x} 10^{18}$$

$$\text{atoms / cm}^2$$

$$\sigma = 3 \text{ x } 10^{-30} \text{ cm}^2$$

$$\phi = (0.5 \text{ x } 10^{-6} \text{ Coulombs/ sec}) / 1.602 \text{ x} 10^{-19} \text{ Coulombs/ ion} = 3012 \text{ x} 10^{12} \text{ ions/ sec}$$

$$t_i = 60 \text{ sec}$$

$$\lambda = (\text{In } 2) / 55 \text{ sec} = 1.26 \text{ x } 10^{-2} \text{ sec}^{-1}$$

$$\mathrm{A} = 7.2 \text{ dis/ sec}$$

Let us consider what we can learn about cross sections from some general considerations. Consider the reaction of an uncharged particle (a neutron) with a nucleus as shown in figure.

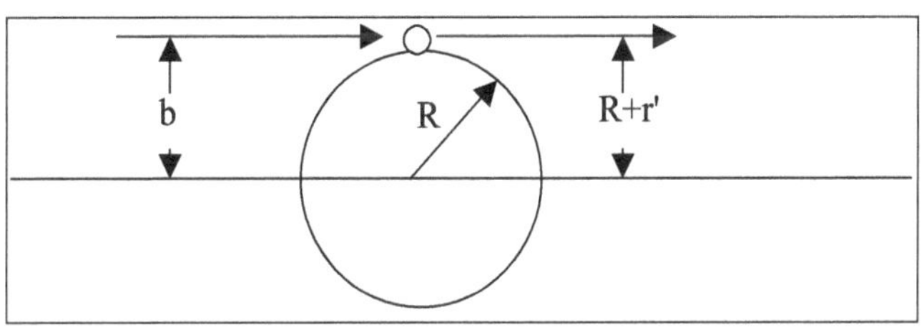

Schematic diagram of a grazing collision of a neutron with a nucleus.

The neutron makes a grazing collision with the nucleus. The impact parameter b is taken to be the sum of the radii of the projectile and target nuclei. Thus the cross section can be written as,

$$\sigma \approx \pi(\mathrm{R} + \mathrm{r'})^2 = \pi r_0^2\,(\mathrm{A_P} + \mathrm{A_T})^2$$

where r' is the radius of the projectile. Applying classical mechanics to this problem, we can write for the orbital angular momentum, ℓ,

$$\ell = \vec{r} \text{ x } \overrightarrow{p} = pb$$

In quantum mechanics, ℓ ℓ° and the momentum, p, is given by:

$$p = \frac{\hbar}{\lambda\!\!\!^{-}}$$

Thus we have,

$$\ell\hbar = \frac{\hbar b}{\lambda\!\!\!^{-}}$$

$$b = \ell\lambda\!\!\!^{-}$$

This is not quite right because ℓ is quantized but b is not. We get around this by associating b with certain rings or zones on the target. Figure suggests that for head-on collisions (ℓ = 0), the range of b is from 0 to λ while for ℓ = 1 collisions, the range of b is from λ to 2λ. Thus the cross section is larger for larger impact parameters and these larger impact parameters are associated with larger angular momenta. We can write the cross section for a specific value of ℓ as,

$$\sigma_\ell = \pi(\ell+1)^2 \lambda\!\!\!^- {}^2 - \pi \ell^2 \lambda\!\!\!^- {}^2$$

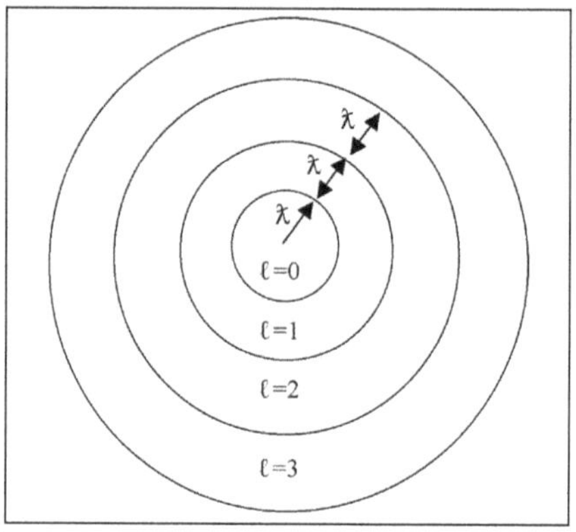

Schematic bulls-eye view of the target nucleus.

$$\sigma_\ell = \pi \lambda\!\!\!^- {}^2(\ell^2 + 2\ell + 1 - \ell^2)$$

$$\sigma_\ell = \pi \lambda\!\!\!^- {}^2 \, (2\ell + 1)$$

The total reaction cross section is obtained by summing over all ℓ values as,

$$\sigma_{total} = \sum_\ell \sigma_\ell = \sum_{\ell=0}^{\ell_{max}} \pi \lambda\!\!\!^- {}^2(2\ell+1) = \pi \lambda\!\!\!^- {}^2 \sum_{\ell=0}^{\ell_{max}} (2\ell+1) = \pi \lambda\!\!\!^- {}^2 \, (\ell_{max}+1)^2$$

We can write for the maximum angular momentum, ℓ_{max},

$$\ell_{max} = \frac{R}{\lambda\!\!\!^-}$$

$$\ell_{max} + 1 = \frac{R + \lambda\!\!\!^-}{\lambda\!\!\!^-}$$

Thus we have for the total cross section,

$$\sigma_{total} = \pi(R + \lambda\!\!\!^-)^2$$

The total cross section is proportional to the size of the target nucleus and the "size" of the projectile nucleus. Since the wave length of the projectile, λ, goes to infinity as the projectile energy goes to zero, the cross sections for neutrons at low energies can be very large. We need to indicate how the problem would look if we used quantum mechanics to treat it.

In quantum mechanics, we can write an similar expression for the total reaction cross section,

$$\sigma_{total} = \pi\lambda\!\!\!^{-2} \sum_{\ell=0}^{\infty} (2\ell+1)\, T_\ell$$

where the transmission coefficient T_ℓ varies between 0 and 1. The transmission coefficient expresses the probability that a given angular momentum transfer ℓ will occur. At high projectile energies, $T_\ell = 1$ for $\ell \le \ell_{max}$ and $T_\ell = 0$ for $\ell \ge \ell_{max}$. At very low projectile energies, $T_\ell = \varepsilon^{1/2}$ for $\ell = 0$ and $T_\ell = 0$ for $\ell > 0$, where ε is the projectile energy. Thus, at very low energies, we have:

$$\sigma_{total} \propto \pi\lambda\!\!\!^{-2} \sqrt{\varepsilon} \propto \pi \frac{\hbar}{2m\varepsilon} \sqrt{\varepsilon} \propto \frac{1}{\sqrt{\varepsilon}}$$

Such behavior of the cross sections for neutron induced reactions is referred to as '1/v" behavior.

Now let us consider the interaction of a charged particle with a nucleus as shown in figure.

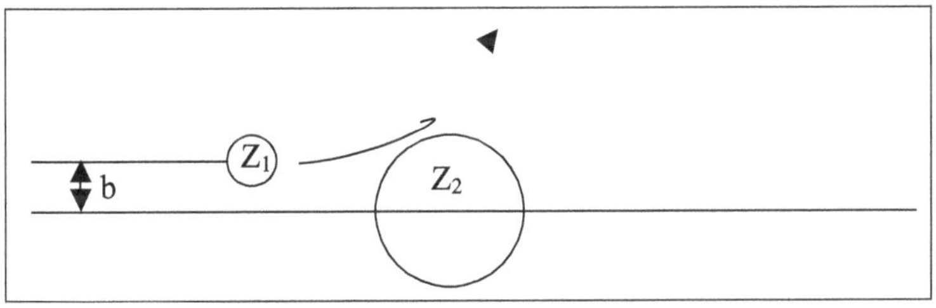

Schematic diagram of a charged particle induced reaction.

As the projectile approaches the target nucleus, it feels the long range Coulomb force and is deflected. As a consequence, the range of collisions corresponds to a smaller range of impact parameters. If the incident projectile has an energy ε at an infinite separation from the target nucleus, at the distance of closest approach R, it has a kinetic energy of ε - B where B, the Coulomb barrier, is given by,

$$B = Z_1 Z_2 e^2 / R$$

At the point of closest approach, the momentum p of the projectile is $(2mT)^{1/2}$. Thus we can write,

$$p = (2\,mT)^{1/2} = (2\mu)^{1/2}(\varepsilon - B)^{1/2} = (2\mu\varepsilon)^{1/2}(1 - B/\varepsilon)^{1/2}$$

where μ is the reduced mass of the system $\left(= A_1A_2/\left(A_1 + A_2\right)\right)$. Classically we have, for the orbital angular momentum,

$$\ell = \vec{r}\ x\ \overrightarrow{p}$$

$$\ell_{max} = R\,(2\mu\varepsilon)^{1/2}\left(1 - \frac{B}{\varepsilon}\right)^{1/2}$$

Quantum mechanically, we have $\ell \rightarrow \ell\,\hbar$. So we can write,

$$\sigma_{total} = \pi\lambda\!\!\!^{-2}(\ell_{\max}+1)^2 \approx \pi\lambda\!\!\!^{-2}\ell_{\max}^2 = \pi\lambda\!\!\!^{-2}R^2\frac{2\mu\varepsilon}{\hbar}\left(1-\frac{B}{\varepsilon}\right) = \pi\lambda\!\!\!^{-2}R^2\left(1-\frac{B}{\varepsilon}\right)$$

$$\sigma_{total} = \pi R^2\left(1-\frac{B}{\varepsilon}\right)$$

Note this last classical expression is valid only when ε > B. The combined general properties of cross sections for charged and uncharged particles are shown in figure.

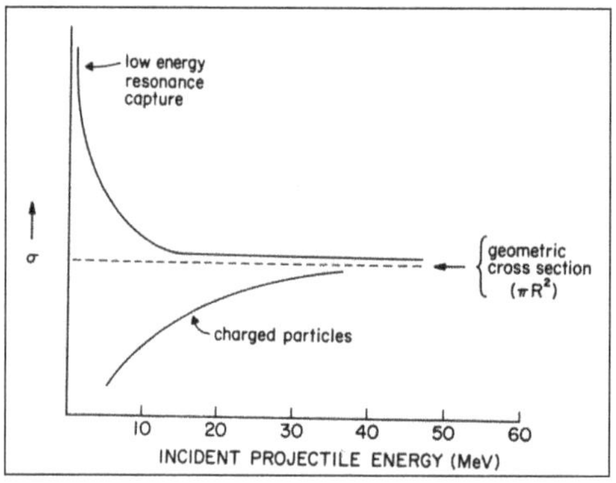

Near threshold behavior of neutron and charged particle induced reactions.

Sample problem: Calculate the energy dependence of the total reaction cross section for the ^{48}Ca + ^{208}Pb reaction.

Solution:

$$\sigma_{total} = \pi R^2\,(1-\frac{B}{\varepsilon})$$

$$R = R_{Pb} + R_{Ca} = 1.2\,(208^{1/3} + 48^{1/3}) = 11.47 \text{ fm}$$

$$B = Z_1Z_2e^2/R = (82)(20)(1.44\,\text{MeV} - \text{fm})/11.47\,\text{fm} = 205.9\,\text{MeV}$$

ε = energy of the projectile in the cm system.

Elastic Scattering

When a beam of particles, represented in quantum mechanics by a plane wave, hits a nucleus the wave function is modified by the presence of a scattering potential V (r), responsible for the appearance of a phase in the outgoing part of the wave. Elastic scattering is just one of the channels for the which the reaction can proceed and is known as elastic channel. Inelastic scattering and all the other channels are grouped in the reaction channel.

The occurrence of a nuclear reaction through a given reaction channel leads to a modification of the outgoing part of the wave function not only by a phase factor, but also by changing its magnitude, indicating that there is a loss of particles in the elastic channel. For a projectile with momentum $p = \hbar k$, this can be expressed by:

$$\Psi \sim \frac{1}{2i}\sum_{l=0}^{\infty}(2l+1)i^l P_l(\cos\theta)\frac{\eta_l e^{i(kr - l\pi/2)} - e^{-i(kr - l\pi/2)}}{kr},$$

where the complex coefficient η_l is the factor mentioned above. If $\eta = 1$ the sum in above equation can be done analytically, leading to $\Psi \sim \exp(\text{ik.r})$ i.e. a plane wave. But if $\eta_l = \exp[i\delta l]$, with δ_l real, the incoming and outgoing waves have the same magnitude, i.e. the scattering is elastic.

The sum in equation $\Psi \sim \frac{1}{2i}\sum_{l=0}^{\infty}(2l+1)i^l P_l(\cos\theta)\frac{\eta_l e^{i(kr - l\pi/2)} - e^{-i(kr - l\pi/2)}}{kr}$, is know as partial wave expansion of the scattering wave. The label l = 0, 1, 2, · · · = (s, p, d, · · · waves) denotes the contribution of a particular angular momentum (in units of $\hbar$) to the total wavefunction. Classically the angular momentum of an incident particle is given by l = kR, where R is known as impact parameter which is the perpendicular distance to the target if the projectile were undeflected. In quantum mechanics l is not continuous, varying in steps of one.

We can rewrite equation $\Psi \sim \frac{1}{2i}\sum_{l=0}^{\infty}(2l+1)i^l P_l(\cos\theta)\frac{\eta_l e^{i(kr - l\pi/2)} - e^{-i(kr - l\pi/2)}}{kr}$, as a sum of a plane wave and a scattering outgoing wave in the form $\Psi \sim \exp(ik.r) + f(\theta)e^{ikr}/r$, where $f(\theta)$ accounts for the distortion of the outgoing wave at the scattering angle θ. It is known as the scattering amplitude:

$$f(\theta) = \frac{1}{2ik}\sum_{l=0}^{\infty}(2l+1)(\eta_l - 1)P_l(\cos\theta)$$

One can now calculate the cross section by counting the number of particles that are scattered through angle θ. This can be done by calculating the particle current associated with the wavefunction Ψ. One gets the differential scattering cross section,

$$\frac{d\sigma_e}{d\Omega} = |f(\theta)|^2 = \frac{1}{4k^2}\left|\sum_{l=0}^{\infty}(2_l+1)(1-\eta_l)P_l(\cos\theta)\right|^2.$$

The total scattering cross section, equation $\sigma_\alpha = \int d\Omega\left[\frac{d\sigma_\alpha(\Omega)}{d\Omega}\right]$ becomes:

$$\sigma_e = \pi\bar{\lambda}^2\sum_{l=0}^{\infty}(2l+1)|1-\eta_l|^2$$

with $\bar{\lambda} = \lambda/2\pi = 1/k$.

Rutherford Scattering

One of the first possible outcomes of the collision of a charged particle with a nucleus is Rutherford or Coulomb scattering. The incident charged particle feels the long-range Coulomb force of the positively charged nucleus and is deflected from its path.

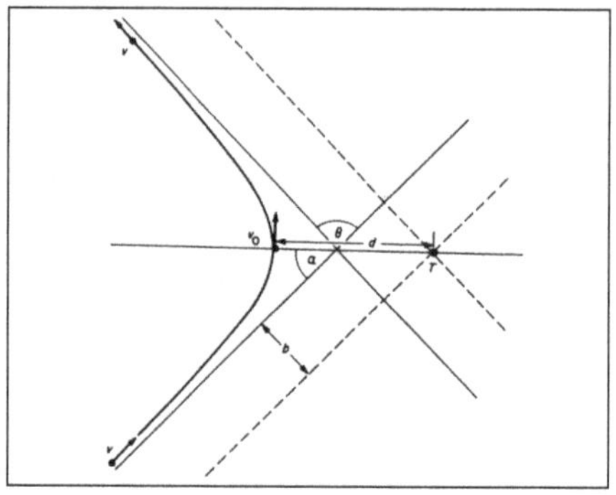

Schematic diagram of Rutherford scattering.

The Coulomb force acting between a projectile of mass m, charge Z_1e and a target nucleus with charge Z_2e is given as,

$$F_{coul} = \frac{Z_1 Z e^2}{r^2}$$

where r is the distance between the projectile and target nuclei. The potential energy in this interaction is given as,

$$PE = \frac{Z_1 Z_2 e^2}{r}$$

Consider a target nucleus that is much heavier than the projectile nucleus so that we can neglect the recoil of the target nucleus in the interaction. The projectile will follow a hyperbolic orbit, as shown in figure where b is the impact parameter, T_p is the kinetic energy of the projectile and d is the distance of closest approach. At infinity, the projectile velocity is v. At r=d, the projectile velocity is v_o.

Conservation of energy gives,

$$\frac{1}{2}mv^2 = \frac{1}{2}mv_0^2 + \frac{Z_1 Z_2 e^2}{d}$$

Rearranging, we have:

$$\left(\frac{v_0}{v}\right)^2 = 1 - \frac{d_0}{d}$$

where d_o is given as,

$$d_0 = \frac{2Z_1 Z_2 e^2}{mv^2} = \frac{Z_1 Z_2 e^2}{T_p}$$

If we now invoke the conservation of angular momentum, we can write:

$$\mathrm{mvb} = \mathrm{mv_0\, d}$$

$$b^2 = \left(\frac{v_0}{v}\right)^2 d^2 = d(d - d_0)$$

It is a property of a hyperbola that,

$$\mathrm{d} = \mathrm{b}\cot(\alpha / 2)$$

Substituting from above, we have:

$$\tan\alpha = 2\mathrm{b}/\,\mathrm{d_0}$$

Since $\theta = \pi - 2\alpha$, we can write,

$$\cot\left(\frac{\theta}{2}\right) = \frac{2b}{d_0}$$

In figure, we show the expected orbits of the projectile nuclei after undergoing Rutherford scattering for a typical case. Note that the most probable grazing trajectories result in projectiles being scattered to forward angles but that some head on collisions result in large angle scattering.

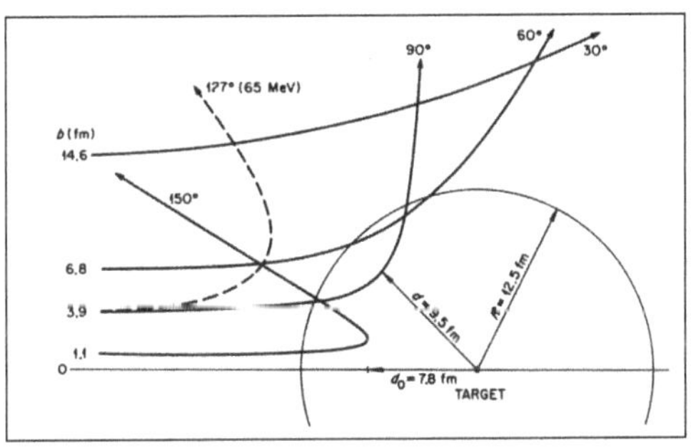

Diagram showing some representative projectile orbits for the interaction of 130 MeV ^{16}O with ^{208}Pb.

It was these latter events that led Rutherford to conclude that there was a massive object at the center of the atom.

We can make these observations more quantitative by considering the situation where a flux of I_0 particles/unit area is incident on a plane normal to the beam direction. The flux of particles passing through a ring of width db and with impact parameters between b and b + db is given as,

$$\mathrm{dI} = (\text{Flux/unit area})(\text{area of ring})$$

$$\mathrm{dI} = \mathrm{I_0}(2\pi\,\mathrm{bdb})$$

Substituting from above, we have:

$$dI = \frac{1}{4}\pi I_0 d_0^2 \frac{\cos\left(\frac{\theta}{2}\right)}{\sin^3\left(\frac{\theta}{2}\right)} d\theta$$

If we want to calculate the number of projectile nuclei that undergo Rutherford scattering into a solid angle $d\Omega$ at a plane angle θ, we can write:

$$\frac{d\sigma}{d\Omega} = \frac{dI}{I_0} = \frac{1}{d\Omega} = \left(\frac{d_0}{4}\right)^2 \frac{1}{\sin^4\left(\frac{\theta}{2}\right)} = \left(\frac{Z_1 Z_2 e^2}{4T_p^{cm}}\right)^2 \frac{1}{\sin^4\left(\frac{\theta}{2}\right)}$$

if we remember that,

$$d\Omega = 2\pi\, sin\theta\, d\theta$$

Note the strong dependence of the Rutherford scattering cross section upon scattering angle. Remember that Rutherford scattering is not a nuclear reaction, as it does not involve the nuclear force, only the Coulomb force between the charged nuclei. Remember that Rutherford scattering will occur to some extent in all studies of charged particle induced reactions and will furnish a "background" of scattered particles at forward angles.

Sample calculation: Calculate the differential cross section for the Rutherford scattering of 215 MeV (lab energy) ^{48}Ca from ^{208}Pb at an angle of 20°.

$$\frac{d\sigma}{d\Omega} = \left(\frac{Z_1 Z_2 e^2}{4T_p^{cm}}\right)^2 \frac{1}{\sin^4\left(\frac{\theta}{2}\right)}$$

$$T_p^{cm} = 215x\frac{208}{256} = 174.7\ MeV$$

$$\frac{d\sigma}{d\Omega} = \left(\frac{20x82x1.44}{4x174.7}\right) \frac{1}{\sin^4\left(\frac{20}{2}\right)} = 12562 f\ m^2\ 125.6b.$$

Direct Reactions

A direct reaction is said to occur if one of the participants in the initial two-body interaction involving the incident projectile leaves the nucleus. Generally speaking, these direct reactions are divided into two classes, the stripping reactions in which part of the incident projectile is "stripped away" and enters the target nucleus and the pickup reactions in which the outgoing emitted particle is a combination of the incident projectile and a few target nucleons.

Let us consider stripping reactions first and in particular, the most commonly encountered stripping reaction, the (d, p) reaction. Formally the result of a (d, p) reaction is to introduce a neutron into the target nucleus and thus this reaction should bear some resemblance to the simple neutron capture reaction. But because of the generally higher angular momenta associated with the (d, p) reaction, there can be differences between the two reactions. Consider the A (d, p) B* reaction where the recoil nucleus B is produced in an excited state B*. We sketch out a simple picture of this reaction and the momentum relations in figure.

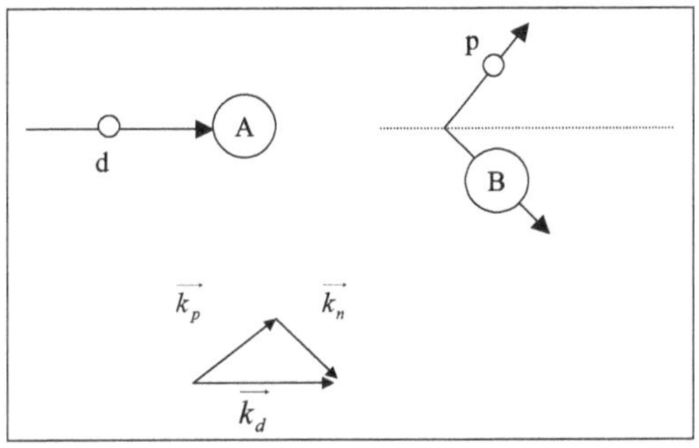

Sketch of a (d, p) reaction and the associated momentum triangle.

The momentum diagram for the reaction shown in figure assumes the momentum of the incident deuteron is $k_d\hbar$, the momentum of the emitted proton is $k_p\hbar$ while $k_n\hbar$ is the momentum of the stripped neutron. From conservation of momentum we have,

$$k_n^2 = k_d^2 + k_p^2 - 2k_dk_p \cos\theta$$

If the neutron is captured at impact parameter R, the orbital angular momentum transferred to the nucleus, $\ell_n\hbar$, is given by,

$$\ell_n\hbar = \vec{r} \times \vec{p} = Rk_n\hbar$$

$$\ell_n = Rk_n$$

Since we have previously shown that k_n is a function of the angle θ, we can now associate each orbital angular momentum transfer in the reaction with a given angle θ corresponding to the direction of motion of the outgoing proton. Thus the (d, p) reaction becomes a very powerful spectroscopic tool. By measuring the energy of the outgoing proton, we can deduce the Q value of the reaction and thus the energy of any excited state of the residual nucleus that is formed. From the direction of motion of the proton, we can deduce the orbital angular momentum transfer in the reaction, ℓ_n. If we know the ground state spin and parity of the residual nucleus, we can deduce information about the spin and parity of the excited states of the residual nucleus using the rules,

$$||(||J_A - \ell_n|| - 1/2|| \le J_{B*} \le J_A + \ell_n + 1/2$$

$$\pi_A\pi_{B*} = (-1)^{\ell_n}$$

Other stripping reactions are reactions like $(\alpha, t), (\alpha, d)$, etc. Typical pickup reactions are (p, d), (p, t), (α, ^{6}Li),etc.

Sample Problem: Calculate the angle at which the (d, p) cross section has a maximum for ℓ = 0,1,2,3 and 4. Assume a deuteron energy of 7 MeV and a proton energy of 13 MeV. Use R = 6 fm.

$$k_d = 0.82\ \text{fm}^{-1}$$

$$k_p = 0.79\ \text{fm}^{-1}$$

$$k_n = \frac{\ell}{R}$$

Thus for $\ell = 0,1,2,3,4, k_n = 0, 0.17\,\text{fm}^{-1}, 0.33\,\text{fm}^{-1}, 0.50\,\text{fm}^{-1}, 0.67\ \text{fm}^{-1}$. Solving the momentum triangle,

$$\cos\vartheta = \frac{-k_n^2 + k_d^2 + k_p}{2\,k_d k_p}$$

$$\theta = 0°, 12°, 24°, 36°, 49°\ \text{for}\ \ell = 0,1,2,3,4$$

Compound Nucleus Reactions

The compound nucleus is a relatively long-lived reaction intermediate that is the result of a complicated set of two body interactions in which the energy of the projectile is distributed among all the nucleons of the composite system. How long does the compound nucleus live? From our definition, we can say the compound nucleus must live for at least several times the time it would take a nucleon to traverse the nucleus (10^{-22} seconds). Thus the time scale of compound nuclear reactions is of the order of $10^{-18} - 10^{-16}$ sec. Lifetimes as long as 10^{-14} sechave been observed. These relatively long times should be compared to the typical time scale of a direct reaction of 10^{-22}sec.

Another important feature of compound nucleus reactions is the mode of decay of the compound nucleus is independent of its mode of formation. (the Bohr independence hypothesis or the amnesia assumption). While this statement is not true in general, it remains a useful tool for understanding certain features of compound nuclear reactions.

Because of the long time scale of the reaction and the "amnesia" of the compound nucleus about its mode of formation, one can show that the angular distribution of the products is symmetric about 90 degrees (in the frame of the moving compound nucleus).

The cross section for a compound nuclear reaction can be written as the product of two factors, the probability of forming the compound nucleus and the probability that the compound nucleus decays in a given way. The probability of forming the compound nucleus can be written as,

$$\sigma = \pi \lambda\!\!\!^{-2} \sum_{\ell=0}^{\infty} (2\ell + 1)\, T_\ell$$

The probability of decay of the compound nucleus into a given set of products β can be written as,

$$\text{probability} = \left[\frac{T_\ell^\beta(E_\beta)}{\sum_{\ell_\gamma, E_\gamma} T_\ell^\gamma(E_\gamma)} \right]$$

where T_I is the transmission coefficient for CN decay into products i. Figure shows a schematic view of the levels of the compound nucleus.

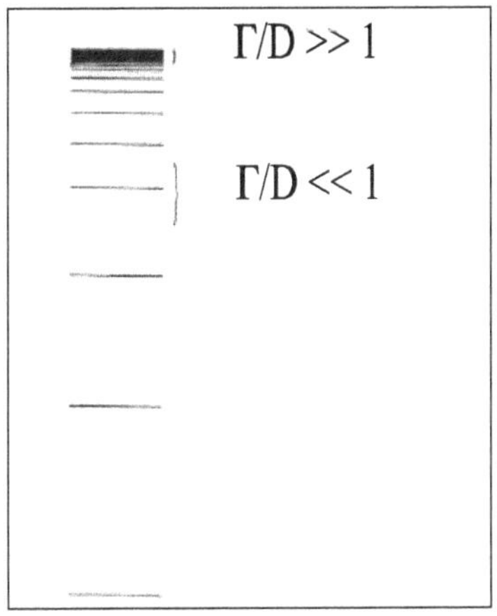

Schematic view of the levels of a compound nucleus. The increasing number of levels as the CN excitation energy increases. Quantitatively, the number of levels per MeV of excitation energy, E, increases approximately exponentially as $E^{1/2}$.

The interesting categories of CN reactions can be defined by the ratio of the width of a compound nucleus level, Γ, to the average spacing between compound nuclear levels, D. (Recall from the Heisenberg uncertainty principle that $\Gamma \cdot \tau \geq \hbar$, where τ is the mean life of a compound nucleus level.) The categories are: (a) $\Gamma / D << 1$, *i.e.*, the case of isolated non-overlapping levels of the compound nucleus and (b) $\Gamma / D >> 1$, the case of many overlapping levels in the compound nucleus. Intuitively category: (a) reactions are those in which the excitation energy of the compound nucleus is low while category, (b) reactions are those in which the excitation energy is high.

Let us first consider the case of $\Gamma / D << 1$. This means that at certain values of the compound nucleus excitation energy, individual levels of the compound nucleus can be excited (*i.e.*, when the excitation energy exactly equals the energy of a given CN level.) When this happens, there will be a sharp rise or resonance in the reaction cross section akin to the absorption of infrared radiation by sodium chloride when the radiation frequency equals the natural crystal oscillation frequency. In this case, the formula for the cross section (the Breit Wigner single level formula) for the reaction $a + A \rightarrow C \rightarrow b + B$ is,

$$\sigma = \pi D^2 \frac{(2J_C + 1)}{(2J_A + 1)(2J_a + 1)} \frac{\Gamma_{aA}\Gamma_{bB}}{(\varepsilon - \varepsilon_0)^2 + \left(\frac{\Gamma}{2}\right)^2}$$

where J_i is the spin of ith nucleus, Γ_{aA}, Γ_{bB}, and Γ are the partial widths for the formation of C, the decay of C into b+B and the total width for the decay of C, respectively. The symbols ε and ε_0 refer to the energy of the projectile nucleus and the projectile energy corresponding to the excitation of a single isolated level.

Applying this formula to the case of (n, γ) reactions gives,

$$\sigma_{n,\gamma} = \pi D^2 \frac{(2J_C + 1)}{(2J_A + 1)(2)} \frac{\Gamma_n \Gamma_\gamma}{(\varepsilon - \varepsilon_0)^2 + \left(\frac{\Gamma}{2}\right)^2}$$

An example of this behavior is shown in figure.

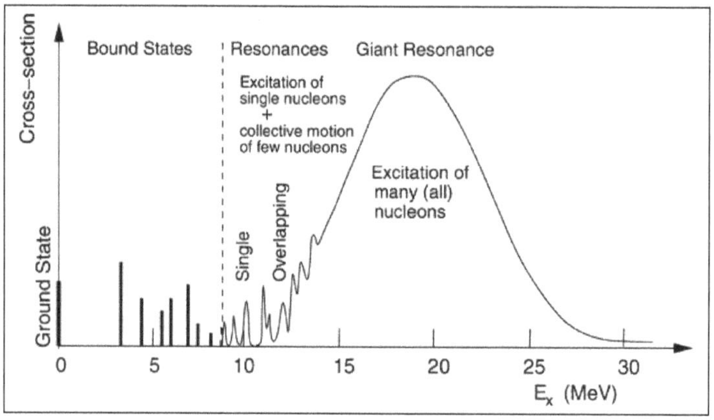

Resonance behavior.

Resonances are seen in low energy neutron induced reactions where one is populating levels in the compound nucleus at excitation energies of the order of the neutron binding energy where the spacing between levels is of the order of eV. For neutron energies well below ε_0, so that $(\varepsilon - \varepsilon_0)^2 \approx\sim \varepsilon_0^2$, then the cross section for the (n, γ) reaction goes as 1/v where v is the neutron velocity.

Let us now consider the case where $\Gamma / D >> 1$, i.e., many overlapping levels of the compound nucleus are populated. (We are also tacitly assuming a large range of compound nuclear excitation energies). The cross section for the reaction $\mathrm{a + A \rightarrow C \rightarrow + B}$ can be written as,

$$\sigma_{\mathrm{ab}} = \sigma_{\mathrm{C}}\, P_{\mathrm{C}}(\mathrm{b})$$

where $\sigma_{\mathrm{C}}(\mathrm{a})$ is the cross section for the formation of the compound nucleus C and $\mathrm{P_C}$ is the probability that C will decay to form b + B. Clearly $\Sigma\, \mathrm{P_C}(\mathrm{b}) = 1$. Now let us consider, in detail, the probability that emitted particle b has an energy ε_{b}. First of all, we can write down that the maximum energy that b can have is $E_C^* - \mathrm{S_b}$ where E_C^* is the excitation energy of the compound nucleus and $\mathrm{S_b}$ is the separation energy of b in the residual nucleus B. But b can be emitted with a variety of energies less than this with the result that the nucleus B will be left in an excited state. By using the arguments of detailed balance from statistical mechanics. We can write for the probability of emitting a particle b with an energy ε_{b} ($< \varepsilon_{\mathrm{max}}$, leaving the nucleus B at an excitation energy E_B^*):

$$W_b(\varepsilon_b)\, d\varepsilon_b = \frac{(2J_b + 1)\mu}{\pi^2 h^3} \varepsilon_{\mathrm{b}} \sigma_{inv} \frac{\rho(E_B^*)}{\rho(E_C^*)} d\varepsilon_b$$

In this equation, μ is the reduced mass of the system and σ_{inv} is the cross section for the inverse process in which the particle b is captured by the nucleus B where b has an energy, ε_b. The symbols $\rho(E^*_B)$ and $\rho(E^*_C)$ refer to the level density in the nucleus B excited to an excitation energy E_B^* and the level density in the compound nucleus C excited to an excitation energy, E_C^*. The inverse cross section can be calculated using the same formulas used to calculate the compound nucleus formation cross section. Using the Fermi gas model, we can calculate the level densities of the excited nucleus as,

$$\rho(E^*) = C\exp\{2\,(aE^*)^{1/2}\}$$

where the level density parameter, a, is A/12 – A/8. The nuclear temperature T is given by the relation,

$$E^* = aT^2 - T$$

The ratio of emission widths for emitted particles x and y is given as,

$$\frac{\Gamma_x}{\Gamma_y} = \frac{g_x\mu_x}{g_y\mu_y}\frac{R_x}{R_y}\frac{a_x}{a_y}\exp\left\{2\left(a_xR_x\right)^{1/2} - 2\left(a_yR_y\right)^{1/2}\right\}$$

where g_i is the spin of the ith particle, a_i and R_i are the level density parameter and maximum excitation energy for the residual nucleus that results from the emission of the ith particle. R is formally $E^* - S - \varepsilon_s$ where ε_s is the threshold for charged particle emission (ε_s for neutrons is 0).

If the emitted particles are neutrons, the emitted neutron energy spectrum has the form,

$$N(\varepsilon)d\varepsilon = \frac{\varepsilon}{T^2}\exp(-\varepsilon / T)d\varepsilon$$

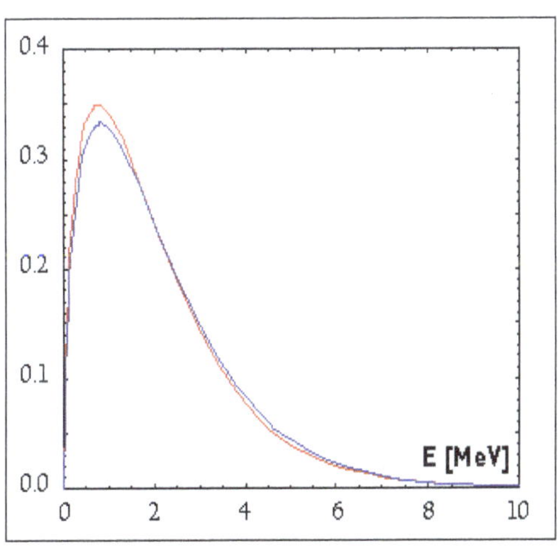

Spectrum of evaporated neutrons.

In other words, the particles are emitted with a Maxwellian energy distribution. The most probable energy is T while the average energy is 2T. What we are saying is that the compound nucleus “evaporates” particles like molecules leaving the surface of a hot liquid. By measuring the energy spectrum of the particles emitted in a compound nuclear reaction, we are using a "nuclear thermometer" in that,

$$\frac{d(\ln N(\varepsilon)d\varepsilon)}{\dfrac{\varepsilon\sigma_{inv}}{d\varepsilon}} = \frac{1}{T}$$

Charged particles may also be evaporated except the minimum kinetic energy is not zero as it is for neutrons. Instead the threshold for charged particle emission ε_s (which is approximately the Coulomb barrier) determines the minimum energy of an evaporated particle. The energy spectrum of evaporated charged particles is,

$$N(\varepsilon)d\varepsilon = \frac{\varepsilon - \varepsilon_s}{T^2}\exp\left(-(\varepsilon - \varepsilon_s)/T\right)d\varepsilon$$

What will be the distribution in space of the reaction products? Let us assume that because the compound nucleus has "forgotten" its mode of formation, there should be no preferential direction for the emission of the decay products.

Thus we might expect that all angles of emission of the particles, θ, to be equally probable. Thus we would expect that P(θ), the probability of emitting a particle at an angle θ, might be a constant. Then we would expect that $d\sigma / d\Omega(\theta)$ would be given as,

$$\frac{d\sigma}{d\Omega}(\theta) = \int P(\vartheta)\frac{d\theta}{d\Omega}$$

This assumes that we are making the measurement of the emitted particle angular distribution in the frame of the moving compound nucleus. In the laboratory frame, there will be appear to be more particles emitted in the forward direction (with higher energies) than are emitted in the backward direction).

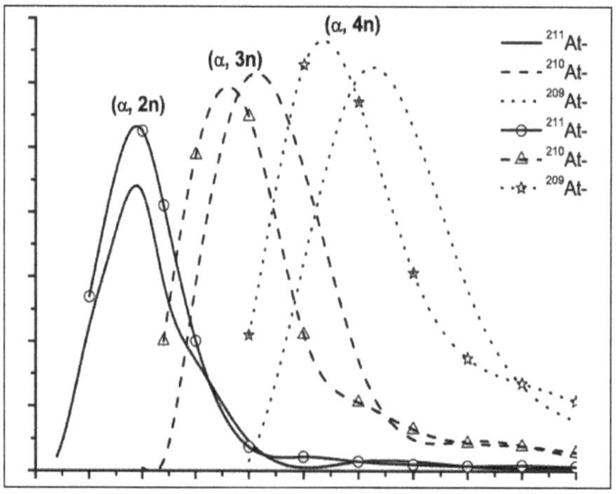

Excitation function the for ^{209}Bi(a, xn) reaction.

The energy variation of the cross section (the *excitation function*) for processes involving evaporation is fairly distinctive as shown in figure, where the excitation function for the ^{209}Bi(α, xn) reaction is shown. Starting from the threshold ε_s, the cross section rises with increasing energy because the formation cross section for the compound nucleus is increasing. Eventually the excitation energy of the compound nucleus becomes large enough that emission of two neutrons is energetically possible. This "2n out" process will dominate over the "1n out" process and the cross section for the "1n out" process will decrease. Eventually the "3n out" process will dominate over the "2n out" process. We expect the peaks for the individual "xn out" processes to be at $S_{n1}+2T, S_{n1}+S_{n2}+4T, S_{n1}+S_{n2}+S_{n3}+6T$, etc. (where we neglect any changes in T during the emission process).

Let us recapitulate what we have said about compound nuclear reactions. We have said that they are basically nuclear reactions with a long-lived reaction intermediate, which is formed by a complicated set of two-body interactions. We can write down a set of equations that describes the overall compound nuclear cross section. We have shown how this general formula simplifies for specific cases, the case of exciting a single level of the compound nucleus where we see spikes or resonances in the cross section and the case of higher excitation energies where the compound nucleus behaves like a hot liquid, evaporating particles. At all excitation energies, the angular distribution of the reaction products is symmetric with respect to a plane perpendicular to the incident particle direction.

Neutron Capture

The neutron capture is one of the possible absorption reactions that may occur. In fact, for non-fissionable nuclei it is the only possible absorption reaction. Capture reactions result in the loss of a neutron coupled with the production of one or more gamma rays. This capture reaction is also referred to as a radiative capture or (n, γ) reaction, and its cross-section is denoted by σ_γ.

Neutron Absorption Reaction

The neutron absorption reaction is the most important type of reactions that take place in a nuclear reactor. The absorption reactions are reactions, where the neutron is completely absorbed and compound nucleus is formed. Therefore a variety of emissions or decays may follow. The most important absorption reactions are divided by the exit channel into two following reactions:

- Radiative Capture: Most absorption reactions result in the loss of a neutron coupled with the production of one or more gamma rays. This is referred to as a capture reaction, and its cross-section is denoted by σ_γ.
- Neutron-induced Fission Reaction: Some nuclei (fissionable nuclei) may undergo a fission event, leading to two or more fission fragments (nuclei of intermediate atomic weight) and a few neutrons. In a fissionable material, the neutron may simply be captured, or it may cause nuclear fission. For fissionable materials we thus divide the absorption cross section as $\sigma_a = \sigma_\gamma + \sigma_f$.

The radiative capture is a reaction, in which the incident neutron is completely absorbed and compound nucleus is formed. The compound nucleus then decays to its ground state by gamma emission. This process can occur at all incident neutron energies, but the probability of the interaction strongly depends on the incident neutron energy and also on the target energy (temperature). In fact the energy in the center-of-mass system determines this probability.

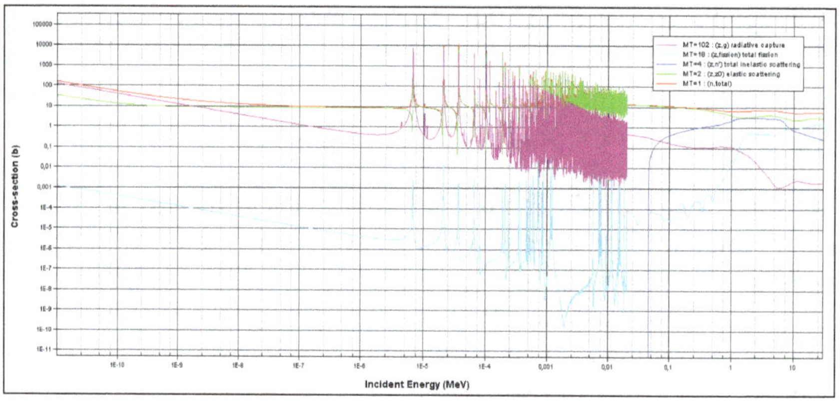

Comparison of cross-sections.

Neutron Capture at Small Neutron Flux

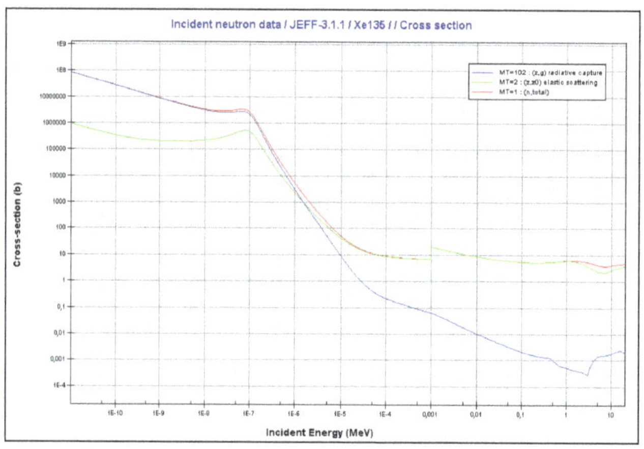

Neutron absorption and scattering: Comparison of cross-sections.

It must be noted we have to distinguish between radiative captures at small neutron flux and at high neutron flux. At small neutron flux, as in a nuclear reactor, the compound nucleus has time to decay between two neutron captures. The most usual capture reactions that occur inside a power reactor are below:

$$
\begin{aligned}
&{}^{1}_{0}n_{th} + {}^{238}_{92}\mathrm{U} \rightarrow [{}^{239}_{92}\mathrm{U}] \xrightarrow{\sim 99.9\%} {}^{239}_{92}\mathrm{U} + \gamma \\
&{}^{1}_{0}n_{th} + {}^{235}_{92}\mathrm{U} \rightarrow [{}^{236}_{92}\mathrm{U}] \xrightarrow{15\%} {}^{236}_{92}\mathrm{U} + \gamma \\
&{}^{1}_{0}n_{th} + {}^{239}_{94}Pu \rightarrow [{}^{240}_{94}Pu] \xrightarrow{27\%} {}^{240}_{94}Pu + \gamma \\
&{}^{1}_{0}n_{th} + {}^{233}_{92}\mathrm{U} \rightarrow [{}^{234}_{92}\mathrm{U}] \xrightarrow{6\%} {}^{234}_{92}\mathrm{U} + \gamma \\
&{}^{1}_{0}n_{th} + {}^{135}_{54}\mathrm{X}e \rightarrow [{}^{136}_{54}\mathrm{X}e] \xrightarrow{100\%} {}^{136}_{54}\mathrm{X}e \\
&{}^{1}_{0}n_{th} + {}^{157}_{64}Gd \rightarrow [{}^{158}_{64}Gd] \xrightarrow{100\%} {}^{158}_{64}Gd
\end{aligned}
$$

Neutron capture reactions which are of importance in power reactors.

Neutron Capture at High Neutron Flux

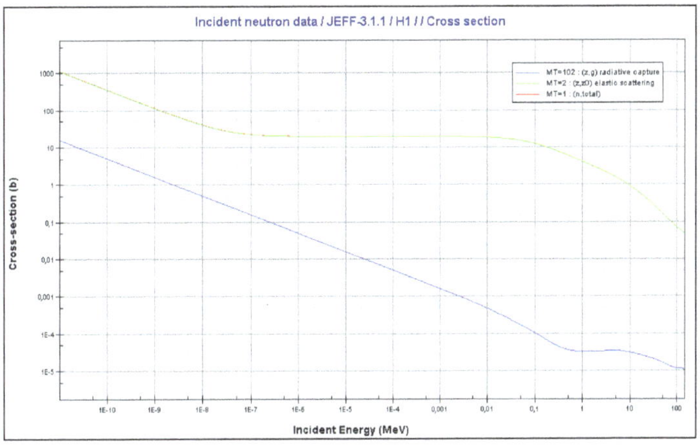

Neutron absorption and scattering: Comparison of cross-sections.

At very high flux the atomic nuclei do not necessarily have enough time to decay via beta particle emission between neutron captures. This happens inside stars, where a really tremendous flux may be reached. As a result of many capture reactions without beta decay the mass number rises by a large amount, while the atomic number stays the same. Only afterwards, the highly unstable nuclei decay via many β–decays to stabilize itself.

Since the process entails a succession of many rapid neutron captures, it is called the r-process. The r-process is a nucleosynthesis process that occurs in core-collapse supernovae and is responsible for the creation of approximately half of the neutron-rich atomic nuclei heavier than iron.

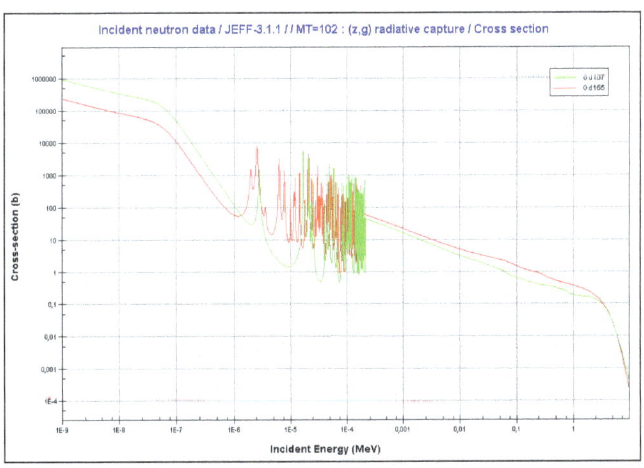

Gadolinium 155 and 157: Comparison of radiative capture cross-sections.

Neutron Capture and Fuel Breeding

In reactor calculations, the neutron capture reaction is as important as the fission reaction. Its impact on the neutron balance is evident. But this reaction is of importance also from another point of view. Fissionable nuclei or even fissile nuclei may capture a neutron, this capture leads to formation of unstable nuclei with higher neutron number. Such unstable nuclei undergo a nuclear decay, which may lead to formation of another fissile nuclei. This process is also referred to as the nuclear transmutation and is responsible for new fuel breeding in nuclear reactors.

Plutonium 239 Breeding

$$n + {}^{238}_{92}\text{U} \rightarrow {}^{239}_{92}\text{U} + \gamma \rightarrow {}^{239}_{93}\text{Np} \rightarrow {}^{239}_{94}\text{Pu}$$

Neutron capture may also be used to create fissile ${}^{239}\text{Pu}$ from ${}^{238}\text{U}$, which is the dominant constituent of naturally occurring uranium (99.28%). Absorption of a neutron in the ${}^{238}\text{U}$ nucleus yields ${}^{239}\text{U}$. The half-life of ${}^{239}\text{U}$ is approximately 23.5 minutes. ${}^{239}\text{U}$ decays (negative beta decay) to ${}^{239}\text{Np}$ (neptunium), whose half-life is 2.36 days. ${}^{239}\text{Np}$ decays (negative beta decay) to ${}^{239}\text{Pu}$.

Uranium 233 Breeding

$$n + {}^{232}_{90}\text{Th} \rightarrow {}^{232}_{90}\text{Th} + \gamma \rightarrow {}^{233}_{91}\text{Pa} \rightarrow {}^{233}_{92}\text{U}$$

${}^{232}\text{Th}$ is the predominant isotope of natural thorium. If this fertile material is loaded in the nuclear reactor, the nuclei of ${}^{232}\text{Th}$ absorb a neutron and become nuclei of ${}^{233}\text{Th}$. The half-life of ${}^{233}\text{Th}$ is approximately 21.8 minutes. ${}^{233}\text{Th}$ decays (negative beta decay) to ${}^{233}\text{Pa}$ (protactinium), whose half-life is 26.97 days. ${}^{233}\text{Pa}$ decays (negative beta decay) to ${}^{233}\text{U}$, that is very good fissile material. On the other hand proposed reactor designs must attempt to physically isolate the protactinium from further neutron capture before beta decay can occur.

Radiative Capture Cross-section

Table of cross-sections.

		Thermal neutron			Fast neutron		
		Scattering	Capture	Fission	Scattering	Capture	Fission
Moderator	H-1	20	0.2	-	4	0.00004	-
	H-2	4	0.0003	-	3	0.000007	-
	C-12	5	0.002	-	2	0.00001	-
Structural materials, others	Zr-90	5	0.006	-	5	0.006	-
	Fe-56	10	2	-	20	0.003	-
	Cr-52	3	0.5	-	3	0.002	-
	Ni-58	20	3	-	3	0.008	-
	O-16	4	0.0001	-	3	0.00000003	-
Absorber	B-10	2	200	-	2	0.4	-
	Cd-113	100	30	-	4	0.05	-
	Xe-135	400	2,000,000	-	5	0.0008	-
	In-115	2	100	-	4	0.02	-
Fuel	U-235	10	99	583	4	0.09	1
	U-238	9	2	0.00002	5	0.07	0.3
	Pu-239	8	269	748	5	0.05	2

The likelihood of a neutron radiative capture is represented by the radiative capture cross section as σ_γ. As usual, the cross-section can be divided into three regions according to the incident neutron energy.

- 1/v Region;
- Resonance Region;
- Fast Neutrons Region.

1/v Region

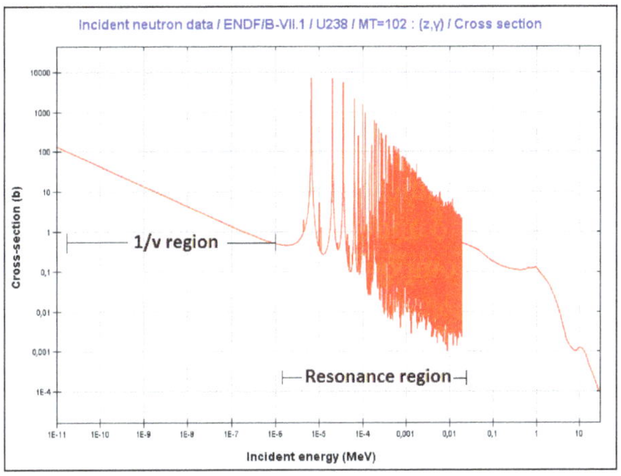

Radiative Capture Cross-section – region of resonances of 238U nuclei.

In the common case, the cross section is usually much larger at low energies than at high energies. For thermal neutrons (in 1/v region), also radiative capture cross-sections increase as the velocity (kinetic energy) of the neutron decreases. Therefore the 1/v Law can be used to determine shift in capture cross-section, if the neutron is in equilibrium with a surrounding medium. This phenomenon is due to the fact the nuclear force between the target nucleus and the neutron has a longer time to interact.

Resonance Region

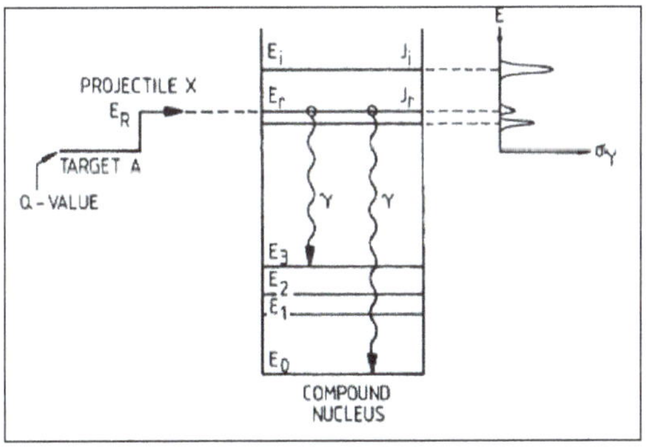

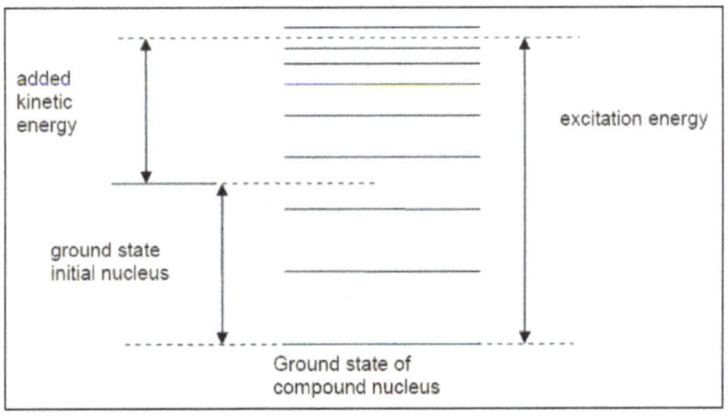

Energy levels of compound state. For neutron absorption reaction on ^{238}U the first resonance E1 corresponds to the excitation energy of 6.67eV. E0 is a base state of ^{239}U.

The largest cross-sections are usually at neutron energies, that lead to long-lived states of the compound nucleus. The compound nuclei of these certain energies are referred to as nuclear resonances and its formation is typical in the resonance region. The widths of the resonances increase in general with increasing energies. At higher energies the widths may reach the order of the distances between resonances and then no resonances can be observed. The narrowest resonances are usually compound states of heavy nuclei (such as fissionable nuclei).

Since the mode of decay of the compound nucleus does not depend on the way the compound nucleus was formed, the nucleus sometimes emits a gamma ray (radiative capture) or sometimes emits a neutron (scattering). In order to understand the way, how a nucleus will stabilize itself, we have to understand the behaviour of compound nucleus.

The compound nucleus emits a neutron only after one neutron obtains an energy in collision with other nucleon greater than its binding energy in the nucleus. It have some delay, because the excitation energy of the compound nucleus is divided among several nucleons. It is obvious the average time that elapses before a neutron can be emitted is much longer for nuclei with large number of nucleons than when only a few nucleons are involved. It is a consequence of sharing the excitation energy among a large number of nucleons.

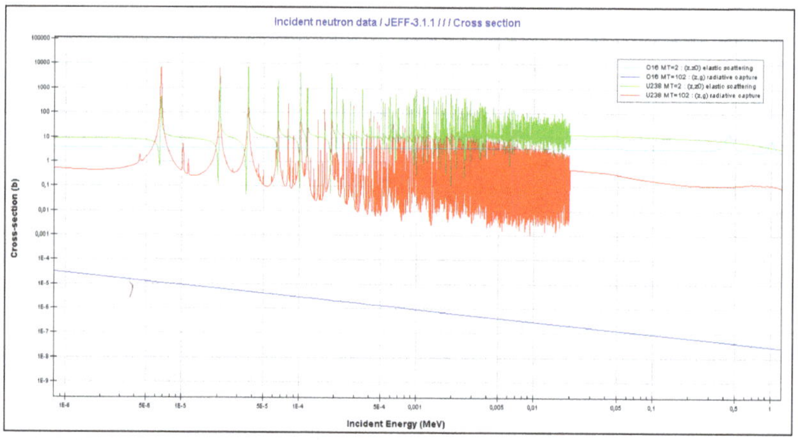

Radiative capture and elastic scattering cross-sections in ^{16}O and ^{238}U.

This is the reason the radiative capture is comparatively unimportant in light nuclei but becomes increasingly important in the heavier nuclei.

The lifetime of a compound nucleus is inversely proportional to its total width. Narrow resonances therefore correspond to capture while the wider resonances are due to scattering.

Example: The first resonance in ^{238}U at 6.67 eV, which corresponds to the first virtual level in ^{239}U, has a total width of only 0.027 eV, and the mean life of this state is 2.4×10^{-14}s. By contrast, the resonance observed at 443 keV in ^{16}O, which corresponds to the first virtual state in ^{17}O, has a total width of 41 keV, giving a mean lifetime of 1.5×10^{-21}s. Thus it is highly likely that the compound state in ^{239}U decays at least to some extent by gamma ray emission, while compound state in ^{17}O must decay primarily by nucleon emission. The 443-keV resonance in ^{16}O is clearly a scattering resonance, whereas the 6.67-eV resonance in ^{238}U is in part a capture resonance.

Doppler Broadening

In general, Doppler broadening is the broadening of spectral lines due to the Doppler effect caused by a distribution of kinetic energies of molecules or atoms. In reactor physics a particular case of this phenomenon is the thermal Doppler broadening of resonances caused by the thermal motion of the target particle in the nuclear fuel.

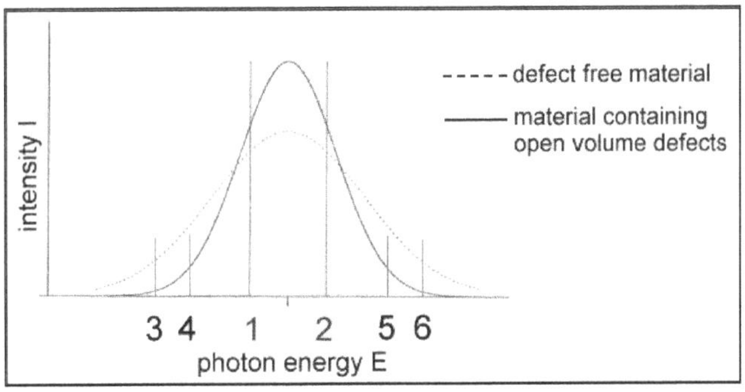

Doppler effect improves reactor stability. Broadened resonance (heating of a fuel) results in a higher probability of absorbtion, thus causes negative reactivity insertion (reduction of reactor power).

The Doppler broadening of resonances is very important phanomenon, which improves reactor stability. The prompt temperature coefficient of most thermal reactors is negative, owing to an nuclear Doppler effect. Although the absorbtion cross-section depends significantly on incident neutron energy, the shape of the cross-section curve depends also on target temperature.

Nuclei are located in atoms which are themselves in continual motion owing to their thermal energy. As a result of these thermal motions neutrons impinging on a target appears to the nuclei in the target to have a continuous spread in energy. This, in turn, has an effect on the observed shape of resonance. The resonance becomes shorter and wider than when the nuclei are at rest.

Although the shape of a resonance changes with temperature, the total area under the resonance remains essentially constant. But this does not imply constant neutron absorbtion. Despite the constant area under resonance, a resonance integral, which determines the absorbtion, increases with increasing target temperature. This, of course, decreases coefficient k (negative reactivity is inserted).

Fast Neutrons Region

The radiative capture cross-section at energies above the resonance region drops rapidly to very small values. This rapid drop is caused by the compound nucleus, which is formed in more highly-excited states. In these highly-excited states it is more likely that one neutron obtains an energy in collision with other nucleon greater than its binding energy in the nucleus. The neutron emission becomes dominant and gamma decay becomes less important. Moreover, at high energies, the inelastic scattering and (n,2n) reaction are highly probable at the expense of both elastic scattering and radiative capture.

Nuclear Fission

Nuclear fission is the subdivision of a heavy atomic nucleus, such as that of uranium or plutonium, into two fragments of roughly equal mass. The process is accompanied by the release of a large amount of energy.

In nuclear fission the nucleus of an atom breaks up into two lighter nuclei. The process may take place spontaneously in some cases or may be induced by the excitation of the nucleus with a variety of particles (e.g., neutrons, protons, deuterons, or alpha particles) or with electromagnetic radiation in the form of gamma rays. In the fission process, a large quantity of energy is released, radioactive products are formed, and several neutrons are emitted. These neutrons can induce fission in a nearby nucleus of fissionable material and release more neutrons that can repeat the sequence, causing a chain reaction in which a large number of nuclei undergo fission and an enormous amount of energy is released. If controlled in a nuclear reactor, such a chain reaction can provide power for society's benefit. If uncontrolled, as in the case of the so-called atomic bomb, it can lead to an explosion of awesome destructive force.

The discovery of nuclear fission has opened a new era—the "Atomic Age." The potential of nuclear fission for good or evil and the risk/benefit ratio of its applications have not only provided the basis of many sociological, political, economic, and scientific advances but grave concerns as well. Even from a purely scientific perspective, the process of nuclear fission has given rise to many puzzles and complexities, and a complete theoretical explanation is still not at hand.

Fundamentals of the Fission Process

Structure and Stability of Nuclear Matter

The fission process may be best understood through a consideration of the structure and stability of nuclear matter. Nuclei consist of nucleons (neutrons and protons), the total number of which is equal to the mass number of the nucleus. The actual mass of a nucleus is always less than the sum of the masses of the free neutrons and protons that constitute it, the difference being the mass equivalent of the energy of formation of the nucleus from its constituents. The conversion of mass to energy follows Einstein's equation, $E = mc^2$, where E is the energy equivalent of a mass, m and c is the velocity of light. This difference is known as the mass defect and is a measure of the total binding energy (and hence, the stability) of the nucleus. This binding energy is released during the formation of a nucleus from its constituent nucleons and would have to be supplied to the nucleus to decompose it into its individual nucleon components.

A curve illustrating the average binding energy per nucleon as a function of the nuclear mass number is shown in figure. The largest binding energy (highest stability) occurs near mass number 56—the mass region of the element iron. Figure indicates that any nucleus heavier than mass number 56 would become a more stable system by breaking into lighter nuclei of higher binding energy, the difference in binding energy being released in the process. (It should be noted that nuclei lighter than mass number 56 can gain in stability by fusing to produce a heavier nucleus of greater mass defect—again, with the release of the energy equivalent of the mass difference. It is the fusion of the lightest nuclei that provides the energy released by the Sun and constitutes the basis of the hydrogen, or fusion, bomb. Efforts to harness fusion reaction for power production have been actively pursued.

On the basis of energy considerations alone, figure would indicate that all matter should seek its most stable configuration, becoming nuclei of mass number near 56. However, this does not happen, because barriers to such a spontaneous conversion are provided by other factors. A good qualitative understanding of the nucleus is achieved by treating it as analogous to a uniformly charged liquid drop. The strong attractive nuclear force between pairs of nucleons is of short range and acts only between the closest neighbours. Since nucleons near the surface of the drop have fewer close neighbours than those in the interior, a surface tension is developed, and the nuclear drop assumes a spherical shape in order to minimize this surface energy. (The smallest surface area enclosing a given volume is provided by a sphere.) The protons in the nucleus exert a long-range repulsive (Coulomb) force on each other because of their positive charge. As the number of nucleons in a nucleus increases beyond about 40, the number of protons must be diluted with an excess of neutrons to maintain relative stability.

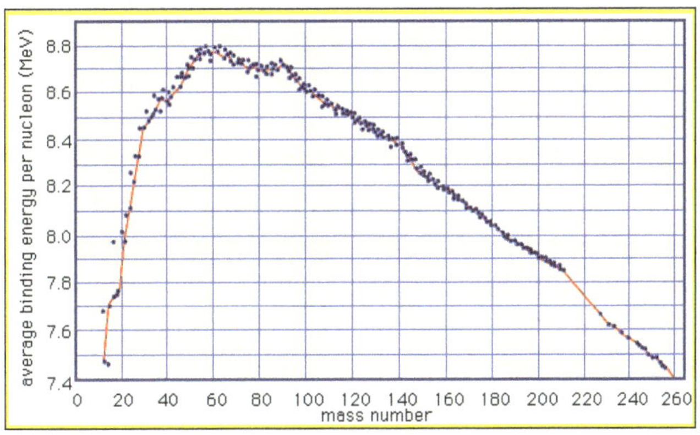

The average binding energy per nucleon as a function of the mass number, *A*. The line connects the odd-*A* points.

If the nucleus is excited by some stimulus and begins to oscillate (i.e., deform from its spherical shape), the surface forces will increase and tend to restore it to a sphere, where the surface tension is at a minimum. On the other hand, the Coulomb repulsion decreases as the drop deforms and the protons are positioned farther apart. These opposing tendencies set up a barrier in the potential energy of the system, as indicated in figure.

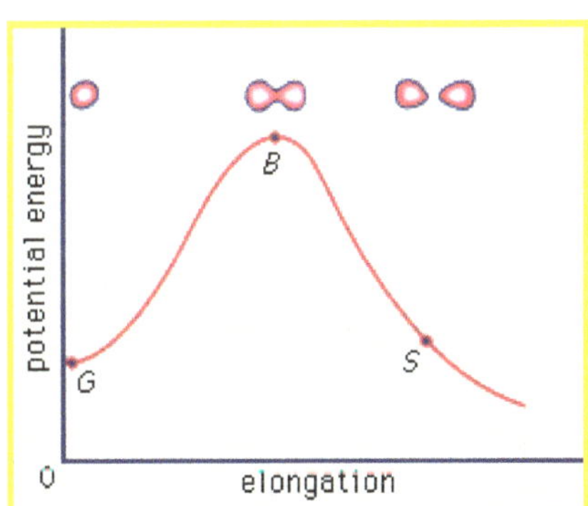

The potential energy as a function of elongation of a fissioning nucleus. G is the ground state of the nucleus; *B* is the top of the barrier to fission (called the saddle point); and *S* is the scission point. The nuclear shape at these points is shown at the top.

The curve in figure rises initially with elongation, since the strong, short-range nuclear force that gives rise to the surface tension increases. The Coulomb repulsion between protons decreases faster with elongation than the surface tension increases, and the two are in balance at point *B*, which represents the height of the barrier to fission. (This point is called the "saddle point" because, in a three-dimensional view of the potential energy surface, the shape of the pass over the barrier resembles a saddle.) Beyond point *B*, the Coulomb repulsion between the protons drives the nucleus into further elongation until at some point, *S* (the scission point), the nucleus breaks in two.

Qualitatively, at least, the fission process is thus seen to be a consequence of the Coulomb repulsion between protons.

Induced Fission

The height and shape of the fission barrier are dependent on the particular nucleus being considered. Fission can be induced by exciting the nucleus to an energy equal to or greater than that of the barrier. This can be done by gamma-ray excitation (photofission) or through excitation of the nucleus by the capture of a neutron, proton, or other particle (particle-induced fission). The binding energy of a particular nucleon to a nucleus will depend on—in addition to the factors considered above—the odd–even character of the nucleus. Thus, if a neutron is added to a nucleus having an odd number of neutrons, an even number of neutrons will result, and the binding energy will be greater than for the addition of a neutron that makes the total number of neutrons odd. This "pairing energy" accounts in part for the difference in behaviour of nuclides in which fission can be induced with slow (low-energy) neutrons and those that require fast (higher-energy) neutrons. Although the heavy elements are unstable with respect to fission, the reaction takes place to an appreciable extent only if sufficient energy of activation is available to surmount the fission barrier. Most nuclei that are fissionable with slow neutrons contain an odd number of neutrons (e.g., uranium-233, uranium-235, or plutonium-239), whereas most of those requiring fast neutrons (e.g., thorium-232 or uranium-238) have an even number. The addition of a neutron in the former case liberates sufficient binding energy to induce fission. In the latter case, the binding energy is less and may be insufficient to surmount the barrier and induce fission. Additional energy must then be supplied in the form of the kinetic energy of the incident neutron. (In the case of thorium-232 or uranium-238, a neutron having about 1 MeV of kinetic energy is required.)

Spontaneous Fission

The laws of quantum mechanics deal with the probability of a system such as a nucleus or an atom being in any of its possible states or configurations at any given time. A fissionable system (uranium-238, for example) in its ground state (i.e., at its lowest excitation energy and with an elongation small enough that it is confined inside the fission barrier) has a small but finite probability of being in the energetically favoured configuration of two fission fragments. In effect, when this occurs, the system has penetrated the barrier by the process of quantum mechanical tunneling. This process is called spontaneous fission because it does not involve any outside influences. In the case of uranium-238, the process has a very low probability, requiring more than 10^{15} years for half of the material to be transformed (its so-called half-life) by this reaction. On the other hand, the probability for spontaneous fission increases dramatically for the heaviest nuclides known and becomes the dominant mode of decay for some—those having half-lives of only fractions of a second. In fact, spontaneous fission becomes the limiting factor that may prevent the formation of still heavier (super-heavy) nuclei.

Stages of Fission

A pictorial representation of the sequence of events in the fission of a heavy nucleus is given in figure. The approximate time elapse between stages of the process is indicated at the bottom of the figure.

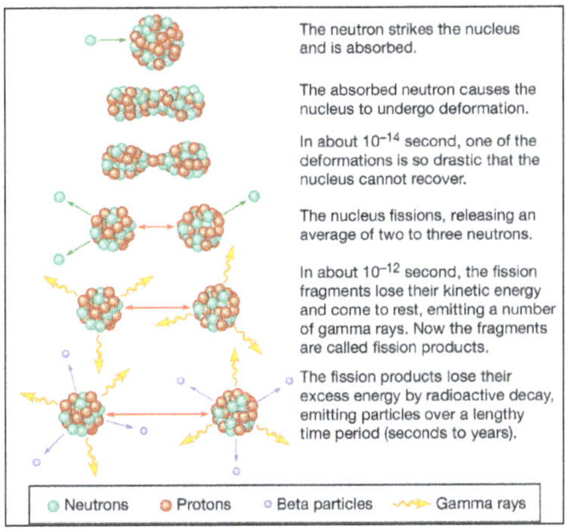

Sequence of events in the fission of a uranium nucleus by a neutron.

Phenomenology of Fission

When a heavy nucleus undergoes fission, a variety of fragment pairs may be formed, depending on the distribution of neutrons and protons between the fragments. This leads to probability distribution of both mass and nuclear charge for the fragments. The probability of formation of a particular fragment is called its fission yield and is expressed as the percentage of fissions leading to it.

The separated fragments experience a large Coulomb repulsion due to their nuclear charges, and they recoil from each other with kinetic energies determined by the fragment charges and the distance between the charge centres at the time of scission. Variations in these parameters lead to a distribution of kinetic energies, even for the same mass split.

The initial velocities of the recoiling fragments are too fast for the outer (atomic) electrons of the fissioning atom to keep pace, and many of them are stripped away. Thus, the nuclear charge of the fragment is not fully neutralized by the atomic electrons, and the fission fragments fly apart as highly charged atoms. As the nucleus of the fragment adjusts from its deformed shape to a more stable configuration, the deformation energy (i.e., the energy required to deform it) is recovered and converted into internal excitation energy, and neutrons and prompt gamma rays (an energetic form of electromagnetic radiation given off nearly coincident with the fission event) may be evaporated from the moving fragment. The fast-moving, highly charged atom collides with the atoms of the medium through which it is moving, and its kinetic energy is transferred to ionization and heating of the medium as it slows down and comes to rest. The range of fission fragments in air is only a few centimetres.

During the slowing-down process, the charged atom picks up electrons from the medium and becomes neutral by the time it stops. At this stage in the sequence of events, the atom produced is called a fission product to distinguish it from the initial fission fragment formed at scission. Since a few neutrons may have been lost in the transition from fission fragment to fission product, the two may not have the same mass number. The fission product is still not a stable species but is radioactive, and it finally reaches stability by undergoing a series of beta decays, which may vary over a time scale of fractions of a second to many years. The beta emission consists of electrons and antineutrinos, often accompanied by gamma rays and X-rays.

The distributions in mass, charge, and kinetic energy of the fragments have been found to be dependent on the fissioning species as well as on the excitation energy at which the fission act occurs. Many other aspects of fission have been observed, adding to the extensive phenomenology of the process and providing an intriguing set of problems for interpretation. These include the systematics of fission cross sections (a measure of the probability for fission to occur); the variation of the number of prompt neutrons emitted as a function of the fissioning species and the particular fragment mass split; the angular distribution of the fragments with respect to the direction of the beam of particles inducing fission; the systematics of spontaneous fission half-lives; the occurrence of spontaneous fission isomers (excited states of the nucleus); the emission of light particles (hydrogen-3, helium-3, helium-4, etc.) in small but significant numbers in some fission events; the presence of delayed neutron emitters among the fission products; the time scale on which the various stages of the process take place; and the distribution of the energy release in fission among the particles and radiations produced.

Fission Fragment Mass Distributions

The distribution of the fragment masses formed in fission is one of the most striking features of the process. It is dependent on the mass of the fissioning nucleus and the excitation energy at which the fission occurs. At low excitation energy, the fission of such nuclides as uranium-235 or plutonium-239 is asymmetric; i.e., the fragments are formed in a two-humped probability (or yield) distribution favouring an unequal division in mass. This is illustrated in figure. As will be noted, the light group of fragment masses shifts to higher mass numbers as the mass of the fissioning nucleus increases, whereas the position of the heavy group remains nearly stationary. As the excitation energy of the fission increases, the probability for a symmetric mass split increases, while that for asymmetric division decreases. Thus, the valley between the two peaks increases in probability (yield of formation), and at high excitations the mass distribution becomes single-humped, with the maximum yield at symmetry. Radium isotopes show interesting triple-humped mass distributions, and nuclides lighter than radium show a single-humped, symmetric mass distribution. (These nuclides, however, require a relatively high activation energy to undergo fission.) For very heavy nuclei in the region of fermium-260, the mass-yield curve becomes symmetric (single-humped) even for spontaneous fission, and the kinetic energies of the fragments are unusually high. An understanding of these mass distributions has been one of the major puzzles of fission, and a complete theoretical interpretation is still lacking, albeit much progress has been made.

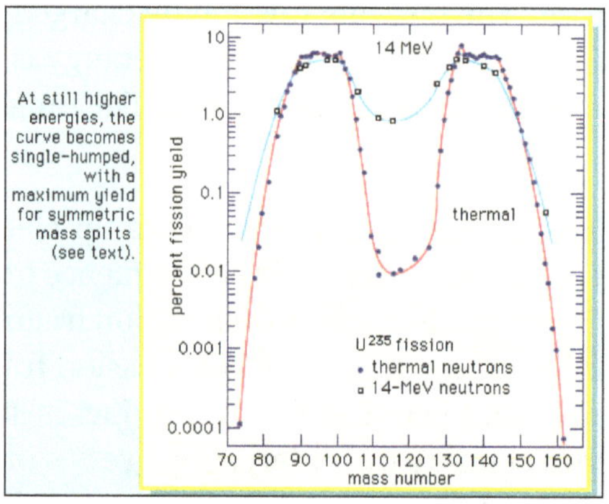

Mass distribution dependence on the energy excitation in the fission of uranium-235. At still higher energies, the curve becomes single-humped, with a maximum yield for symmetric mass splits.

Fission Decay Chains and Charge Distribution

In order to maintain stability, the neutron-to-proton (n/p) ratio in nuclei must increase with increasing proton number. The ratio remains at unity up to the element calcium, with 20 protons. It then gradually increases until it reaches a value of about 1.5 for the heaviest elements. When a heavy nucleus fissions, a few neutrons are emitted; however, this still leaves too high an n/p ratio in the fission fragments to be consistent with stability for them. They undergo radioactive decay and reach stability by successive conversions of neutrons to protons with the emission of a negative electron (called a beta particle, β^-) and an antineutrino. The mass number of the nucleus remains the same, but the nuclear charge (atomic number) increases by one, and a new element is formed for each such conversion. The successive beta decays constitute an isobaric, fission-product decay chain for each mass number. The half-lives for the decay of the radioactive species generally increase as they approach the stable isobar of the chain. (Species of the same element characterized by the same nuclear charge, Z [number of protons], but differing in their number of neutrons [and therefore in mass number A] are called isotopes. Species that have the same mass number, A, but differ in Z are known as isobars.)

For a typical mass split in the neutron-induced fission of uranium-235, the complementary fission-product masses of 93 and 141 may be formed following the emission of two neutrons from the initial fragments. The division of charge (i.e., protons) between the fragments represents an important parameter in the fission process. Thus, for the mass numbers 93 and 141, the following isobaric fission-product decay chains would be formed (the half-lives for the beta-decay processes are indicated above the arrows):

$$^{93}_{36}\mathrm{Kr} \xrightarrow[\beta^-]{1.3\ \mathrm{sec}} {}^{93}_{37}\mathrm{Rb} \xrightarrow[\beta^-]{6\ \mathrm{sec}} {}^{93}_{38}\mathrm{Sr} \xrightarrow[\beta^-]{7.5\ \mathrm{min}} {}^{93}_{39}\mathrm{Y} \xrightarrow[\beta^-]{10\ \mathrm{hr}} {}^{93}_{40}\mathrm{Zr} \xrightarrow{1.5\times10^{6}\ \mathrm{yr}} {}^{93m}_{41}\mathrm{Nb} \xrightarrow{13\ \mathrm{yr}} {}^{93}_{41}\mathrm{Nb}\ (\mathrm{stable})$$

$$^{93}_{40}\mathrm{Zr} \xrightarrow{\beta^-} {}^{93}_{41}\mathrm{Nb}\ (\mathrm{stable})$$

$$^{141}_{54}\mathrm{Xe} \xrightarrow[\beta^-]{1.7\ \mathrm{sec}} {}^{141}_{55}\mathrm{Cs} \xrightarrow[\beta^-]{25\ \mathrm{sec}} {}^{141}_{56}\mathrm{Ba} \xrightarrow[\beta^-]{18\ \mathrm{min}} {}^{141}_{57}\mathrm{La} \xrightarrow[\beta^-]{3.9\ \mathrm{hr}} {}^{141}_{58}\mathrm{Ce} \xrightarrow[\beta^-]{32.5\ \mathrm{d}} {}^{141}_{59}\mathrm{Pr}\ (\mathrm{stable})$$

(The left subscript on the element symbol denotes Z, while the superscript denotes A.) The 92 protons of the uranium nucleus must be conserved, and complementary fission-product pairs—such as krypton-36 with barium-56, rubidium-37 with cesium-55, or strontium-38 with xenon-54—would be possible.

The percentage of fissions in which a particular nuclide is formed as a primary fission product (i.e., as the direct descendant of an initial fragment following its de-excitation) is called the independent yield of that product. The total yield for any nuclide in the isobaric decay chain is the sum of its independent yield and the independent yields of all of its precursors in the chain. The total yield for the entire chain is called the cumulative yield for that mass number.

Extensive radiochemical investigations have suggested that the most probable charge division is one that is displaced from stability about the same distance in both chains. This empirical observation is called the equal charge displacement (ECD) hypothesis, and it has been confirmed by several physical measurements. In the above example the ECD would predict the most probable charges at about rubidium-37 and cesium-55. A strong shell effect modifies the ECD expectations for fragments having 50 protons. The dispersion of the charge formation probability about the most probable charge (*Zp*) is rather narrow and approximately Gaussian in shape and is nearly independent of the mass split as well as of the fissioning species. The most probable charge for an isobaric chain is a useful concept in the description of the charge dispersion, and it need not have an integral value. As the energy of fission increases, the charge division tends toward maintaining the n/p ratio in the fragments the same as that in the fissioning nucleus. This is referred to as an unchanged charge distribution.

Prompt Neutrons in Fission

The average number of neutrons emitted per fission (represented by the symbol $\bar{\upsilon}$) varies with the fissioning nucleus. It is about 2.0 for the spontaneous fission of uranium-238 and 4.0 for that of fermium-257. In the thermal-neutron induced fission of uranium-235, $\bar{\upsilon}$= 2.4. The actual number of neutrons emitted, however, varies with each fission event, depending on the mass split. Although there is still controversy regarding the number of neutrons emitted at the instant of scission, it is generally agreed that most of the neutrons are given off by the recoiling fission fragments soon after scission occurs. The number of neutrons emitted from each fragment depends on the amount of energy the fragment possesses. The energy can be in the form of internal excitation (heat) energy or stored as energy of deformation of the fragment to be released when the fragment returns to its stable equilibrium shape.

Delayed Neutrons in Fission

A few of the fission products have beta-decay energies that exceed the binding energy of a neutron in the daughter nucleus. This is likely to happen when the daughter nucleus contains one or two neutrons more than a closed shell of 50 or 82 neutrons, since these "extra" neutrons are more loosely bound. The beta decay of the precursor may take place to an excited state of the daughter from which a neutron is emitted. The neutron emission is "delayed" by the beta-decay half-life of the precursor. Six such delayed neutron emitters have been identified, with half-lives varying from about 0.5 to 56 seconds. The yield of the delayed neutrons is only about 1 percent of that of the prompt neutrons, but they are very important for the control of the chain reaction in a nuclear reactor.

Energy Release in Fission

The total energy release in a fission event may be calculated from the difference in the rest masses of the reactants (e.g., $^{235}U + n$) and the final stable products (e.g., $^{93}Nb + {}^{141}Pr + 2n$). The energy equivalent of this mass difference is given by the Einstein relation, $E = mc^2$. The total energy release depends on the mass split, but a typical fission event would have the total energy release distributed approximately as follows for the major components in the thermal neutron-induced fission of uranium-235:

Energy component	Number per fission	Total energy
Kinetic energy of fission fragments	2	170 Mev
Kinetic energy of prompt neutrons	2.5	5

Binding energy from capture of prompt neutrons	2.5	~ 12
Prompt gamma rays	8	8
	Total	195 Mev

The energy release from the capture of the prompt neutrons depends on how they are finally stopped, and some will escape the core of a nuclear reactor.

This energy is released on a time scale of about 10^{-12} second and is called the prompt energy release. It is largely converted to heat within an operating reactor and is used for power generation. Also, there is a delayed release of energy from the radioactive decay of the fission products varying in half-life from fractions of a second to many years. The shorter-lived species decay in the reactor, and their energy adds to the heat generated; however, the longer-lived species remain radioactive and pose a problem in the handling and disposition of the reactor fuel elements when they need to be replaced. The antineutrinos that accompany the beta decay of the fission products are unreactive, and their kinetic energy (about 10 MeV per fission) is not recovered. Overall, about 200 MeV of energy per fission may be recovered for power applications.

Fission Theory

Nuclear fission is a complex process that involves the rearrangement of hundreds of nucleons in a single nucleus to produce two separate nuclei. A complete theoretical understanding of this reaction would require a detailed knowledge of the forces involved in the motion of each of the nucleons through the process. Since such knowledge is still not available, it is necessary to construct simplified models of the actual system to simulate its behaviour and gain as accurate a description as possible of the steps in the process. The successes and failures of the models in accounting for the various observations of the fission process can provide new insights into the fundamental physics governing the behaviour of real nuclei, particularly at the large nuclear deformations encountered in a nucleus undergoing fission.

The framework for understanding nuclear reactions is analogous to that for chemical reactions and involves the concept of a potential-energy surface on which the reaction occurs. The driving force for physical or chemical reactions is the tendency to lower the potential energy and increase the stability of the system. Thus, for example, a stone at the top of a hill will roll down the hill, converting its potential energy at the top to kinetic energy of motion, and will come to rest at the bottom in a more stable state of lower potential energy. The potential energy is calculated as a function of various parameters of the system being studied. In the case of fission, the potential energy may be calculated as a function of the shape of the system as it proceeds over the barrier to the scission point, and the path of lowest potential energy may be determined.

As has been pointed out, an exact calculation of the nuclear potential energy is not yet possible, and it is to approximate this calculation that various models have been constructed to simulate the real system. Some of the models were developed to address aspects of nuclear structure and spectroscopy as well as features of nuclear reactions, and they also have been employed in attempts to understand the complexity of nuclear fission. The models are based on different assumptions and approximations of the nature of the nuclear forces and the dynamics of the path to scission. No one model can account for all of the extensive phenomenology of fission, but each addresses different aspects of the process and provides a foundation for further development toward a complete theory.

Nuclear Models and Nuclear Fission

The nucleus exhibits some properties that reflect the collective motion of all its constituent nucleons as a unit, as well as other properties that are dependent on the motion and state of the individual nucleons.

The analogy of the nucleus to a drop of an incompressible liquid was first suggested by George Gamow in 1935 and later adapted to a description of nuclear reactions. Bohr proposed the so-called compound nucleus description of nuclear reactions, in which the excitation energy of the system formed by the absorption of a neutron or photon, for example, is distributed among a large number of degrees of freedom of the system. This excited state persists for a long time relative to the periods of motion of nucleons across the nucleus and then decays by emission of radiation, the evaporation of neutrons or other particles, or by fission. The liquid-drop model of the nucleus accounts quite well for the general collective behaviour of nuclei and provides an understanding of the fission process on the basis of the competition between the cohesive nuclear force and the disruptive Coulomb repulsion between protons. It predicts, however, a symmetric division of mass in fission, whereas an asymmetric mass division is observed. Moreover, it does not provide an accurate description of fission barrier systematics or of the ground-state masses of nuclei. The liquid-drop model is particularly useful in describing the behaviour of highly excited nuclei, but it does not provide an accurate description for nuclei in their ground or low-lying excited states. Many versions of the liquid-drop model employing improved sets of parameters have been developed. However, investigators have found that mass asymmetry and certain other features in fission cannot be adequately described on the basis of the collective behaviour posited by such models alone.

A preference for the formation of unequal masses (i.e., an asymmetric division) was observed early in fission research, and it has remained the most puzzling feature of the process to account for. Investigators have invoked various models other than that of the liquid drop in an attempt to address this question. Dealing with the mutual interaction of all the nucleons in a nucleus has been simplified by treating it as if it were equivalent to the interaction of one particle with an average spherical static potential field that is generated by all the other nucleons. The methods of quantum mechanics provide the solution for the motion of a nucleon in such a potential. A characteristic set of energy levels for neutrons and protons is obtained, and, analogous to the set of levels of the electrons in an atom, the levels group themselves into shells at certain so-called magic numbers of nucleons. (For both neutrons and protons, these numbers are 2, 8, 20, 28, 50, 82, and 126.) Shell closures at these nuclear numbers are marked by especially strong binding, or extra stability. This constitutes the essence of the spherical-shell model (sometimes called the independent-particle, or single-particle, model), as developed by Maria Goeppert Mayer and J. Hans D. Jensen and their colleagues. It accounts well for ground-state masses and spins and for the existence of isomeric nuclear states (excited states having measurable half-lives) that occur when nuclear levels of widely differing spins lie relatively close to each other. The agreement with observations is excellent for spherical nuclei with nucleon numbers near the magic shell numbers. The spherical-shell model, however, does not agree well with the properties of nuclei that have other nucleon numbers—e.g., the nuclei of the lanthanide and actinide elements, with nucleon numbers between the magic numbers.

In the lanthanide and actinide nuclei, the ground state is not spherical but rather deformed into a prolate spheroidal shape—that of a football or watermelon. For such nuclei, the allowed states of motion of a nucleon must be calculated in a potential having a symmetry corresponding to a spheroid rather than a sphere. This was first done by Aage Bohr, Ben R. Mottelson, and Sven G. Nilsson in 1955, and the level structure was calculated as a function of the deformation of the nucleus. A spheroid has three axes of symmetry, and it can rotate in space as a unit about any one of them. The rotation can occur independent of the internal state of excitation of the individual nucleons. Various modes of vibration of the spheroid also may take place. Since this deformed shell model has components of both the independent-particle motion and the collective motion of the nucleus as a whole (i.e., rotations and vibrations), it is sometimes referred to as the unified model.

In Aage Bohr's application of the unified model to the fission process, the sequence of potential-energy surfaces for the excited states of the system are considered to be functions of a deformation parameter (i.e., elongation) characterizing the motion toward fission and evaluated at the saddle point. As the system passes over the saddle point, most of its excitation energy is used up in deforming the nucleus, and the system remains "cold" i.e., it manifests little excitation, or heat, energy. Thus, only the low-lying excited states are available to the system. The spin and parity of the particular state (or channel) in which the system exists as it passes over the saddle point are then expected to determine the fission properties. In this channel (or transition-state) analysis of fission, a number of characteristics of the process are qualitatively accounted for. Hence, fission thresholds would depend on the spin and parity of the compound nuclear state, the fission fragment angular distribution would be governed by the collective rotational angular momentum of the state, and asymmetry in the mass distribution would result from passage over the barrier in a state of negative parity (which does not possess reflection symmetry). This model gives a good qualitative interpretation of many fission phenomena, but it must assume that at least some of the properties of the transition state at the saddle point are not altered by dynamical considerations in the descent of the system to the scission point. It is the only model that provides a satisfactory interpretation of the angular distributions of fission fragments, and it has attractive features that must be included in any complete theory of fission.

The first application of the spherical-shell model to fission was the recognition that the positions of the peaks in the fission mass distribution correlated fairly well with the magic numbers and suggested a qualitative interpretation of the asymmetric mass division. Thus, a preference for the formation of nuclei with neutron numbers close to 82 would favour the formation of nuclides near the peak in the heavy group and would thus determine the mass split for the fissioning system. Some extra stability for nuclear configurations of 50 protons would also be expected, but this is not particularly evident. In fact, the so-called doubly magic nucleus tin-132, with 50 protons and 82 neutrons, has a rather low yield in low-energy fission.

A more quantitative application of the spherical-shell model to fission was undertaken by Peter Fong in the United States in 1956. He related the probability of formation of a given pair of fragments to the available density of states for that pair of fragments at the scission point in a statistical-model approach. A model of this sort predicts that the system, in its random motions, will experience all possible configurations and so will have a greater probability of being in the region where the greatest number of such configurations (or states) is concentrated. The model assumes that the potential energy at the saddle point is essentially all converted to excitation energy and that a statistical equilibrium among all possible states is established at the scission point. The extra binding energy for closed-shell nuclei leads to a higher density of states at a given excitation energy than is present for other nuclei and, hence, leads to a higher probability of formation. An asymmetric mass distribution in good agreement with that observed for the neutron-induced fission of uranium-235 is obtained. Moreover, the changes in the mass distribution with an increased excitation energy of fission (e.g., an increase in the probability of symmetric fission relative to asymmetric fission) are accounted for by the decrease in importance of the shell effects as the excitation energy increases. Other features of the fission process also are qualitatively explained; however, extensive changes in the parameters of the model are required to obtain agreement with experiments for other fissionable nuclides. Then, too, there are fundamental problems concerning the validity of some of the basic assumptions of the model.

The fundamental question as to the validity of models that evaluate the properties of the system at the scission point (the so-called scission-point models of fission) is whether the system remains long enough at this point on the steep decline of the potential-energy surface for a quasi-equilibrium condition to be established. There is some evidence that such a condition may indeed prevail, but it is not clearly established. Nonetheless, such models have proved quite useful in interpreting observations of mass, charge, and kinetic energy distributions, as well as of neutron emission dependence on fragment mass. It seems very likely that the fragment shell structure plays a significant role in determining the course of the fission process.

Although the single-particle models provide a good description of various aspects of nuclear structure, they are not successful in accounting for the energy of deformation of nuclei (i.e., surface energy), particularly at the large deformations encountered in the fission process. A major breakthrough occurred when a hybrid model incorporating shell effects as a correction to the potential energy of the liquid-drop model was proposed by the Russian physicist V.M. Strutinskii in 1967. This approach retains the dominant collective surface and Coulomb effects while adding shell and pairing corrections that depend on deformation. Shell corrections of several million electron volts are calculated, and these can have a significant effect on a liquid-drop barrier of about 5 MeV. The nucleon numbers at which the shells appear depend on the deformation and may differ from the spherical model magic numbers. In the vicinity of the fission barrier, the shells introduce structure in the liquid-drop potential-energy curve, as illustrated in figure. The relative heights and widths of the two peaks vary with the mass and charge of the fissioning system.

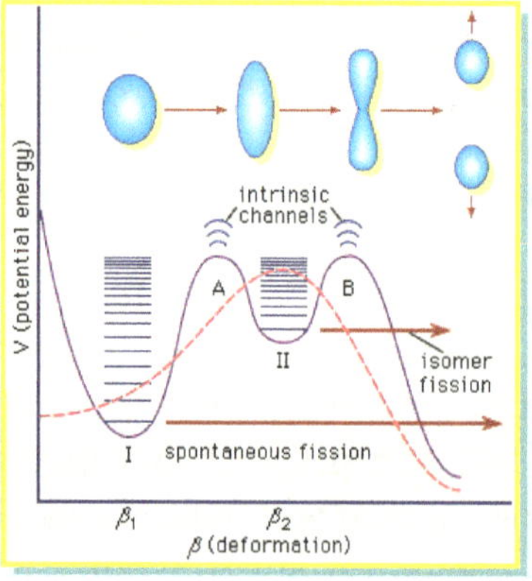

Above figure schematic illustrations of single-humped and double-humped fission barriers. The former are represented by the dashed line and the latter by the continuous line. Intrinsic excitations in the first and second wells at deformations β_1 and β_2 are designated class I and class II states, respectively. Intrinsic channels at the two barriers also are illustrated. The transition in the shape of the nucleus as a function of deformation is schematically represented in the upper part of the figure. Spontaneous fission of the ground state and isomeric state occurs from the lowest energy class I and class II states, respectively.

The double-humped barrier provides a satisfactory explanation for a number of puzzling observations in fission. The existence of short-lived, spontaneous fission isomers, for example, is understood as the consequence of the population of states in the second well (class II). These isomers have a much smaller barrier to penetrate and so exhibit a much shorter spontaneous fission half-life. The change in shape associated with these states, as compared to class I states, also hinders a rapid return to the ground state by gamma emission. (Class II states are also called shape isomers.) The systematics of neutron-induced fission cross sections and structure in some fission-fragment angular distributions also find an interpretation in the implications of the double-humped barrier.

The Strutinskii procedure provided a strong stimulus for calculations of the potential-energy surfaces appropriate to fissioning systems, since it provided a consistent and useful prescription for treating both the macroscopic (liquid-drop) and microscopic (single-particle) effects in deformed nuclei. Many calculations of the potential-energy surface employing different model potentials and parameters have been carried out as functions of the shapes of the system. The work of the American nuclear physicists W.J. Swiatecki, James R. Nix, and their collaborators has been particularly noteworthy in such studies, which also include some attempts to treat the dynamical evolution of the fission process.

Calculations for the actinide elements indicate that, at deformations corresponding to the second barrier, the potential energy for asymmetric mass splits is lower than that for symmetric ones; hence, the former are favoured at that stage of the process. For larger deformations, however, a single potential does not represent the incipient formation of two fragments very well. In fact, a discontinuity occurs at the scission point, and the results of the calculation depend on whether the scission configuration is treated as one nucleus or as two separate nuclei.

A two-centre potential may also be used to represent the nature of the forces at work in a fissioning nucleus. In such a model, the potential energy surfaces are represented by two overlapping spheres or spheroids. It is equivalent to a one-centre potential when there is a complete overlap at small deformations, and it has the correct asymptotic behaviour as the nascent fragments separate. This approach indicates a preformation of the final shell structure of the fragments early in the process.

Although the validity of the assumptions inherent in scission-point models may be in question, the results obtained with them are in excellent agreement with observation. Representative of such a model is the Argonne Scission-Point model, which uses a macroscopic-microscopic calculation with deformed fragment shell and pairing corrections to determine the potential energy of a system of two nearly touching spheroids and which includes their interaction in terms of a neck connecting them. Models of this kind provide a simple approach to a highly quantitative and detailed study of the dependence of the probability of formation of a given fragment pair on the neutron and proton number and on the deformation in each fragment. They account very well for the mass, charge, and kinetic-energy distributions and the neutron-emission dependence on mass number for a broad range of fissioning nuclei. The scission-point models, however, do not address questions of fission probability or the angular distributions of the fragments. As the fission-excitation energy increases, the shell correction diminishes and the macroscopic (liquid-drop) behaviour dominates.

Nuclides in the region of fermium-264 have been observed to undergo symmetric fission with unusually high fragment kinetic energies. This appears to be the consequence of the stability for the magic number configurations of 50 protons and 82 neutrons. The formation of two doubly magic fragments of tin-132 is strongly favoured energetically, whereas the formation of only one such fragment in the low-energy fission of uranium or plutonium isotopes is not. The fragments of tin-132 are spherical rather than deformed, and a more compact configuration at the scission point (with the charge centres closer together) leads to higher fragment kinetic energies.

It is evident that shell effects, both in the fissioning system at the saddle point and in the deformed fragments near the scission point, are important in interpreting many of the features of the fission process. The stage of the process at which the various fragment distributions are determined is, however, not clearly established. All the components of a reasonable understanding of fission seem to be at hand, but they have yet to be synthesized into a complete, dynamic theory.

Considerations of the dynamics of the descent of the system on the potential-energy surface from the saddle point to the scission point involve two extreme points of view. An "adiabatic" approximation may be valid if the collective motion of the system is considered to be so slow—or the coupling between the collective and internal single-particle degrees of freedom (i.e., between macroscopic and microscopic behaviour) so weak—that the fast single-particle motions can readily adjust to the changes in shape of the fissioning nucleus as it progresses toward scission. In this case, the changes in the system take place without the gain or loss of heat energy. The decrease in potential energy between the saddle and scission points will then appear primarily in the collective degrees of freedom at scission and be associated with the kinetic energy of the relative motion of the nascent fragments (referred to as pre-scission kinetic energy). On the other hand, if the collective motion toward scission is relatively fast or the coupling-to-particle motion stronger, collective energy can be transformed into internal excitation (heat) energy of the nucleons. (This is analogous to heating in the motion of a viscous fluid.) In such a "non-adiabatic" process the mixing among the single-particle degrees of freedom may be sufficiently complete that a statistical model may be applicable at the scission point. Either extreme represents an approximation of complex behaviour, and some experimental evidence in support of either interpretation may be advanced. As in most such instances in nature, the truth probably lies somewhere between the extremes, with both playing some role in the fission process.

Fission Chain Reactions and their Control

The emission of several neutrons in the fission process leads to the possibility of a chain reaction if at least one of the fission neutrons induces fission in another fissile nucleus, which in turn fissions and emits neutrons to continue the chain. If more than one neutron is effective in inducing fission in other nuclei, the chain multiplies more rapidly. The condition for a chain reaction is usually expressed in terms of a multiplication factor, k, which is defined as the ratio of the number of fissions produced in one step (or neutron generation) in the chain to the number of fissions in the preceding generation. If k is less than unity, a chain reaction cannot be sustained. If $k = 1$, a steady-state chain reaction can be maintained; and if k is greater than 1, the number of fissions increases at each step, resulting in a divergent chain reaction. The term critical assembly is applied to a configuration of fissionable material for which $k = 1$; if $k > 1$, the assembly is said to be supercritical. A critical assembly might consist of the fissile material in the form of a metal or oxide, a moderator to slow the fission neutrons, and a reflector to scatter neutrons that would otherwise be lost back into the assembly core.

In a fission bomb it is desirable to have k as large as possible and the time between steps in the chain as short as possible so that many fissions occur and a large amount of energy is generated within a brief period ($\sim 10^{-7}$ second) to produce a devastating explosion. If one kilogram of uranium-235 were to fission, the energy released would be equivalent to the explosion of 20,000 tons of the chemical explosive trinitrotoluene (TNT). In a controlled nuclear reactor, k is kept equal to unity for steady-state operation. A practical reactor, however, must be designed with k somewhat greater than unity. This permits power levels to be increased if desired, as well as allowing for the following: the gradual loss of fuel by the fission process; the buildup of "poisons" among the fission products being formed that absorb neutrons and lower the k value; and the use of some of the neutrons produced for research studies or the preparation of radioactive species for various applications. The value of k is controlled during the operation of a reactor by the positioning of movable rods made of a material that readily absorbs neutrons (i.e., one with a high neutron-capture cross section), such as boron, cadmium, or hafnium. The delayed-neutron emitters among the fission products increase the time between successive neutron generations in the chain reaction and make the control of the reaction easier to accomplish by the mechanical movement of the control rods.

Fission reactors can be classified by the energy of the neutrons that propagate the chain reaction. The most common type, called a thermal reactor, operates with thermal neutrons (those having the same energy distribution as gas molecules at ordinary room temperatures). In such a reactor, the fission neutrons produced (with an average kinetic energy of more than 1 MeV) must be slowed down to thermal energy by scattering from a moderator, usually consisting of ordinary water, heavy water (D_2O), or graphite. In another type, termed an intermediate reactor, the chain reaction is maintained by neutrons of intermediate energy, and a beryllium moderator may be used. In a fast reactor, fast fission neutrons maintain the chain reaction, and no moderator is needed. All of the reactor types require a coolant to remove the heat generated; water, a gas, or a liquid metal may be used for this purpose, depending on the design needs.

Uses of Fission Reactors and Fission Products

A nuclear reactor is essentially a furnace used to produce steam or hot gases that can provide heat directly or drive turbines to generate electricity. Nuclear reactors are employed for commercial electric-power generation throughout much of the world and as a power source for propelling submarines and certain kinds of surface vessels. Another important use for reactors is the utilization of their high neutron fluxes for studying the structure and properties of materials and for producing a broad range of radionuclides, which, along with a number of fission products, have found many different applications. Heat generated by radioactive decay can be converted into electricity through the thermoelectric effect in semiconductor materials and thereby produce what is termed an atomic battery. When powered by either a long-lived beta-emitting fission product (e.g., strontium-90 or promethium-147) or one that emits alpha particles (plutonium-238 or curium-244), these batteries are a particularly useful source of energy for cardiac pacemakers and for instruments employed in remote unmanned facilities, such as those in outer space, the polar regions of the Earth, or the open seas.

Nuclear Fusion

Nuclear fusion is the process by which multiple atomic particles join together to form a heavier nucleus. It is accompanied by the release or absorption of energy. Iron and nickel nuclei have the largest binding energies per nucleon of all nuclei and therefore are the most stable. The fusion of two nuclei lighter than iron or nickel generally releases energy, while the fusion of nuclei heavier than iron or nickel absorbs energy. The opposite is true for nuclear fission. Nuclear fusion is naturally found in stars.

Fusion reactions power the stars and produce all but the lightest elements in a process called nucleosynthesis. Whereas the fusion of light elements in the stars releases energy, production of the heaviest elements absorbs energy, so it can only take place in the extremely high-energy conditions of supernova explosions.

When the fusion reaction is a sustained uncontrolled chain, it can result in a thermonuclear explosion, such as what is generated by a hydrogen bomb. Reactions that are not self-sustaining can still release considerable energy, as well as large numbers of neutrons.

Research into controlled fusion, with the aim of producing fusion power for the production of electricity, has been conducted for over 50 years. It has been accompanied by extreme scientific and technological difficulties, and as of yet has not been successful in producing workable designs. As of the present, the only self-sustaining fusion reactions produced by humans have been produced in hydrogen bombs, where the extreme power of a fission bomb is necessary to begin the process. While some plans have been put forth to attempt to use the explosions of hydrogen bombs to generate electricity (e.g. PACER), none of these have ever moved far past the design stage.

It takes considerable energy to force nuclei to fuse, even those of the lightest element, hydrogen. This is because all nuclei have a positive charge (due to their protons), and as like charges repel, nuclei strongly resist being put too close together. Accelerated to high speeds (that is, heated to thermonuclear temperatures), however, they can overcome this electromagnetic repulsion and get close enough for the strong nuclear force to be active, achieving fusion. The fusion of lighter nuclei, creating a heavier nucleus and a free neutron, will generally release more energy than it took to force them together—an exothermic process that can produce self-sustaining reactions.

The energy released in most nuclear reactions is much larger than that in chemical reactions, because the binding energy that holds a nucleus together is far greater than the energy that holds electrons to a nucleus. For example, the ionization energy gained by adding an electron to a hydrogen nucleus is 13.6 electron volts—less than one-millionth of the 17 MeV released in the D-T (deuterium-tritium) reaction shown to the top right. Fusion reactions have an energy density many times greater than nuclear fission—that is, per unit of mass the reactions produce far greater energies, even though individual fission reactions are generally much more energetic than individual fusion reactions—which are themselves millions of times more energetic than chemical reactions.

Only the direct conversion of mass into energy, such as with collision of matter and antimatter, is more energetic per unit of mass than nuclear fusion.

Building upon the nuclear transmutation experiments of Ernest Rutherford done a few years earlier, fusion of light nuclei (hydrogen isotopes) was first observed by Mark Oliphant in 1932, and the steps of the main cycle of nuclear fusion in stars were subsequently worked out by Hans Bethe throughout the remainder of that decade. Research into fusion for military purposes began in the early 1940s, as part of the Manhattan Project, but was not successful until 1952. Research into controlled fusion for civilian purposes began in the 1950s, and continues to this day.

Requirements

A substantial energy barrier must be overcome before fusion can occur. At large distances two naked nuclei repel one another because of the repulsive electrostatic force between their positively charged protons. If two nuclei can be brought close enough together, however, the electrostatic repulsion can be overcome by the nuclear force which is stronger at close distances.

When a nucleon such as a proton or neutron is added to a nucleus, the nuclear force attracts it to other nucleons, but primarily to its immediate neighbors due to the short range of the force. The nucleons in the interior of a nucleus have more neighboring nucleons than those on the surface. Since smaller nuclei have a larger surface area-to-volume ratio, the binding energy per nucleon due to the strong force generally increases with the size of the nucleus but approaches a limiting value corresponding to that of a fully surrounded nucleon.

The electrostatic force, on the other hand, is an inverse-square force, so a proton added to a nucleus will feel an electrostatic repulsion from *all* the other protons in the nucleus. The electrostatic energy per nucleon due to the electrostatic force thus increases without limit as nuclei get larger.

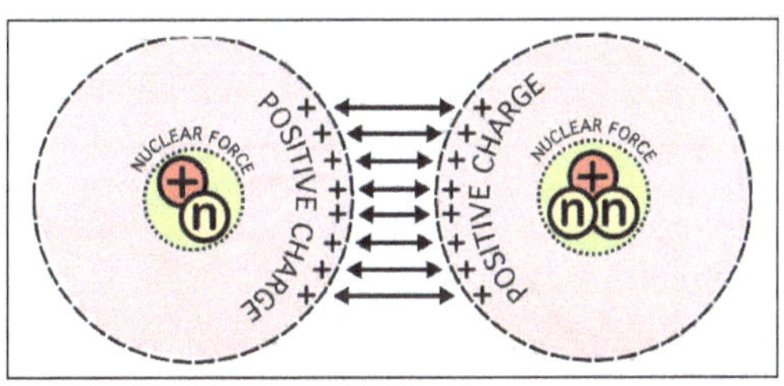

The electrostatic force caused by positively charged nuclei are very strong over long distances, but at short distances the nuclear force is stronger. As such, the main technical difficulty for fusion is getting the nuclei close enough to fuse (distances not to scale).

The net result of these opposing forces is that the binding energy per nucleon generally increases with increasing size, up to the elements iron and nickel, and then decreases for heavier nuclei. Eventually, the binding energy becomes negative and very heavy nuclei are not stable. The four most tightly bound nuclei, in decreasing order of binding energy, are ^{62}Ni, ^{58}Fe, ^{56}Fe, and ^{60}Ni. Even though the nickel isotope ^{62}Ni is more stable, the iron isotope ^{56}Fe is an order of magnitude more common. This is due to a greater disintegration rate for ^{62}Ni in the interior of stars driven by photon absorption.

A notable exception to this general trend is the helium-4 nucleus, whose binding energy is higher than that of lithium, the next heavier element. The Pauli exclusion principle provides an explanation for this exceptional behavior—it says that because protons and neutrons are fermions, they cannot exist in exactly the same state. Each proton or neutron energy state in a nucleus can accommodate both a spin up particle and a spin down particle. Helium-4 has an anomalously large binding energy because its nucleus consists of two protons and two neutrons; so all four of its nucleons can be in the ground state. Any additional nucleons would have to go into higher energy states.

The situation is similar if two nuclei are brought together. As they approach each other, all the protons in one nucleus repel all the protons in the other. Not until the two nuclei actually come in contact can the strong nuclear force take over. Consequently, even when the final energy state is lower, there is a large energy barrier that must first be overcome. It is called the Coulomb barrier.

The Coulomb barrier is smallest for isotopes of hydrogen—they contain only a single positive charge in the nucleus. A bi-proton is not stable, so neutrons must also be involved, ideally in such a way that a helium nucleus, with its extremely tight binding, is one of the products.

Using deuterium-tritium fuel, the resulting energy barrier is about 0.01 MeV. In comparison, the energy needed to remove an electron from hydrogen is 13.6 eV, about 750 times less energy. The (intermediate) result of the fusion is an unstable ^{5}He nucleus, which immediately ejects a neutron with 14.1 MeV. The recoil energy of the remaining ^{4}He nucleus is 3.5 MeV, so the total energy liberated is 17.6 MeV. This is many times more than what was needed to overcome the energy barrier.

If the energy to initiate the reaction comes from accelerating one of the nuclei, the process is called beam-target fusion; if both nuclei are accelerated, it is beam-beam fusion. If the nuclei are part of a plasma near thermal equilibrium, one speaks of thermonuclear fusion. Temperature is a measure of the average kinetic energy of particles, so by heating the nuclei they will gain energy and eventually have enough to overcome this 0.01 MeV. Converting the units between electron-volts and Kelvin shows that the barrier would be overcome at a temperature in excess of 120 million Kelvin—a very high temperature.

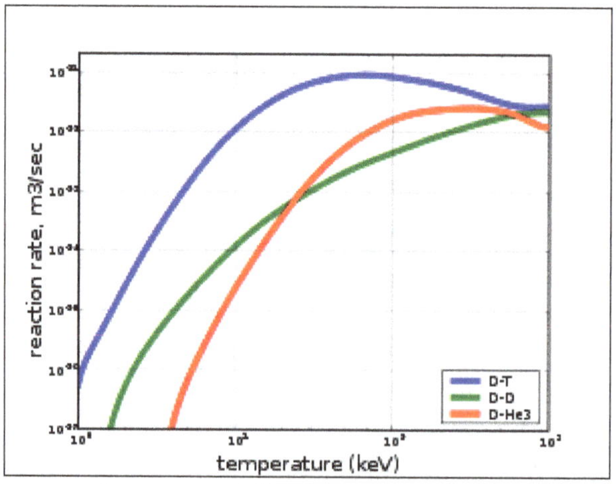

The fusion reaction rate increases rapidly with temperature until it maximizes and then gradually drops off. The DT rate peaks at a lower temperature (about 70 keV, or 800 million Kelvin) and at a higher value than other reactions commonly considered for fusion energy.

There are two effects that lower the actual temperature needed. One is the fact that temperature is the average kinetic energy, implying that some nuclei at this temperature would actually have much higher energy than 0.01 MeV, while others would be much lower. It is the nuclei in the high-energy tail of the velocity distribution that account for most of the fusion reactions. The other effect is quantum tunneling. The nuclei do not actually have to have enough energy to overcome the Coulomb barrier completely. If they have nearly enough energy, they can tunnel through the remaining barrier. For this reason fuel at lower temperatures will still undergo fusion events at a lower rate.

The reaction cross section σ is a measure of the probability of a fusion reaction as a function of the relative velocity of the two reactant nuclei. If the reactants have a distribution of velocities, e.g. a thermal distribution with thermonuclear fusion, then it is useful to perform an average over the distributions of the product of cross section and velocity. The reaction rate (fusions per volume per time) is $\langle \sigma v \rangle$ times the product of the reactant number densities:

$$f = n_1 n_2 \langle \sigma v \rangle$$

If a species of nuclei is reacting with itself, such as the DD reaction, then the product $n_1 n_2$ must be replaced by $(1/2)n^2$.

$\langle \sigma v \rangle$ increases from virtually zero at room temperatures up to meaningful magnitudes at temperatures of 10–100 keV. At these temperatures, well above typical ionization energies (13.6 eV in the hydrogen case), the fusion reactants exist in a plasma state.

The significance of $\langle \sigma v \rangle$ as a function of temperature in a device with a particular energy confinement time is found by considering the Lawson criterion.

Fusion Reaction

Fusion reactions constitute the fundamental energy source of stars, including the Sun. The evolution of stars can be viewed as a passage through various stages as thermonuclear reactions and nucleosynthesis cause compositional changes over long time spans. Hydrogen (H) "burning" initiates the fusion energy source of stars and leads to the formation of helium (He). Generation of fusion energy for practical use also relies on fusion reactions between the lightest elements that burn to form helium. In fact, the heavy isotopes of hydrogen—deuterium (D) and tritium (T)—react more efficiently with each other, and, when they do undergo fusion, they yield more energy per reaction than do two hydrogen nuclei. (The hydrogen nucleus consists of a single proton. The deuterium nucleus has one proton and one neutron, while tritium has one proton and two neutrons).

Fusion reactions between light elements, like fission reactions that split heavy elements, release energy because of a key feature of nuclear matter called the binding energy, which can be released through fusion or fission. The binding energy of the nucleus is a measure of the efficiency with which its constituent nucleons are bound together. Take, for example, an element with *Z* protons and *N* neutrons in its nucleus. The element's atomic weight *A* is *Z* + *N*, and its atomic number is *Z*. The binding energy *B* is the energy associated with the mass difference between the *Z* protons and *N* neutrons considered separately and the nucleons bound together (*Z* + *N*) in a nucleus of mass *M*. The formula is:

$$\mathrm{B} = \left(\mathrm{Zm_p} + \mathrm{Nm_n} - \mathrm{M}\right)C^2,$$

where m_p and m_n are the proton and neutron masses and c is the speed of light. It has been determined experimentally that the binding energy per nucleon is a maximum of about 1.4×10^{-12} joule at an atomic mass number of approximately 60—that is, approximately the atomic mass number of iron. Accordingly, the fusion of elements lighter than iron or the splitting of heavier ones generally leads to a net release of energy.

Two Types of Fusion Reactions

Fusion reactions are of two basic types:

1. Those that preserve the number of protons and neutrons,

2. Those that involve a conversion between protons and neutrons. Reactions of the first type are most important for practical fusion energy production, whereas those of the second type are crucial to the initiation of star burning. An arbitrary element is indicated by the notation ${}^{A}_{Z}X$, where Z is the charge of the nucleus and A is the atomic weight. An important fusion reaction for practical energy generation is that between deuterium and tritium (the D-T fusion reaction). It produces helium (He) and a neutron (n) and is written:

$$\mathrm{D + T \rightarrow He + n.}$$

To the left of the arrow (before the reaction) there are two protons and three neutrons. The same is true on the right.

The other reaction, that which initiates star burning, involves the fusion of two hydrogen nuclei to form deuterium (the H-H fusion reaction):

$$\mathrm{H + H \rightarrow D +} \beta^{+} + \mathrm{v},$$

where β^{+} represents a positron and v stands for a neutrino. Before the reaction there are two hydrogen nuclei (that is, two protons). Afterward there are one proton and one neutron (bound together as the nucleus of deuterium) plus a positron and a neutrino (produced as a consequence of the conversion of one proton to a neutron).

Both of these fusion reactions are exoergic and so yield energy. The German-born physicist Hans Bethe proposed in the 1930s that the H-H fusion reaction could occur with a net release of energy and provide, along with subsequent reactions, the fundamental energy source sustaining the stars. However, practical energy generation requires the D-T reaction for two reasons: first, the rate of reactions between deuterium and tritium is much higher than that between protons; second, the net energy release from the D-T reaction is 40 times greater than that from the H-H reaction.

Energy Released in Fusion Reactions

Energy is released in a nuclear reaction if the total mass of the resultant particles is less than the mass of the initial reactants. To illustrate, suppose two nuclei, labeled X and a, react to form two other nuclei, Y and b, denoted:

$$X + a \rightarrow Y + b.$$

The particles a and b are often nucleons, either protons or neutrons, but in general can be any nuclei. Assuming that none of the particles is internally excited (i.e., each is in its ground state), the energy quantity called the Q-value for this reaction is defined as,

$$Q = (m_x + m_a - m_b - m_y)c^2,$$

where the m-letters refer to the mass of each particle and c is the speed of light. When the energy value Q is positive, the reaction is exoergic; when Q is negative, the reaction is endoergic (i.e., absorbs energy). When both the total proton number and the total neutron number are preserved before and after the reaction (as in D-T reactions), then the Q-value can be expressed in terms of the binding energy B of each particle as,

$$Q = B_y + B_b - B_x - B_a.$$

The D-T fusion reaction has a positive Q-value of 2.8×10^{-12} joule. The H-H fusion reaction is also exoergic, with a Q-value of 6.7×10^{-14} joule. To develop a sense for these figures, one might consider that one metric ton (1,000 kg, or almost 2,205 pounds) of deuterium would contain roughly 3×10^{32} atoms. If one ton of deuterium were to be consumed through the fusion reaction with tritium, the energy released would be 8.4×10^{20} joules. This can be compared with the energy content of one ton of coal—namely, 2.9×10^{10} joules. In other words, one ton of deuterium has the energy equivalent of approximately 29 billion tons of coal.

Rate and Yield of Fusion Reactions

The energy yield of a reaction between nuclei and the rate of such reactions are both important. These quantities have a profound influence in scientific areas such as nuclear astrophysics and the potential for nuclear production of electrical energy.

When a particle of one type passes through a collection of particles of the same or different type, there is a measurable chance that the particles will interact. The particles may interact in many ways, such as simply scattering, which means that they change direction and exchange energy, or they may undergo a nuclear fusion reaction. The measure of the likelihood that particles will interact is called the cross section, and the magnitude of the cross section depends on the type of interaction and the state and energy of the particles. The product of the cross section and the atomic density of the target particle is called the macroscopic cross section. The inverse of the macroscopic cross section is particularly noteworthy as it gives the mean distance an incident particle will travel before interacting with a target particle; this inverse measure is called the mean free path. Cross sections are measured by producing a beam of one particle at a given energy, allowing the beam to interact with a (usually thin) target made of the same or a different material, and measuring deflections or reaction products. In this way it is possible to determine the relative likelihood of one type of fusion reaction versus another, as well as the optimal conditions for a particular reaction.

The cross sections of fusion reactions can be measured experimentally or calculated theoretically, and they have been determined for many reactions over a wide range of particle energies. They are well known for practical fusion energy applications and are reasonably well known, though with gaps, for stellar evolution. Fusion reactions between nuclei, each with a positive charge of one or more, are the most important for both practical applications and the nucleosynthesis of the light elements in the burning stages of stars. Yet, it is well known that two positively charged nuclei repel each other electrostatically—i.e., they experience a repulsive force inversely proportional to the square of the distance separating them. This repulsion is called the Coulomb barrier. It is highly unlikely that two positive nuclei will approach each other closely enough to undergo a fusion reaction unless they have sufficient energy to overcome the Coulomb barrier. As a result, the cross section for fusion reactions between charged particles is very small unless the energy of the particles is high, at least 10^4 electron volts (1 eV ≅ 1.602×10^{-19} joule) and often more than 10^5 or 10^6 eV. This explains why the centre of a star must be hot for the fuel to burn and why fuel for practical fusion energy systems must be heated to at least 50,000,000 kelvins (K; 90,000,000 °F). Only then will a reasonable fusion reaction rate and power output be achieved.

The phenomenon of the Coulomb barrier also explains a fundamental difference between energy generation by nuclear fusion and nuclear fission. While fission of heavy elements can be induced by either protons or neutrons, generation of fission energy for practical applications is dependent on neutrons to induce fission reactions in uranium or plutonium. Having no electric charge, the neutron is free to enter the nucleus even if its energy corresponds to room temperature. Fusion energy, relying as it does on the fusion reaction between light nuclei, occurs only when the particles are sufficiently energetic to overcome the Coulomb repulsive force. This requires the production and heating of the gaseous reactants to the high temperature state known as the plasma state.

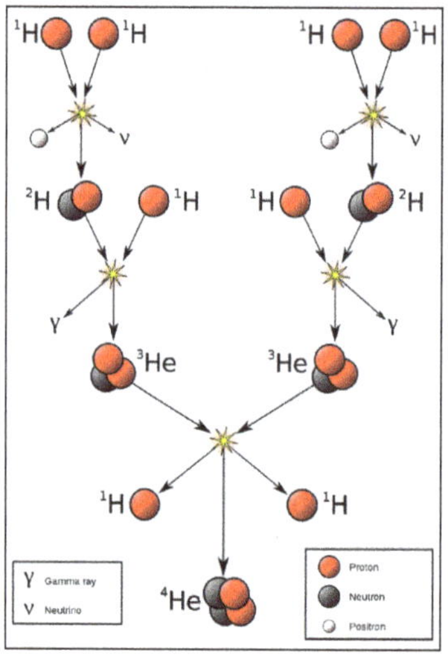

Plasma State

Typically, a plasma is a gas that has had some substantial portion of its constituent atoms or molecules ionized by the dissociation of one or more of their electrons. These free electrons enable plasmas to conduct electric charges, and a plasma is the only state of matter in which thermonuclear reactions can occur in a self-sustaining manner. Astrophysics and magnetic fusion research, among other fields, require extensive knowledge of how gases behave in the plasma state. The stars, the solar wind, and much of interstellar space are examples where the matter present is in the plasma state. Very high-temperature plasmas are fully ionized gases, which means that the ratio of neutral gas atoms to charged particles is small. For example, the ionization energy of hydrogen is 13.6 eV, while the average energy of a hydrogen ion in a plasma at 50,000,000 K is 6,462 eV. Thus, essentially all of the hydrogen in this plasma would be ionized.

A reaction-rate parameter more appropriate to the plasma state is obtained by accounting for the fact that the particles in a plasma, as in any gas, have a distribution of energies. That is to say, not all particles have the same energy. In simple plasmas this energy distribution is given by the Maxwell-Boltzmann distribution law, and the temperature of the gas or plasma is, within a proportionality constant, two-thirds of the average particle energy; i.e., the relationship between the average energy E and temperature T is $E = 3kT/2$, where k is the Boltzmann constant, 8.62×10^{-5} eV per kelvin. The intensity of nuclear fusion reactions in a plasma is derived by averaging the product of the particles' speed and their cross sections over a distribution of speeds corresponding to a Maxwell-Boltzmann distribution. The cross section for the reaction depends on the energy or speed of the particles. The averaging process yields a function for a given reaction that depends only on the temperature and can be denoted $f(T)$. The rate of energy released (i.e., the power released) in a reaction between two species, a and b, is:

$$P_{ab} = n_a n_b f_{ab}(T) U_{ab},$$

where n_a and n_b are the density of species a and b in the plasma, respectively, and U_{ab} is the energy released each time a and b undergo a fusion reaction. The parameter P_{ab} properly takes into account both the rate of a given reaction and the energy yield per reaction.

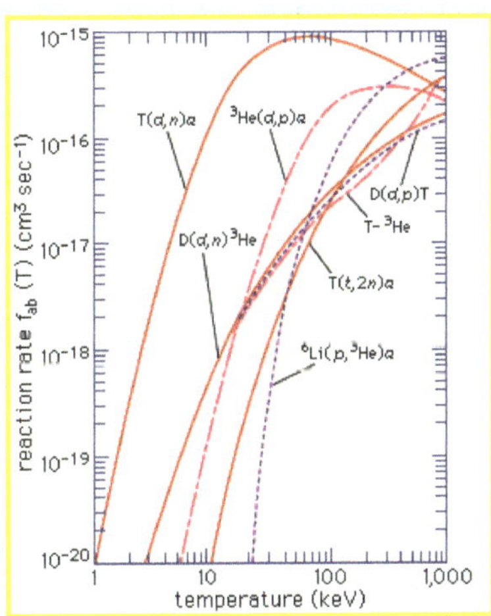

One keV is equivalent to a temperature of 11,000,000 K. The rate of reaction between deuterium and tritium is seen to be higher than all others and is very substantial, even at temperatures in the 5- to 10-keV range (see text).

The reaction rate as a function of plasma temperature, expressed in kiloelectron volts (keV; 1 keV is equivalent to a temperature of 11,000,000 K). The rate of reaction between deuterium and tritium is seen to be higher than all others and is very substantial, even at temperatures in the 5-to-10-keV range.

Fusion Reactions in Stars

Fusion reactions are the primary energy source of stars and the mechanism for the nucleosynthesis of the light elements. In the late 1930s Hans Bethe first recognized that the fusion of hydrogen nuclei to form deuterium is exoergic (i.e., there is a net release of energy) and, together with subsequent nuclear reactions, leads to the synthesis of helium. The formation of helium is the main source of energy emitted by normal stars, such as the Sun, where the burning-core plasma has a temperature of less than 15,000,000 K. However, because the gas from which a star is formed often contains some heavier elements, notably carbon (C) and nitrogen (N), it is important to include nuclear reactions between protons and these nuclei. The reaction chain between protons that ultimately leads to helium is the proton-proton cycle. When protons also induce the burning of carbon and nitrogen, the CN cycle must be considered; and, when oxygen (O) is included, still another alternative scheme, the CNO bi-cycle, must be accounted for.

The proton-proton nuclear fusion cycle in a star containing only hydrogen begins with the reaction,

$$\mathrm{H} + \mathrm{H} \rightarrow \mathrm{D} + \beta^{+} + \nu;\ \mathrm{Q} = 1.44\ \mathrm{MeV},$$

where the *Q*-value assumes annihilation of the positron by an electron. The deuterium could react with other deuterium nuclei, but, because there is so much hydrogen, the D/H ratio is held to very low values, typically 10^{-18}. Thus, the next step is:

$$\mathrm{H} + \mathrm{D} \rightarrow {}^{3}\mathrm{He} + \gamma; Q = 5.49\ \mathrm{MeV},$$

where γ indicates that gamma rays carry off some of the energy yield. The burning of the helium-3 isotope then gives rise to ordinary helium and hydrogen via the last step in the chain:

$${}^{3}\mathrm{He} + {}^{3}\mathrm{He} \rightarrow {}^{4}\mathrm{He} + 2(\mathrm{H});\ \mathrm{Q} = 12.86\ \mathrm{MeV}.$$

At equilibrium, helium-3 burns predominantly by reactions with itself because its reaction rate with hydrogen is small, while burning with deuterium is negligible due to the very low deuterium concentration. Once helium-4 builds up, reactions with helium-3 can lead to the production of still-heavier elements, including beryllium-7, beryllium-8, lithium-7, and boron-8, if the temperature is greater than about 10,000,000 K.

The stages of stellar evolution are the result of compositional changes over very long periods. The size of a star, on the other hand, is determined by a balance between the pressure exerted by the hot plasma and the gravitational force of the star's mass. The energy of the burning core is transported toward the surface of the star, where it is radiated at an effective temperature. The effective temperature of the Sun's surface is about 6,000 K, and significant amounts of radiation in the visible and infrared wavelength ranges are emitted.

Fusion Reactions for Controlled Power Generation

Reactions between deuterium and tritium are the most important fusion reactions for controlled power generation because the cross sections for their occurrence are high, the practical plasma temperatures required for net energy release are moderate, and the energy yield of the reactions are high—17.58 MeV for the basic D-T fusion reaction.

It should be noted that any plasma containing deuterium automatically produces some tritium and helium-3 from reactions of deuterium with other deuterium ions. Other fusion reactions involving elements with an atomic number above 2 can be used, but only with much greater difficulty. This is because the Coulomb barrier increases with increasing charge of the nuclei, leading to the requirement that the plasma temperature exceed 1,000,000,000 K if a significant rate is to be achieved. Some of the more interesting reactions are:

a. $\mathrm{H} + {}^{11}\mathrm{B} \rightarrow 3({}^{4}\mathrm{He});\ \mathrm{Q} = 8.68\ \mathrm{MeV};$

b. $\mathrm{H} + {}^{6}\mathrm{Li} \rightarrow {}^{3}\mathrm{He} + {}^{4}\mathrm{He};\ \mathrm{Q} = 4.023\ \mathrm{MeV};$

c. $^{3}He + ^{6}Li \rightarrow H + 2(^{4}He)$; $Q = 16.88$ MeV; and

d. $^{3}He + ^{6}Li \rightarrow D + ^{7}Be$; $Q = 0.113$ MeV.

Reaction(b) converts lithium-6 to helium-3 and ordinary helium. Interestingly, if reaction (b) is followed by reaction (c), then a proton will again be produced and be available to induce reaction (b), thereby propagating the process. Unfortunately, it appears that reaction is 10 times more likely to occur than reaction (c).

Methods of Achieving Fusion Energy

Practical efforts to harness fusion energy involve two basic approaches to containing a high-temperature plasma of elements that undergo nuclear fusion reactions: magnetic confinement and inertial confinement. A much less likely but nevertheless interesting approach is based on fusion catalyzed by muons; research on this topic is of intrinsic interest in nuclear physics. These three methods are described in some detail in this section. In addition, the processes popularly dubbed cold fusion.

Magnetic Confinement

In magnetic confinement the particles and energy of a hot plasma are held in place using magnetic fields. A charged particle in a magnetic field experiences a Lorentz force that is proportional to the product of the particle's velocity and the magnetic field. This force causes electrons and ions to spiral about the direction of the magnetic line of force, thereby confining the particles. When the topology of the magnetic field yields an effective magnetic well and the pressure balance between the plasma and the field is stable, the plasma can be confined away from material boundaries. Heat and particles are transported both along and across the field, but energy losses can be prevented in two ways. The first is to increase the strength of the magnetic field at two locations along the field line. Charged particles contained between these points can be made to reflect back and forth, an effect called magnetic mirroring. In a basically straight system with a region of intensified magnetic field at each end, particles can still escape through the ends due to scattering between particles as they approach the mirroring points. Such end losses can be avoided altogether by creating a magnetic field in the topology of a torus (i.e., configuration of a doughnut or inner tube).

External magnets can be arranged to create a magnetic field topology for stable plasma confinement, or they can be used in conjunction with magnetic fields generated by currents induced to flow in the plasma itself. The late 1960s witnessed a major advance by the Soviet Union in harnessing fusion reactions for practical energy production. Soviet scientists achieved a high plasma temperature (about 3,000,000 K), along with other physical parameters, in a machine referred to as a tokamak. A tokamak is a toroidal magnetic confinement system in which the plasma is kept stable both by an externally generated, doughnut-shaped magnetic field and by electric currents flowing within the plasma. Since the late 1960s the tokamak has been the major focus of magnetic fusion research worldwide, though other approaches such as the stellarator, the compact torus, and the reversed field pinch (RFP) have also been pursued. In these approaches, the magnetic field lines follow a helical, or screwlike, path as the lines of magnetic force proceed around the torus. In the tokamak the pitch of the helix is weak, so the field lines wind loosely around the poloidal direction (through the central hole) of the torus. In contrast, RFP field lines wind much tighter, wrapping many times in the poloidal direction before completing one loop in the toroidal direction (around the central hole).

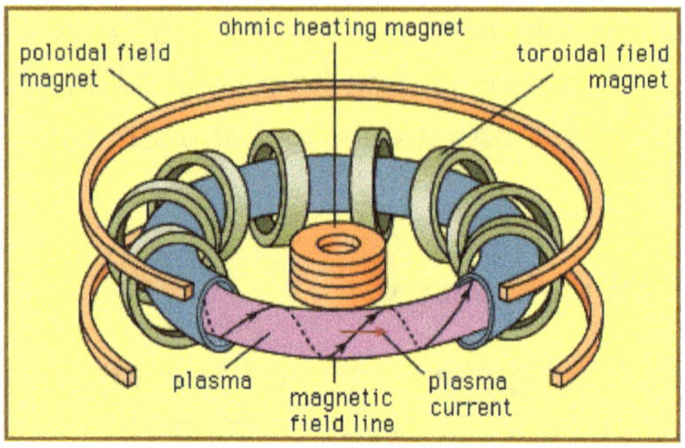

Tokamak magnetic confinement.

Magnetically confined plasma must be heated to temperatures at which nuclear fusion is vigorous, typically greater than 75,000,000 K (equivalent to an energy of 4,400 eV). This can be achieved by coupling radio-frequency waves or microwaves to the plasma particles, by injecting energetic beams of neutral atoms that become ionized and heat the plasma, by magnetically compressing the plasma, or by the ohmic heating (also known as Joule heating) that occurs when an electric current passes through the plasma.

Employing the tokamak concept, scientists and engineers in the United States, Europe, and Japan began in the mid-1980s to use large experimental tokamak devices to attain conditions of temperature, density, and energy confinement that now match those necessary for practical fusion power generation. The machines employed to achieve these results include the Joint European Torus (JET) of the European Union, the Japanese Tokamak-60 (JT-60), and, until 1997, the Tokamak Fusion Test Reactor (TFTR) in the United States. Indeed, in both the TFTR and the JET devices, experiments using deuterium and tritium produced more than 10 megawatts of fusion power and essentially energy breakeven conditions in the plasma itself. Plasma conditions approaching those achieved in tokamaks were also achieved in large stellarator machines in Germany and Japan during the 1990s.

Inertial Confinement Fusion

In this approach, a fuel mass is compressed rapidly to densities 1,000 to 10,000 times greater than normal by generating a pressure as high as 10^{17} pascals (10^{12} atmospheres) for periods as short as a nanosecond (10^{-9} second). Near the end of this time period, the implosion speed exceeds about 3×10^5 metres per second. At maximum compression of the fuel, which is now in a cool plasma state, the energy in converging shock waves is sufficient to heat the very centre of the fuel to temperatures high enough to induce fusion reactions (greater than an equivalent energy of about 4,400 eV). If the mass of this highly compressed fuel material is large enough, energy will be generated through fusion reactions before this hot plasma ball disassembles. Under proper conditions, much more energy can be released than is required to compress and shock heat the fuel to thermonuclear burning conditions.

The physical processes in ICF bear a relationship to those in thermonuclear weapons and in star formation—namely, collapse, compression heating, and the onset of nuclear fusion. The situation in star formation differs in one respect: gravity is the cause of the collapse, and a collapsed star begins to expand again due to heat from exoergic nuclear fusion reactions. The expansion is ultimately arrested by the gravitational force associated with the enormous mass of the star, at which point a state of equilibrium in both size and temperature is achieved. In contrast, the fuel in a thermonuclear weapon or ICF completely disassembles. In the ideal ICF case, however, this does not occur until about 30 percent of the fusion fuel has burned.

Over the decades, very significant progress has been made in developing the technology and systems for high-energy, short-time-pulse drivers that are necessary to implode the fusion fuel. The most common driver is a high-power laser, though particle accelerators capable of producing beams of high-energy ions are also used. Lasers that produce more than 100,000 joules in pulses of about one nanosecond are now used in experiments, and the power available in short bursts exceeds 10^{14} watts.

Two lasers capable of delivering up to 5,000,000 joules in equally short bursts, generating a power level on the fusion targets in excess of 5×10^{14} watts, are operational. One facility is the Laser MegaJoule in Bordeaux, France. The other is the National Ignition Facility at the Lawrence Livermore National Laboratory in Livermore, California, U.S.

Muon-catalyzed Fusion

The need in traditional schemes of nuclear fusion to confine very high-temperature plasmas has led some researchers to explore alternatives that would permit fusion reactants to approach each other more closely at much lower temperatures. One method involves substituting muons (μ) for the electrons that ordinarily surround the nucleus of a fuel atom. Muons are negatively charged subatomic particles similar to electrons, except that their mass is a little more than 200 times the electron mass and they are unstable, having a half-life of about 2.2×10^{-6} second. In fact, fusion has been observed in liquid and gas mixtures of deuterium and tritium at cryogenic temperatures when muons were injected into the mixture.

Muon-catalyzed fusion is the name given to the process of achieving fusion reactions by causing a deuteron (deuterium nucleus, D^+), a triton (tritium nucleus, T^+), and a muon to form what is called a muonic molecule. Once a muonic molecule is formed, the rate of fusion reactions is approximately 3×10^{-8} second. However, the formation of a muonic molecule is complex, involving a series of atomic, molecular, and nuclear processes.

In schematic terms, when a muon enters a mixture of deuterium and tritium, the muon is first captured by one of the two hydrogen isotopes in the mixture, forming either atomic D^+-μ or T^+-μ, with the atom now in an excited state. The excited atom relaxes to the ground state through a cascade collision process, in which the muon may be transferred from a deuteron to a triton or vice versa. More important, it is also possible that a muonic molecule (D^+-μ-T^+) will be formed. Although a much rarer reaction, once a muonic molecule does form, fusion takes place almost immediately, releasing the muon in the mixture to be captured again by a deuterium or tritium nucleus and allowing the process to continue. In this sense the muon acts as a catalyst for fusion reactions within the mixture. The key to practical energy production is to generate enough fusion reactions before the muon decays.

The complexities of muon-catalyzed fusion are many and include generating the muons (at an energy expenditure of about five billion electron volts per muon) and immediately injecting them into the deuterium-tritium mixture. In order to produce more energy than what is required to initiate the process, about 300 D-T fusion reactions must take place within the half-life of a muon.

Cold Fusion and Bubble Fusion

Two disputed fusion experiments merit mention. In 1989 two chemists, Martin Fleischmann of the University of Utah and Stanley Pons of the University of Southampton in England, announced that they had produced fusion reactions at essentially room temperature. Their system consisted of electrolytic cells containing heavy water (deuterium oxide, D_2O) and palladium rods that absorbed the deuterium from the heavy water. Efforts to give a theoretical explanation of the results failed, as did worldwide efforts to reproduce the claimed cold fusion.

In 2002 Rusi Taleyarkhan and colleagues at Purdue University in Lafayette, Indiana, claimed to have observed a statistically significant increase in nuclear emissions of products of fusion reactions (neutrons and tritium) during acoustic cavitation experiments with chilled deuterated (bombarded with deuterium) acetone. Their experimental setup was based on the known phenomenon of sonoluminescence. In sonoluminescence a gas bubble is imploded with high-pressure sound waves. At the end of the implosion process, and for a short time afterward, conditions of high density and temperature are achieved that lead to light emission. By starting with larger, millimetre-sized cavitations (bubbles) that had been deuterated in the acetone liquid, the researchers claimed to have produced densities and temperatures sufficient to induce fusion reactions just before the bubbles broke up. As with cold fusion, most attempts to replicate their results have failed.

Conditions for Practical Fusion Yield

Two conditions must be met to achieve practical energy yields from fusion. First, the plasma temperature must be high enough that fusion reactions occur at a sufficient rate. Second, the plasma must be confined so that the energy released by fusion reactions, when deposited in the plasma, maintains its temperature against loss of energy by such phenomena as conduction, convection, and radiation. When these conditions are achieved, the plasma is said to be ignited. In the case of stars, or some approaches to fusion by magnetic confinement, a steady state can be achieved, and no energy beyond what is supplied from fusion reactions is needed to sustain the system. In other cases, such as the ICF approach, there is a large temperature excursion once fuel ignition is achieved. The energy yield can far exceed the energy required to attain plasma ignition conditions, but this energy is released in a burst, and the process has to be repeated roughly once every second for practical power to be produced.

The conditions for plasma ignition are readily derived. When fusion reactions occur in a plasma, the power released is proportional to the square of plasma ion density, n^2. The plasma loses energy when electrons scatter from positively charged ions, accelerating and radiating in the process. Such radiation is called bremsstrahlung and is proportional to $n^2T^{1/2}$, where T is the plasma temperature. Other mechanisms by which heat can escape the plasma lead to a characteristic energy-loss time denoted by τ. The energy content of the plasma at temperature T is $3nkT$, where k is the Boltzmann constant. The rate of energy loss by mechanisms other than bremsstrahlung is thus simply $3nkT/\tau$. The energy balance of the plasma is the balance between the fusion energy heating the plasma and the energy-loss rate, which is the sum of $3nkT/\tau$ and the bremsstrahlung. The condition satisfying this balance is called the ignition condition. An equation relates the product of density and energy confinement time, denoted $n\tau$, to a function that depends only on the plasma temperature and the type of fusion reaction. For example, when the plasma is composed of deuterium and tritium, the smallest value of $n\tau$ required to achieve ignition is about 2×10^{20} particles per cubic metre times seconds, and the required temperature corresponds to an energy of about 25,000 eV. If the only energy losses are due to bremsstrahlung escaping from the plasma (meaning τ is infinite), the ignition temperature decreases to an energy level of 4,400 eV. Hence, the keys to generating usable amounts of fusion energy are to attain a sufficient plasma temperature and a sufficient confinement quality, as measured by the product $n\tau$. At a temperature equivalent to 10,000 eV, the $n\tau$ product must be about 3×10^{20} particles per cubic metre times seconds.

Magnetic fusion energy generally creates plasmas with a density of about 3×10^{20} particles per cubic metre, which is about 10^{-8} of normal density. Hence, the characteristic time for heat to escape must be greater than about one second. This is a measure of the required degree of magnetic insulation for the heat content. Under these conditions the plasma remains in energy balance and can operate continuously if the ash of the nuclear fusion, namely helium, is removed (otherwise it will quench the plasma) and fuel is replenished.

ICF creates plasmas of much higher density, generally between 10^{31} and 10^{32} particles per cubic metre, or 1,000 to 10,000 times the normal density. As such, the confinement time, or minimum burn time, can be as short as 20×10^{-12} second. The objective in ICF is to achieve a temperature equivalent of 4,400 eV at the centre of the highly compressed fuel mass, while still having sufficient mass left around the centre so that the disassembly time will exceed the minimum burn time.

References

- Nuclear-reactions, nuclear-power: nuclear-power.net, Retrieved 17 March, 2019
- Neutron-capture-radiative-capture, neutron-nuclear-reactions, nuclear-engineering-fundamentals, reactor-physics, nuclear-power: nuclear-power.net, Retrieved 13 February, 2019
- Nuclear-fission, science: britannica.com, Retrieved 3 May, 2019
- Nuclear-fusion, entry: newworldencyclopedia.org, Retrieved 15 January, 2019
- Nuclear-fusion, science: britannica.com, Retrieved 28 April, 2019

4

Nuclear Decay

The process through which an unstable nucleus loses energy through radiation is termed as nuclear decay. It is also known as radioactivity. The different particles which can undergo nuclear decay are alpha particles, beta particles and gamma particles. This chapter has been carefully written to provide an easy understanding of the varied facets of the nuclear decay of these particles.

Nuclear decay occurs when the nucleus of an atom is unstable and spontaneously emits energy in the form of radiation. The result is that the nucleus changes into the nucleus of one or more other elements. These daughter nuclei have a lower mass and are more stable (lower in energy) than the parent nucleus. Nuclear decay is also called radioactive decay, and it occurs in a series of sequential reactions until a stable nucleus is reached.

Neutron to Proton Ratio

The neutron-to-proton ratio $\left(n:p\,or\,\frac{n}{p}\right)$ determines whether or not an atom is stable. Elements with a ratio close to or equal to one are considered stable. Regardless of the neutron-proton ratio, all elements with an atomic number greater than 83 are unstable and therefore radioactive.

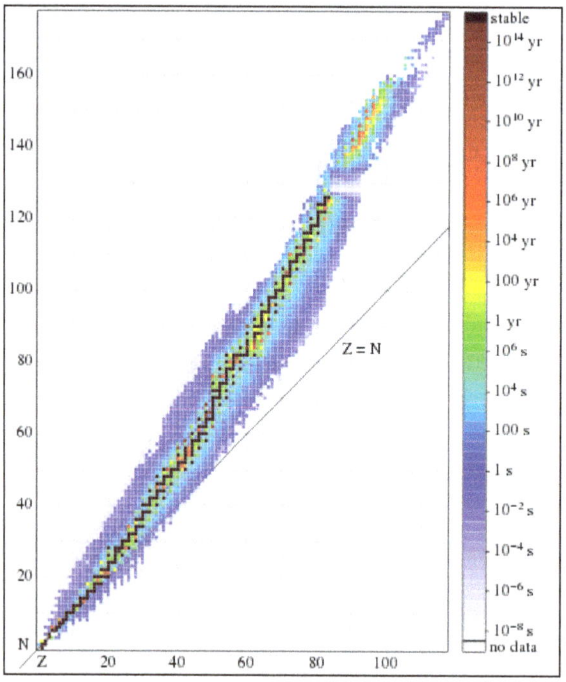

The black line in the picture above shows the stability band, where the ratio is equal to 1:1. As the atomic number increases, the graph curves upward. As atomic mass increases, a larger number of neutrons are required to overcome the strong electrostatic repulsion between the protons. These atoms do not have a stable $\frac{n}{p}$ ratio. To overcome this instability, the nucleus of the atom emits energy. The emissions are called radioactivity.

Radiation

Radiation is a form of energy that is released as electromagnetic waves or particles, moves through space, and may be able to penetrate or interact with different materials. Radiation-caused changes in materials depend on the origin, type of radiation, and the deposited energy.

Classification of Radiations

Radiation is classified into ionizing and nonionizing radiation. Ionizing radiation is divided into direct ionizing and indirect ionizing.

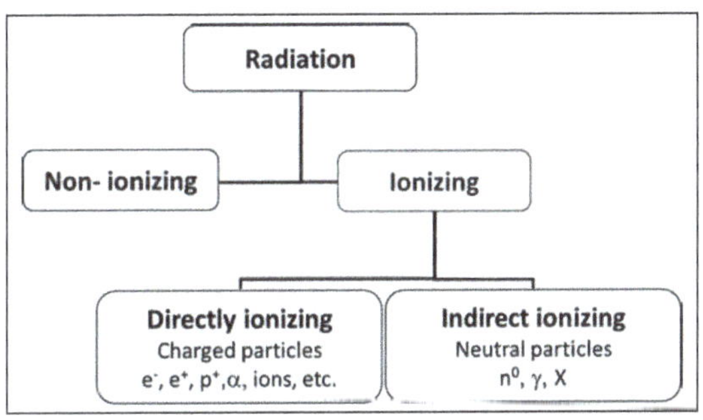

Ionizing Radiation

Ionization radiation refers to the capability of ionizing material either directly or indirectly because of their higher energy such as X-rays, γ-rays, energetic neutrons, electrons, protons, and heavier particles.

- If the energy of electromagnetic waves is high, the frequency will be high with short wavelength such as those of gamma rays or heavy particles (beta and alpha).
- Enough high energy to pull electron from orbit.

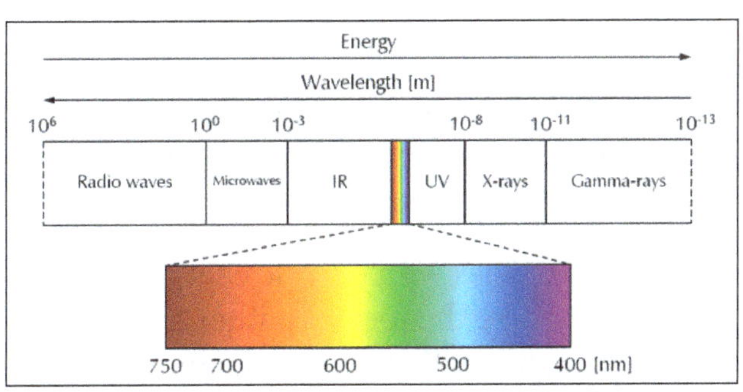

The schematic representation of the different regions of the electromagnetic spectrum.

Nonionizing Radiation

Nonionizing radiation refers to the inability of ionizing materials because of their lower energy, Such as ultraviolet radiation, visible light, infrared photons, microwaves and radio waves.

- If the energy of electromagnetic waves is low, the frequency will be low with long wavelength such as those of radio waves and microwaves.
- Not enough energy to pull electron from orbit, but the electron can exit.

Classification of Ionizing Radiation

Ionizing radiation is classified into two types:

Directly Ionizing Radiation

- This consists of charged particles, such as electrons, protons, α particles, and heavy ions.
    - Energy can be deposited by directly ionizing radiation in the medium through direct coulomb interaction between the directly ionizing charged particles and orbital electrons of atoms.

Indirectly Ionizing Radiation

- This consists of neutral particles, photons (X-ray and γ-rays), and neutrons.
    - Energy can be deposited by indirectly ionizing radiation (photons or neutrons) in the medium through two steps:
        - First step: A charged molecule is transmitted in the medium (photons discharge electrons or positrons; neutrons discharge protons or heavier particles).
        - The second step: The produced charged particles store vitality to the medium through direct Coulomb interaction with the orbital electrons of the atoms.

Ionizing Radiation takes a Few Forms

Alpha, beta, neutron particles, gamma rays and X-rays are each caused by unstable atoms, either through the overabundance of vitality or mass (or both of them). To reach a steady state, they must discharge that additional vitality or mass within the frame of radiation.

Alpha particles (α particle): They are positive charged particles (+2), which are released in the radioactive decay of some nuclei. An alpha is a particle which is emitted from the nucleus of an atom, which consists of (+2) protons and neutrons with mass number (Helium atom). It is strong ionizing with low penetration power and short range.

Beta particles (β+ or β-): They are particles with electric charge ((+) or (−)) emitted from the nucleus during radioactive decay. They take the form of either an electron or a positron (a particle with the size and mass of an electron, but with a positive charge).

Electrons or positrons have small mass and variable energy. Electrons are formed when a neutron transforms into a proton and an electron.

Gamma rays: They are different from alpha or beta rays, because they do not contain any particles, as they are used in electromagnetic radiation. Instead, they consist of a photon of energy, which is released from an unstable nucleus of an atom.

Isomeric transition: It occurs when the excited atomic nucleus changes from a higher to a lower state of the energy by emitting gamma ray.

Internal conversion electron: This process occurs when the gamma rays are not released sometimes, so they provide their exceed energy to the electron in the atomic orbit; this process usually happens to the nearest nucleus.

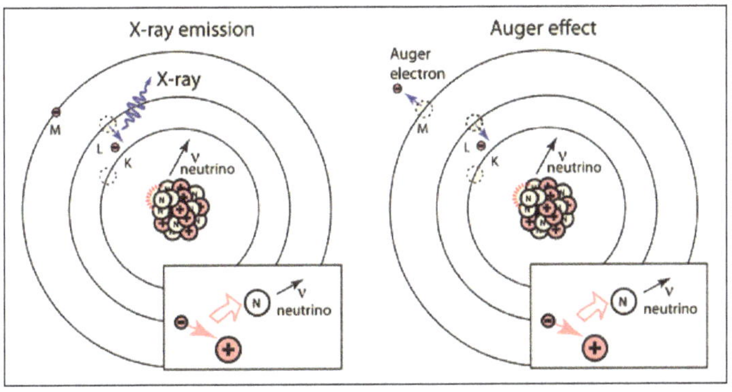

Emission of an auger electron as an alternative to X-ray emission. No X-ray is emitted.

X-rays: They are similar to gamma radiation; the only one primary difference is that they originate from the electron shell. This is generally caused by energy changes in an electron, such as moving from a higher energy level to a lower one. This causes the excess energy to be released. X-rays are called characteristic X-ray. It (X-ray) has longer wavelength and possess (usually) lower energy than gamma radiation, as well. The emission of high-energy waves came from the electron of an atom.

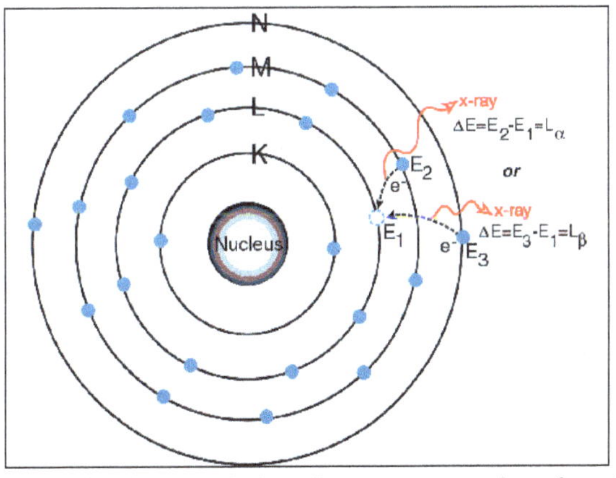

How the emission of the characteristics of X-ray occurs when the orbital electrons move from an outer shell so as to fill in the inner shell vacancy.

A neutron particle is an uncharged element particle with a mass that is slightly greater than that of the proton, and it is found in the nucleus. It is usually released because of spontaneous or induced nuclear fission.

Radioactive Decay

Radioactive decay is the process by which an excited, unstable atomic nucleus loses energy by emitting radiation in the form of particles or electromagnetic waves, thereby transitioning toward a more stable state.

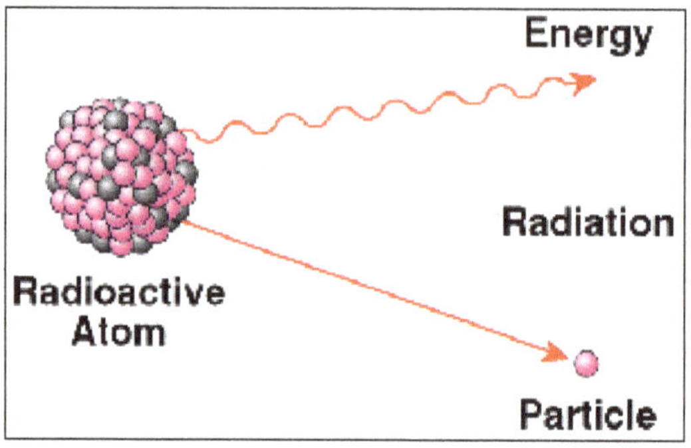

Radioactive decay modes of an atomic nucleus.

The atomic nucleus comprises certain combinations of protons and neutrons held in a stable configuration through a precise balance of powerful forces: the strong force holding the protons and neutrons together is powerful but very short range; the electrostatic repulsion of the positively charged protons is less powerful but long range; the weak force makes the neutron inherently unstable and will turn it into a proton if given the chance. This balance is very delicate: A uranium-238 nucleus has a half-life of 4.5 billion years while uranium-237 with just one less neutron has a half-life of 1.3 minutes.

If there is an imbalance in these forces, the system will eventually shed the excess by ejecting radiation in some combination of particles and wave energy. The most common radioactive decays occur in response to one of three possible types of imbalance. If the nucleus has too many neutrons, one of its neutrons decays (through beta decay) into one proton plus two fragments ejected from the nucleus, a neutrino and an electron (called a a beta particle). If the nucleus has too many protons, it undergoes alpha decay by ejecting two protons and two neutrons as an alpha particle. If the nucleus is excited (has too much energy) it ejects a gamma ray.

Materials exhibiting radioactive decay have yielded widespread application to enhance human welfare. The various applications take advantage of the different decay properties, different decay products, and different chemical properties of the many elements having some isotopes that are radioactive. Major types of applications use the radiation either for diagnosing a problem or for treating a problem by killing specific harmful cells. Areas of application include human and veterinary medicine, nutrition research, basic research in genetics and metabolism, household smoke detectors, industrial and mining inspection of welds, security inspection of cargo, tracing and analyzing pollutants in studies of runoff, and dating materials in geology, paleontology, and archaeology.

Nuclides

Radioactive decay results in an atom of one type, called the *parent nuclide,* being transformed to an atom of a different type, called the *daughter nuclide*. For example, a carbon-14 atom (the "parent") emits radiation and transforms to a nitrogen-14 atom (the "daughter"). This transformation involves quantum probability, so it is impossible to predict when a particular atom will decay. Given a large number of atoms, however, the decay rate is predictable and measured by the "half-life"—the time it takes for 50 percent of the atoms to undergo the change. The half-life of radioactive atoms varies enormously; from fractions of a millisecond to billions of years.

The trefoil symbol is used to indicate radioactive material.

The SI unit of radioactive decay (the phenomenon of natural and artificial radioactivity) is the becquerel (Bq). One Bq is defined as one transformation (or decay) per second. Since any reasonably-sized sample of radioactive material contains many atoms, a Bq is a tiny measure of activity; amounts on the order of TBq (terabecquerel) or GBq (gigabecquerel) are commonly used. Another unit of radio activity is the curie, Ci, which was originally defined as the activity of one gram of pure radium, isotope Ra-226. At present, it is equal (by definition) to the activity of any radionuclide decaying with a disintegration rate of 3.7×10^{10} Bq. The use of Ci is presently discouraged by SI.

The neutrons and protons that constitute nuclei, as well as other particles that may approach them, are governed by several interactions. The strong nuclear force, not observed at the familiar macroscopic scale, is the most powerful force over subatomic distances. The electrostatic force is also significant, while the weak nuclear force is responsible for Beta decay.

The interplay of these forces is simple. Some configurations of the particles in a nucleus have the property that, should they shift ever so slightly, the particles could fall into a lower-energy arrangement with the extra energy moving elsewhere. One might draw an analogy with a snowfield on a mountain. While friction between the snow crystals can support the snow's weight, the system is inherently unstable with regards to a lower-potential-energy state, and a disturbance may facilitate the path to a greater entropy state (that is, towards the ground state where heat will be produced, and thus total energy is distributed over a larger number of quantum states). Thus, an avalanche results. The total energy does not change in this process, but because of entropy effects, avalanches only happen in one direction, and the end of this direction, which is dictated by the largest number of chance-mediated ways to distribute available energy, is what we commonly refer to as the "ground state."

Such a collapse (a *decay event*) requires a specific activation energy. In the case of a snow avalanche, this energy classically comes as a disturbance from outside the system, although such disturbances can be arbitrarily small. In the case of an excited atomic nucleus, the arbitrarily small disturbance comes from quantum vacuum fluctuations. A nucleus (or any excited system in quantum mechanics) is unstable, and can thus *spontaneously stabilize* to a less-excited system. This process is driven by entropy considerations: The energy does not change, but at the end of the process, the total energy is more diffused in spacial volume. The resulting transformation alters the structure of the nucleus. Such a reaction is thus a nuclear reaction, in contrast to chemical reactions, which also are driven by entropy, but which involve changes in the arrangement of the outer electrons of atoms, rather than their nuclei.

Some nuclear reactions *do* involve external sources of energy, in the form of collisions with outside particles. However, these are not considered *decay*. Rather, they are examples of induced nuclear reactions. Nuclear fission and fusion are common types of induced nuclear reactions.

Modes of Decay

Radionuclides can undergo a number of different reactions. These are summarized in the following table. A nucleus with atomic weight *A* and a positive charge *Z* (called atomic number) is represented as (*A*, *Z*).

Mode of decay	Participating particles	Daughter nucleus
Decays with emission of nucleons		
Alpha decay	An alpha particle (A=4, Z=2) emitted from nucleus	(A-4, Z-2)
Proton emission	A proton ejected from nucleus	(A-1, Z-1)
Neutron emission	A neutron ejected from nucleus	(A-1, Z)
Double proton emission	Two protons ejected from nucleus simultaneously	(A-2, Z-2)
Spontaneous fission	Nucleus disintegrates into two or more smaller nuclei and other particles	-
Cluster decay	Nucleus emits a specific type of smaller nucleus (A_1, Z_1) larger than an alpha particle	($A-A_1$, $Z-Z_1$) + (A_1,Z_1)
Different modes of beta decay		
Beta-Negative decay	A nucleus emits an electron and an antineutrino	(A, Z+1)
Positron emission, also Beta-Positive decay	A nucleus emits a positron and a neutrino	(A, Z-1)
Electron capture	A nucleus captures an orbiting electron and emits a neutrino - The daughter nucleus is left in an excited and unstable state	(A, Z-1)
Double beta decay	A nucleus emits two electrons and two antineutrinos	(A, Z+2)
Double electron capture	A nucleus absorbs two orbital electrons and emits two neutrinos - The daughter nucleus is left in an excited and unstable state	(A, Z-2)
Electron capture with positron emission	A nucleus absorbs one orbital electron, emits one positron and two neutrinos	(A, Z-2)
Double positron emission	A nucleus emits two positrons and two neutrinos	(A, Z-2)
Transitions between states of the same nucleus		
Gamma decay	Excited nucleus releases a high-energy photon (gamma ray)	(A, Z)
Internal conversion	Excited nucleus transfers energy to an orbital electron and it is ejected from the atom	(A, Z)

Radioactive decay results in a reduction of summed rest mass, which is converted to energy (the disintegration energy) according to the formula $E = mc^2$. This energy is released as kinetic energy of the emitted particles. The energy remains associated with a measure of mass of the decay system invariant mass, inasmuch the kinetic energy of emitted particles contributes also to the total invariant mass of systems. Thus, the sum of rest masses of particles is not conserved in decay, but the *system* mass or system invariant mass (as also system total energy) is conserved.

Radioactive Series

In a simple, one-step radioactive decay, the new nucleus that emerges is stable. C-14 undergoing beta decay to N-14 and K-40 undergoing electron capture to Ar-40 are examples.

On the other hand, the daughter nuclide of a decay event can be unstable, sometimes even more unstable than the parent. If this is the case, it will proceed to decay again. A sequence of several decay events, producing in the end a stable nuclide, is a *decay chain*. Ultrapure uranium, for instance, is hardly radioactive at all. After a few weeks, however, the unstable daughter nucleides accumulate—such as radium—and it is their radioactivity that becomes noticeable.

Of the commonly occurring forms of radioactive decay, the only one that changes the number of aggregate protons and neutrons *(nucleons)* contained in the nucleus is alpha emission, which reduces it by four. Thus, the number of nucleons modulo 4 is preserved across any decay chain. This leads to the four radioactive decay series with atomic weights 4n+0, 4n+1, 4n+2, and 4n+3.

In an alpha decay, the atomic weight decreases by 4 and the atomic number decreases by 2. In a beta decay, the atomic weight stays the same and the atomic number increases by 1. In a gamma decay, both atomic weight and number stay the same. A branching path occurs when there are alternate routes to the same stable destination. Onc branch is usually highly favored over the other.

These are the four radioactive decay series.

Uranium-235 Series (4n+3):

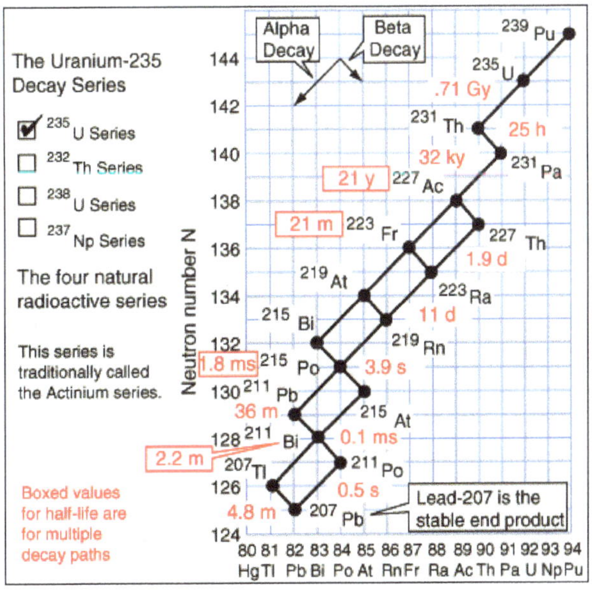

Thorium-232 Series (4n+0):

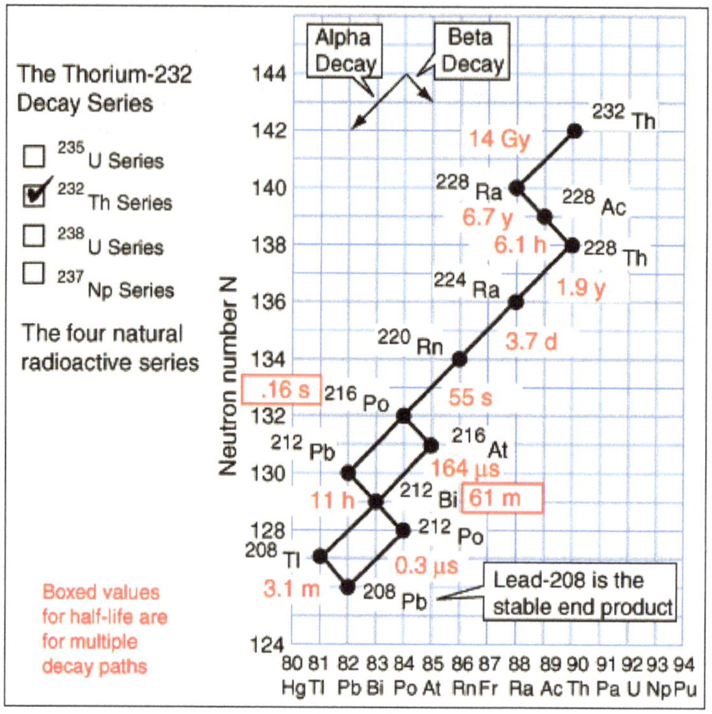

Uranium-238 Series (4n+2):

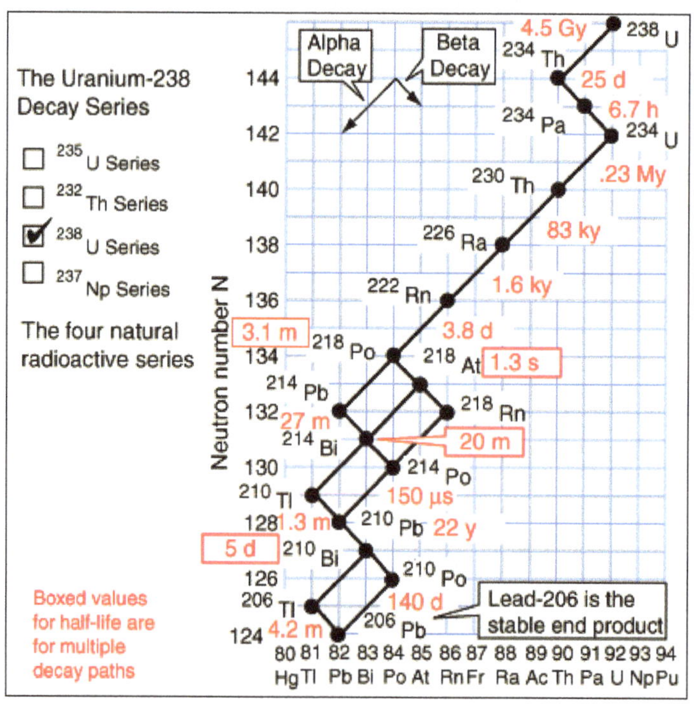

Neptunium-237 Series (4n+1):

The members of this series are not presently found in nature because the half-life of the longest lived isotope in the series is short compared to the age of the earth.

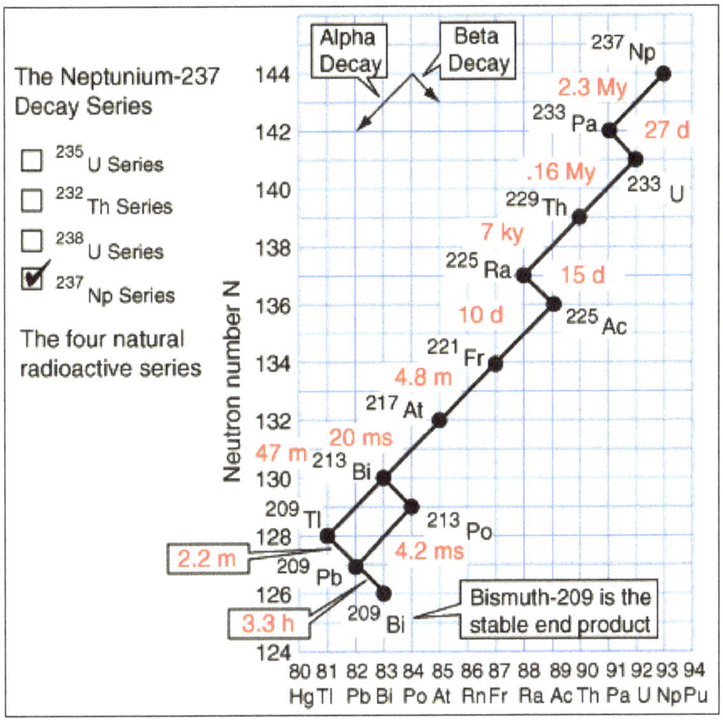

Occurrence

According to the widely accepted Big Bang theory, the universe began as a mixture of hydrogen-1 (75 percent) and helium-4 (25 percent) with only traces of other light atoms. All the other elements, including the radioactive ones, were generated later during the thermonuclear burning of stars—the fusion of the lighter elements into the heavier ones. Stable isotopes of the lightest five elements (H, He, and traces of Li, Be, and B) were produced very shortly after the emergence of the universe, in a process called Big Bang nucleosynthesis. These lightest stable nuclides (including deuterium) survive to today, but any radioactive isotopes of the light elements produced in the Big Bang (such as tritium) have long since decayed. Isotopes of elements heavier than boron were not produced at all in the Big Bang, and these first five elements do not have any long-lived radioisotopes. Thus, all radioactive nuclei are, therefore, relatively young with respect to the birth of the universe, having formed later in various other types of nucleosynthesis in stars (in particular, supernovae), and also during ongoing interactions between stable isotopes and energetic particles. For example, carbon-14, a radioactive nuclide with a half-life of only 5,730 years, is constantly produced in Earth's upper atmosphere due to interactions between cosmic rays and nitrogen.

Applications

Radioactive materials and their decay products—alpha particles (2 protons plus 2 neutrons), beta particles (electrons or positrons), gamma radiation, and the daughter isotopes—have been put to the service of humanity in a great number of ways. At the same time, high doses of radiation from radioactive materials can be toxic unless they are applied with medical precision and control. Such exposures are unlikely except for the unlikely cases of a nuclear weapon detonation or an accident or attack on a nuclear facility.

In medicine, some radioactive isotopes, such as iron-59 and iodine-131, are usable directly in the body because the isotopes are chemically the same as stable iron and iodine respectively. Iron-59, steadily announcing its location by emitting beta-decay electrons, is readily incorporated into blood cells and thereby serves as an aid in studying iron deficiency, a nutritional deficiency affecting more than 2 billion people globally. Iron-59 is an important tool in the effort to understand the many factors affecting a person's ability to metabolize iron in the diet so that it becomes part of the the blood. Iodine-131 administered in the blood to people suffering from hyperthyroidism or thyroid cancer concentrates in the thyroid where gamma radiation emitted by the iodine-131 kills many of the thyroid cells. Hyperthyroidism in cats is treated effectively by one dose of iodine-131.

Radioactive isotopes whose chemical nature does not permit them to be readily incorporated into the body, are delivered to targeted areas by attaching them to a particular molecule that does tend to concentrate in a particular bodily location—just as iodine naturally concentrates in the thyroid gland. For studying activity in the brain, the radioactive isotope fluorine-18 is commonly attached to an analog of the sugar glucose which tends to concentrate in the active regions of the brain within a short time after the molecule is injected into the blood. Fluorine-18 decays by releasing a positron whose life is soon ended as it meets an electron and the two annihilate yielding gamma radiation that is readily detected by the Positron Emission Tomography (PET) technology. Similar techniques of radioisotopic labeling, have been used to track the passage of a variety of chemical substances through complex systems, especially living organisms.

Three gamma emitting radioisotopes are commonly used as a source of radiation. Technetium-99m, a metastable form with a half-life of 6 hours, emits a relatively low frequency gamma radiation that is readily detected. It has been widely used for imaging and functional studies of the brain, myocardium, thyroid, lungs, liver, gallbladder, kidneys, skeleton, blood, and tumors. Gamma radiation from cobalt-60 is used for sterilizing medical equipment, treating cancer, pasteurizing certain foods and spices, gauging the thickness of steel as it is being produced, and monitoring welds. Cesium-137 is used as a source of gamma radiation for treating cancer, measuring soil density at construction sites, monitoring the filling of packages of foods and pharmaceuticals, monitoring fluid flows in production plants, and studying rock layers in oil wells.

Americanium-241, which decays by emitting alpha particles and low energy gamma radiation, is commonly used in smoke detectors as the alpha particles ionize air in a chamber permitting a small current to flow. Smoke particles entering the chamber activate the detector by absorbing alpha particles without being ionized, thereby reducing the current.

On the premise that radioactive decay is truly random (rather than merely chaotic), it has been used in hardware random-number generators. Because the process is not thought to vary significantly in mechanism over time, it is also a valuable tool in estimating the absolute ages of certain materials. For geological materials, the radioisotopes (parents) and certain of their decay products (daughters) become trapped when a rock solidifies, and can then later be used to estimate the date of the solidification (subject to such uncertainties as the possible number of daughter elements present at the time of solidification and the possible number of parent or daughter atoms added or removed over time).

For dating organic matter, radioactive carbon-14 is used because the atmosphere contains a small percentage of carbon-14 along with the predominance of stable carbons 12 and 13. Living plants incorporate the same ratio of carbon-14 to carbon-12 into their tissues and the animals eating the plants have a similar ratio in their tissues. After organisms die, their carbon-14 decays to nitrogen at a certain rate while the carbon-12 content remains constant. Thus, in principle, measuring the ratio of carbon-14 to carbon-12 in the dead organism provides an indication of how long the organism has been dead. This dating method is limited by the 5730 year half-life of carbon-14 to a maximum of 50,000 to 60,000 years. The accuracy of carbon dating has been called into question primarily because the concentration of carbon-14 in the atmosphere varies over time and some plants have the capacity to exclude carbon-14 from their intake.

Radioactive Decay Rates

The decay rate, or activity, of a radioactive substance are characterized by:

Constant quantities:

- Half life—Symbol $t_{1/2}$—The time for half of a substance to decay.
- Mean lifetime—Symbol τ—The average lifetime of any given particle.
- Decay constant—Symbol λ—The inverse of the mean lifetime.

Although these are constants, they are associated with statistically random behavior of substances, and predictions using these constants are less accurate for a small number of atoms.

Time-variable Quantities

- Total activity—Symbol A—Number of decays an object undergoes per second.
- Number of particles—Symbol N—The total number of particles in the sample.
- Specific activity—Symbol S_A—Number of decays per second per amount of substance. The “amount of substance” can be the unit of either mass or volume.

These are related as follows:

$$t_{1/2} = \frac{\ln(2)}{\lambda} = \tau \ln(2)$$

$$A = -\frac{dN}{dt} = \lambda N$$

$$S_A a_0 = -\left.\frac{dN}{dt}\right|_{t=0} = \lambda N_0$$

where, a_0 is the initial amount of active substance—substance that has the same percentage of unstable particles as when the substance was formed.

Activity Measurements

The units in which activities are measured are: Becquerel (symbol Bq) = number of disintegrations per second; curie (Ci) = 3.7×10^{10} disintegrations per second. Low activities are also measured in disintegrations per minute (dpm).

Decay Timing

The decay of an unstable nucleus is entirely random and it is impossible to predict when a particular atom will decay. However, it is equally likely to decay at any time. Therefore, given a sample of a particular radioisotope, the number of decay events $-dN$ expected to occur in a small interval of time dt is proportional to the number of atoms present. If N is the number of atoms, then the probability of decay $(-dN/N)$ is proportional to dt:

$$\left(-\frac{dN}{N}\right)=\lambda\cdot dt$$

Particular radionuclides decay at different rates, each having its own decay constant (λ). The negative sign indicates that N decreases with each decay event. The solution to this first-order differential equation is the following function:

$$N(t)=N_0e^{-M}=N_0e^{-t/\tau}$$

This function represents exponential decay. It is only an approximate solution, for two reasons. Firstly, the exponential function is continuous, but the physical quantity N can only take non-negative integer values. Secondly, because it describes a random process, it is only statistically true. However, in most common cases, N is a very large number and the function is a good approximation.

In addition to the decay constant, radioactive decay is sometimes characterized by the mean lifetime. Each atom "lives" for a finite amount of time before it decays, and the mean lifetime is the arithmetic mean of all the atoms' lifetimes. It is represented by the symbol τ, and is related to the decay constant as follows:

$$\tau=\frac{1}{\lambda}$$

A more commonly used parameter is the half-life. Given a sample of a particular radionuclide, the half-life is the time taken for half the radionuclide's atoms to decay. The half-life is related to the decay constant as follows:

$$t_{1/2}=\frac{\ln 2}{\lambda}=\tau\ln 2$$

This relationship between the half-life and the decay constant shows that highly radioactive substances are quickly spent, while those that radiate weakly endure longer. Half-lives of known radionuclides vary widely, from more than 10^{19} years (such as for very nearly stable nuclides, for example, ^{209}Bi), to 10^{-23} seconds for highly unstable ones.

Radioactive Decay Law

The rate of decay (number of disintegrations per unit time) is proportional to N, the number of radioactive nuclei in the sample:

$$dN/dt\propto-\lambda N$$

The negative sign signifies that N is decreasing with time. λ is called the decay constant - probability per unit time that a given radioactive nucleus will decay.

Large $\lambda \Rightarrow$ rapid decay; $\lambda \Rightarrow$ small slow decay.

Equation ($dN/dt \propto -\lambda N$) can be integrated to give:

$$N(t) = N_0 \exp(-\lambda t)$$

Where N_o = number of radioactive nuclei at t = 0.

Activity and Half-life

Activity: Number of disintegrations per unit time:

$$A(t) = \lambda N(t) \;\; = \lambda N_0 \; \exp(-\lambda t) \;\; = \; A_0 \; \exp(-\lambda t)$$

This has the same exponential fall off with time as $N(t)$.

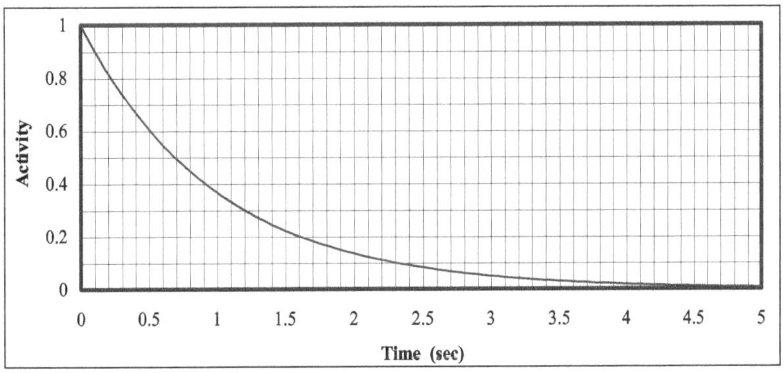

Half-life: Time for half the radioactive nuclei in the sample to decay.

Substituting $N_0 = N_0 / 2$ and $t = t_{1/2}$ into equation ($N(t) = N_0 \exp(-\lambda t)$) gives $t_{1/2} = \ln 2 / \lambda$.

The figure above shows the activity of a sample decaying at a rate of exp (-t). The half-life of this sample = ln2 ($\approx$0.7 s).

Dccay Chains

When nuclei A decay into stable nuclei B, the number of each present at time t is,

$$N_A(t) \; = \; N_A(0)e^{-\lambda t} \; \text{and} \; N_B(t) = N_A(0)\left(1 - e^{-\lambda t}\right)$$

where only nuclei A are present initially.

The number of nuclei A (parent nuclei) decreases with time.

The number of nuclei B (daughter nuclei) increases from zero, initially, and approaches N_A(0) as t $\rightarrow\infty$, i.e. all the parent nuclei eventually become daughter nuclei.

The total number of nuclei is constant: $N_A(t) + N_B(t) = N_A(0)$

If nuclei B are also radioactive, the above equations do not apply, since, as nuclei B are produced, they also decay. The daughter nuclei of B may also be radioactive and a decay chain is set up:

$A \rightarrow B \rightarrow C \rightarrow$....etc.

Alpha Decay

The process of alpha decay is a nuclear reaction that can be written as:

$$^{A}_{Z}(z)_{N} \rightarrow {}^{A-4}_{Z-2}(X)^{2-}_{N-2} + {}^{4}_{2}He^{2+}_{2} + Q_{\alpha}$$

where we have chosen to write out all of the superscripts and subscripts. Thus the α-decay of ^{238}U can be written,

$$^{238}U \rightarrow {}^{234}Th^{2-} + {}^{4}He^{2+} + Q$$

The Q_α-value is positive (exothermic) for spontaneous alpha decay. The helium nucleus emerges with a substantial velocity and is fully ionized, and the atomic electrons on the daughter are disrupted by the sudden change but the whole process conserves electrical charge. We can rewrite the equation in terms of the masses of the neutral atoms:

$$^{A}_{Z}(z)_{N} \rightarrow {}^{A-4}_{Z-2}(X)_{N-2} + {}^{4}_{2}He_{2} + Q_{\alpha}$$

and then calculate the Q_α-value because the *net* change in the atomic binding energies is,

$\left(\sim 65.3Z^{7/5} - 80Z^{2/5}eV\right)$ very small compared to the nuclear decay energy.

What causes α-decay? (Or, what causes Q_α to be positive?) In the language of the semi-empirical mass equation, the emission of an α-particle lowers the Coulomb energy of the nucleus, which increases the stability of heavy nuclei while not affecting the overall binding energy per nucleon because the tightly bound α-particle has approximately the same binding energy/nucleon as the original nucleus.

Two important features of alpha decay are that the energies of the alpha particles are known to generally increase with the atomic number of the parent but yet the kinetic energy of the emitted particle is less than that of the Coulomb barrier in the reverse reaction between the α-particle and the daughter nucleus. In addition, all nuclei with mass numbers greater than A >> 150 are thermodynamically unstable against alpha emission (Q_α is positive) but alpha emission is the dominant decay process only for the heaviest nuclei, A >210. The energies of the emitted α-particles can range from 1.8 MeV (^{144}Nd) to 11.6 MeV ($^{212}Po^m$) with the half-life of ^{144}Nd being $5x10^{29}$ times as long as that of $^{212}Po^m$. Typical heavy element alpha decay energies are in the range from 4-9 MeV.

In general, alpha decay leads to the ground state of the daughter nucleus so that the emitted particle carries away as much energy as possible and as little angular momentum as possible. The ground state spins of even-even parents and daughters (including the alpha particle, of course) are zero which makes l=0 alpha particle emission the most likely process for these nuclei. Small branches are seen to higher excited states but such processes are strongly suppressed. Some decays of odd- A heavy nuclei populate low-lying excited states that match the spin of the parent so that the orbital angular momentum of the α particle can be zero. For example, the strongest branch (83%) of the alpha decay of ^{249}Cf goes to the 9th excited state of ^{245}Cm because this is the lowest lying state with the same spin and parity as that of the parent. Alpha decay to several different excited states of a daughter nucleus is called *fine structure*; α-decay from an excited state of a parent nucleus to the ground state of the daughter nucleus is said to be *long range alpha emission* because these α-particles are more energetic and thus have longer ranges in matter. The most famous case of long range α-emission is that of ^{212}Po^{m} where a 45 s isomeric level at 2.922 MeV decays to the ground state of ^{208}Pb by emitting a 11.65 MeV α-particle.

Energetics of α Decay

The alpha particle, or ^{4}He nucleus, is an especially strongly bound particle. This combined with the fact that the binding energy per nucleon has a maximum value near A >> 56 and systematically decreases for heavier nuclei, creates the situation that nuclei with A³150 have positive Q_α-values for the emission of alpha particles. For example, one of the heaviest naturally occurring isotopes, ^{238}U (with a mass excess, Δ, of +47.3070 MeV) decays by alpha emission to ^{234}Th (Δ = +40.612 MeV) giving a Q_α-value of:

$$Q_\alpha = 47.3070 - (40.612 + 2.4249) = 4.270 \text{ MeV}$$

Note that the decay energy will be divided between the α particle and the heavy recoiling daughter so that the kinetic energy of the alpha particle will be slightly less. (The kinetic energy of the recoiling ^{234}Th nucleus produced in the decay of ^{238}U is ~0.070 MeV.) Conservation of momentum and energy in this reaction requires that the kinetic energy of the α-particle, T_α, is:

$$T_\alpha = \frac{234}{238} Q_á = 4.198 \text{ MeV}$$

The kinetic energies of the emitted alpha particles can be measured very precisely so we should be careful to distinguish between the Q_α-value and the kinetic energy, T_α. The very small recoil energy of the heavy daughter is very difficult to measure but it is still large compared to chemical bond energies and can lead to interesting chemistry. For example, the daughter nuclei may recoil out of the original α-source. This can cause serious contamination problems if the daughters are themselves radioactive.

The Q_α-values generally increase with increasing atomic number but the variation in the mass surface due to shell effects can overwhelm the systematic increase. The sharp peaks near A=214 are due to the effects of the N=126 shell. When ^{212}Po decays by α-emission, the daughter nucleus is doubly magic ^{208P}b (very stable) with a large energy release. The α decay of ^{211}Pb and ^{213}Po will not lead to such a large Q_α because the products are not doubly magic. Similarly, the presence of the 82 neutron closed shell in the rare earth region causes an increase in Q_α, allowing observable α-decay half-lives for several of these nuclei (with N=84). Also one has observed short lived α-emitters near doubly magic ^{100}Sn, including ^{107}Te, ^{108}Te, and ^{111}Xe. And, in addition, alpha emitters have been identified along the proton dripline above A=100. For a set of isotopes (nuclei with a constant atomic number) the decay energy generally decreases with increasing mass. For example, the kinetic energy of alpha particles from the decay of uranium isotopes is typically 4 to 5 MeV, those for californium isotopes are >6 MeV, and those for rutherfordium isotopes are >8 MeV. However, the kinetic energy from the decay Of $^{212}_{84}$Po to the doubly magic $^{208}_{84}$Pb daughter is 8.78 MeV.

The generally smooth variation of Q_α with Z, A of the emitting nucleus and the two body nature of alpha decay can be used to deduce masses of unknown nuclei. One tool in this effort is the concept of closed decay cycles. Consider the α-l and β-l decays connecting $^{237}_{93}$Np, $^{241}_{95}$Am and $^{237}_{92}$U. By conservation of energy, one can state that the sum of the decay energies around the cycle connecting these nuclei must be zero (within experimental uncertainty). In those cases where experimental data or reliable estimates are available for three branches of the cycle, the fourth can be calculated by difference.

Even though the energies released by the decay of a heavy nucleus into an alpha particle and a lighter daughter nucleus are quite substantial, the energies are paradoxically small compared to the energy necessary to bring the alpha particle back into nuclear contact with the daughter. The electrostatic potential energy between the two positively charged nuclei, called the Coulomb potential, can be written as:

$$Vc = \frac{2Z}{R}\frac{e^2}{4\pi\varepsilon_0}$$

where Z is the atomic number of the daughter and R is the separation between the centers of the two nuclei $\frac{e^2}{4\pi\varepsilon_0}$ is 1.440 MeV × fm. To obtain a rough estimate of the Coulomb energy we can take R to be 1.2($A^{1/3}$ + $4^{1/3}$) fm, where A is the mass number of the daughter. For the decay of ^{238}U we get:

$$\text{Vc} - \frac{(2)(90)(1.440\,MeV\,\text{fm})}{1.2\left(234^{1/3}+4^{1/3}\right)\text{fm}} = \frac{259\,\text{MeV fm}}{9.3\,\text{fm}} - 28\,\text{MeV}$$

Which is 6 to 7 times the decay energy. This factor is typical of the ratio of the Coulomb barrier to the Q-value. If we accept for the moment the large difference between the Coulomb barrier and the observed decay energy, then we can attribute the two general features of increasing decay energy with increasing atomic number, Z, and decreasing kinetic energy with increasing mass among a set of isotopes to the Coulomb potential. The higher nuclear charge accelerates the products apart and the larger mass allows the daughter and alpha particle to start further apart.

Example of Decay Energies

Calculate the Q_α-value, kinetic energy, T_α, and the Coulomb barrier, V_C, for the primary branch of the alpha decay of ^{212}Po to the ground state of ^{208}Pb. Using tabulated mass excesses we have:

$$Q_\alpha = -10.381-(-21.759+2.4249) = 8.953\text{MeV}$$

$$T_\alpha = \frac{208}{212}Q = 8.784\,\text{MeV}$$

and

$$Vc - \frac{(2)(82)(1.440\,\text{MeV fm})}{1.2\left(208^{1/3}+4^{\,1/3}\right)\text{fm}} \approx 26\,\text{MeV}$$

The ^{212}Po parent also decays with a 1% branch to the first excited state of ^{208}Pb at an excitation energy of 2.6146 MeV. What is the kinetic energy of this alpha particle?

$$Q_\alpha = 8.953 - 2.6146 = 6.339\,\text{MeV}$$

$$T_\alpha = \frac{208}{212}\bullet = 6.339 = 6.22\text{MeV}$$

Many heavy nuclei (A≥150) are unstable with respect to α-decay. Some of them also undergo β-decay of β- and α-decays until they form one of the stable isotopes of lead or bismuth, $^{206,207,208}Pb$ or ^{209}Bi. We are now in a position to understand why a particular sequence occurs. Each of the mass parabolas can be thought of as a cut through the nuclear mass surface at constant A. ^{235}U decays to ^{231}Th. ^{231}Th then decays to ^{231}Pa by β-decay. This nucleus, being near the bottom of the mass parabola, cannot undergo further β-decay, but decays by β-emission to ^{227}Ac. This nucleus decays by β-emission to ^{227}Th, which must α-decay to ^{223}Ra, etc.

Theory of α-decay

The allowed emission of alpha particles could not be understood in classical pictures of the nucleus. Using simple estimates we have drawn a one dimensional potential energy curve for this system as a function of radius. At the smallest distances, inside the parent nucleus, we have drawn a flat-bottomed potential with a depth of ~-30 MeV. The potential rapidly rises at the nuclear radius and comes to the Coulomb barrier height of V_C >> +28MeV at 9.3 fm. At larger distances the mpotential falls as $\frac{1}{r}$ according to Coulomb's Law.

Starting from a separated alpha particle and the daughter nucleus, we can determine that the distance of closest approach during the scattering of a 4.2 MeV alpha particle will be ~62 fm. This is the distance at which the alpha particle stops moving towards the daughter and turns around because its kinetic energy has been converted into potential energy of repulsion. Now the paradox should be clear: the alpha particle should not get even remotely near to the nucleus; or from the decay standpoint, the alpha particle should be trapped behind a potential energy barrier that it can not get over.

The solution to this paradox was found in quantum mechanics. A general property of quantum mechanical wave functions is that they are only completely confined by potential energy barriers that are infinitely high. Whenever the barrier has a finite size the wave function solution will have its main component inside the potential well plus a small but finite part *inside* the barrier (generally exponentially decreasing with distance) and another finite piece outside the barrier. This phenomenon is called tunneling because the classically trapped particle has a component of its wave function outside the potential barrier and has some probability to go through the barrier to the outside. Some features of tunneling should be obvious: the closer the energy of the particle to the top of the barrier, the more likely that the particle will get out. Also, the more energetic the particle is relative to a given barrier height, the more frequently the particle will "assault" the barrier and the more likely that the particle will escape.

It has been known for some time that half-life for α-decay, $t_{1/2}$, can be written in terms of the square root of the alpha particle decay energy, Q_α, as follows:

$$\log\left(t_{1/2}\right) = A + \frac{B}{\sqrt{Q_\alpha}}$$

where the constants A and B have a Z dependence. This relationship, is known as the Geiger-Nuttall law of α-decay due to the fact that Geiger and Nuttall found a linear relationship between the logarithm of the decay constant and the logarithm of the range of alpha particles from a given natural radioactive decay series. This simple relationship describes the data on α-decay, which span over 20 orders of magnitude in decay constant or half-life. Note that a 1 MeV change in α-decay energy results in a change of 105 in the half-life. A modern representation of this relationship due to Hatsukawa, Nakahara and Hoffman has the form,

$$\log\left(t_{1/2}\right) = A(Z) \times \left[\frac{A_d}{A_p Q_\alpha}\right]^{1/2} \times [\arccos\sqrt{X} - \sqrt{X(1-X)}] - 20.446 + C(Z,N)$$

where A(Z) = 1.40Z + 1710/Z - 47.7 and where C(Z, N) = 0 for ordinary regions outside closed shells and

$$\mathrm{C}(\mathrm{Z},\mathrm{N}) = \left[1.94 - 0.020(82\text{-Z}) - 0.070(126\text{-N})\right]$$
$$\text{for } 78 \le \mathrm{Z} \le 82, 100 \le \mathrm{N} \le 126,$$
$$\mathrm{C}(\mathrm{Z},\mathrm{N}) = \left[1.42 - 0.105(\mathrm{Z}\text{-}82) - 0.067(126\text{-N})\right]$$
$$\text{for } 82 \le \mathrm{Z} \le 90, 110 \le \mathrm{N} \le 126$$

In these equations, A_p, Z refer to the parent nuclide, A_d, and Z_d refer to the daughter nuclide, and X is defined as,

$$X = 1.2249\left(A_d^{1/3} + 4^{1/3}\right) \times \frac{Q_\alpha}{2Z_d e^2}$$

This relationship is useful for predicting the expected α-decay half-lives for unknown nuclei.

The theoretical description of alpha emission relies on calculating the rate in terms of two factors. The overall rate of emission consists of the product of the rate at which an alpha particle appears at the inside wall of the nucleus times the (independent) probability that the alpha particle tunnels through the barrier. Thus, the rate of emission, or the partial decay constant λ_α, is written as the product of a frequency factor, *f*, and a transmission coefficient, T, through the barrier:

$$\lambda_\alpha = fT$$

Some investigators have suggested that this expression should be multiplied by an additional factor to describe the probability of preformation of an alpha particle inside the parent nucleus. Unfortunately, there is no clear way to calculate such a factor but empirical estimates have been made. The theoretical estimates of the emission rates are higher than the observed rates and the preformation factor can be estimated for each measured case. However, there are other uncertainties in the theoretical estimates that contribute to the differences.

The frequency with which an alpha particle reaches the edge of a nucleus can be estimated as the velocity divided by the distance across the nucleus. We can take the distance to be twice the radius (something of a maximum value) but the velocity is a little more subtle to estimate. A lower limit for the velocity could be obtained from the kinetic energy of emitted alpha particle, but the particle is moving inside a potential energy well and its velocity should be larger and correspond to the well depth plus the external energy. Therefore, the frequency can be written:

$$f = \frac{v}{2R} = \frac{\sqrt{2(v_0 + Q)/\mu}}{2R}$$

where we have assumed that the alpha particle is non-relativistic, V_0 is the well depth of approximately 30 MeV, μ is the reduced mass, and R is the radius of the daughter nucleus (because the α-particle needs only reach this distance before it is emitted). We use the reduced mass because the alpha particle is moving inside the nucleus and the total momentum of the nucleus must be zero. The frequency of assaults on the barrier is quite large, usually on the order of 10^{21}/s.

Generalizing the results summarized in equation E-48 to a three dimensional barrier we have:

$$T = e^{-2G}$$

where the Gamow factor (2G) can be written as:

$$2G = \frac{2}{\hbar}\int_R^b 2\mu\left[\left(\frac{Z_\alpha Z_D e^2}{r} - Q_\alpha\right)\right]^{1/2} dr$$

where,

$$Q_\alpha = \frac{Z_\alpha Z_D e^2}{b}$$

and the classical distance of closest approach, b, is given as,

$$b=\frac{Z_{\alpha}Z_{D}e^{2}}{Q_{\alpha}}$$

In these equations, e^2 = 1.440 MeV-fm, Q_α is given in MeV, Z_α, Z_D are the atomic numbers of the α-particle and daughter nucleus, respectively. Rearranging we have,

$$2G=\frac{2}{\hbar}\sqrt{2\mu Q_{\alpha}}\int_{R}^{b}\left(\frac{b}{r}\right)^{1/2}dr$$

This can be integrated to give,

$$2G=\frac{2b}{\hbar}\sqrt{2\mu Q_{\alpha}}\left(\cos^{-1}\sqrt{\frac{R}{b}}-\sqrt{\frac{R}{b}}\sqrt{1-\frac{R}{b}}\right)$$

Substituting back for b and simplifying we have,

$$2G=2\sqrt{\frac{2\mu}{\hbar^{2}Q_{\alpha}}}\left(Z_{\alpha}Z_{D}e^{2}\right)\left(\cos^{-1}\sqrt{\frac{R}{b}}-\sqrt{\frac{R}{b}}\sqrt{1-\frac{R}{b}}\right)$$

For thick barriers $\frac{R}{b}<<1\left(\frac{Q_{\alpha}}{Vc}<<1\right)$ we can approximate,

$$\cos^{-1}\sqrt{\frac{R}{b}}\approx\frac{\pi}{2}-\sqrt{\frac{R}{b}}$$

We get,

$$2G=2\sqrt{\frac{2\mu}{\hbar^{2}Q_{\alpha}}}\left(Z_{\alpha}Z_{D}e^{2}\right)\left(\frac{\pi}{2}\right)$$

Where B is the "effective" Coulomb barrier, *i.e.*,

$$B-\frac{z\alpha Z_{D}e^{2}}{r_{\alpha}+R_{D}}$$

Typically, the Gamow factor is large (2G~60-120) which makes the transmission coefficient T extremely small (~10^{-55} - 10^{-27}). Combining the various equations we have,

$$t_{1/2}=\frac{\ln 2}{\lambda}=\frac{\ln 2}{fT}=\frac{\ln 2}{\frac{\left(2\left(V_{0}+Q_{\alpha}\right)/\mu\right)^{1/2}}{2R}e^{-20}}$$

Or,

$$\log t_{1/2} = a + \frac{b}{\sqrt{Q_\alpha}}$$

i.e., we get the Geiger-Nuttall law of α decay, where a + b are constants, that depend on Z, etc.

This simple estimate tracks the general behavior of the observed emission rates over the very large range in nature. The calculated emission rate is typically one order of magnitude larger than that observed, meaning that *the observed half-lives are longer than predicted.* This has led some researchers to suggest that the probability to find a 'preformed' alpha particle inside a heavy nucleus is on the order of 10^{-1} or less. One estimate of the "preformation factor" is to plot, for even-even nuclei undergoing $\ell = 0$ decay, the ratio of the calculated half-life to the measured half-life. The average preformation factor is $\sim 10^{-2}$.

Example of Emission Rate Calculation

Calculate the emission rate and half-life for ^{238}U decay from the simple theory of alpha decay. Compare this to the observed half-life.

$$\lambda = fT$$

$$f = \frac{\sqrt{2(V_0 + Q)/\mu}}{2R}$$

where,

$$R = r_0\left(A_1^{1/3} + A_1^{1/3}\right) = 1.2\left(4^{1/3} + 234^{1/3}\right) = 9.3\,\text{fm}$$

Note that since we previously calculated b >> 62 fm, R/b = $\frac{8.63}{62} << 1$

$$\mu = 4 \times 234/238 = 3.933\,\text{amu}$$

$$f = \frac{c^* \sqrt{2(30 + 4.2)/(3.933 \times 93.5)}}{2 \times 9.3} = \left(2.26 \times 10^{21}\right)/\text{sec}$$

We know that,

$$T = e^{-2G}$$

where,

$$2G \approx 2\left(\frac{2\mu}{\hbar^2 Q_\alpha}\right)^{1/2}\left({}_{Z\alpha}Z_D e^2\right)\left(\frac{\pi}{2} - 2\sqrt{\frac{Q}{B}}\right)$$

$$2\left(\frac{2\mu}{\hbar^2 Q_\alpha}\right) = 2\left(\frac{2\mu(\text{amu})931.5\left(\frac{\text{MeV}}{\text{amu}}\right)}{(\hbar c)^2(\text{MeV-fm})^2 Q_\alpha(\text{MeV})}\right)^{1/2} = 2\left(\frac{(2)(3.933)(931.5)}{(197.3)^2(4.27)}\right)^{1/2} = 0.420(\text{MeV-fm})^{-1}$$

$$\left(Z_\alpha Z_{De}{}^2\right) = (2)(90)(1.440) = 259.2\text{MeV} - \text{fm}$$

$$\left(\frac{\pi}{2} - 2\sqrt{\frac{Q}{B}}\right) = \frac{\pi}{2} - 2\left(\frac{4.27}{29}\right)^{1/2} = 0.788$$

$$T = e^{-85.8} = 5.43\times10^{-38}$$

$$\lambda = \text{fT} = \left(2.26\text{x}10^{21}\right)\left(5.43\text{x}10^{-38}\right) = 1.23\text{x}10^{-16}s^{-1}$$

$$t_{1/2} = \frac{\ln 2}{\lambda} = 5.65\text{x}10^{15}S = 1.8\text{x}10^{8}y$$

The observed half-life of ^{238}U is 4.47 x 10^9 years which is a factor of ~25 times longer than the calculated value. Note the qualitative aspects of this calculation. The α-particle must hit the border of the parent nucleus ~10^{38} times before it can escape. Also note the extreme sensitivity of this calculation to details of the nuclear radius. A 2% change in R changes λ by a factor of 2. In our example, we approximated R as R_{Th} + Rα. In reality, the α-particle has not fully separated from the daughter nucleus when they exit the barrier. One can correct for this by approximating R >> $1.4A^{1/3}$.

The theory presented above neglects the effects of angular momentum in that it assumes the α-particle carries off no orbital angular momentum (ℓ = 0). If α decay takes place to or from an excited state, some angular momentum may be carried off by the α-particle with a resulting change in the decay constant. In quantum mechanics, we say that the α-particle has to tunnel through a barrier that is larger by an amount called the centrifugal potential,

$$V\ell = \frac{\ell(\ell=1)\hbar^2}{2\mu R^2}$$

where ℓ is the orbital angular momentum of the α-particle, μ is the reduced mass and R is the appropriate radius. This centrifugal potential is added to the potential energy $V_{(r)}$ resulting in a thicker and higher barrier, increasing the half-life. One can evaluate the effect of this centrifugal potential upon α-decay half-lives by simply adding this energy to the Coulomb barrier height. If we define,

$$\sigma = \frac{\text{centrifugal barrier height}}{\text{coulomb barrier height}}$$

We can say,

$$\sigma = \frac{\ell(\ell+1)\hbar^2}{2\mu R^2}\frac{R}{Z_\alpha Z_D e^2} = \frac{\ell(\ell+1)\hbar^2}{2\mu R\, {}_{Z\alpha} Z_D e^2}$$

Then all we need to do is to replace all occurrences of B by B (1+σ). A simple pocket formula that does this is,

$$\lambda_{\ell\neq0} \approx \lambda_{\ell\neq0}\exp[-2.027\ell(\ell+1)Z^{-1/2}A^{-1/6}]$$

This centrifugal barrier correction is a very small effect compared to the effect of Q_α or R upon the decay rate.

We should also note that conservation of angular momentum and parity during the α decay process places some constraints on the daughter states that can be populated. Since the α-particle has no intrinsic spin, the total angular momentum of the α-particle must equal its orbital angular momentum ℓ and the α-particle parity must be $(-1)^\ell$. If parity is conserved in α-decay, the final states are restricted. If the parent nucleus has $J^\pi = 0^+$, then the allowed values of Jπ of the daughter nucleus are 0^+ (ℓ=0), 1^-(ℓ=1), 2^+ (ℓ=2), etc. These rules only specify the required spin and parity of the state in the daughter, while the energy of the state is a separate quantity. the heaviest elements are strongly deformed and are good rotors. The low lying excited states of even-even nuclei form a low-lying rotational band with spins of 2, 4, 6, etc., while odd angular momenta states tend to lie higher in energy. Because of the decrease in the energy of the emitted α-particle when populating these states, decay to these states will be inhibited. Thus the lower available energy suppresses these decays more strongly than the centrifugal barrier.

Example of Angular Momentum in Alpha Decay

^{241}Am is a long-lived alpha emitter that is used extensively as an ionization source in smoke detectors. The parent state has a spin and parity of $5/2^-$ and cannot decay to the $5/2^+$ ground state of ^{237}Np because that would violate parity conservation. Rather it decays primarily to a $5/2^-$ excited state (85.2%, E* = 59.5 keV) and to a $7/2^-$ higher lying excited state (12.8%, E* = 102.9 keV). Estimate these branching ratios and compare them to the observed values.

$$Q_\alpha\left(5/2^-\right)=5.578\ MeV, Q\alpha\left(7/2^-\right)=5.535 MeV$$

$$f\left(5/2^-\right)=2.29x10^{21}\ /\sec, f\left(7/2^-\right)=2.29x10^{21}\ /\sec$$

$$G\left(5/2^-\right)=33.01, G\left(7/2^-\right)=33.84$$

$$\lambda\left(5/2^-\right)=4.89x10^{-8}\ /\sec, \lambda\left(7/2^-\right)=9.2x10^{-9}\ /\sec$$

Assuming that the branches to other states are small and do not contribute to the sum of partial half-lives we can write:

$$\text{Branching ratio } \left(5/2^-\right)\approx\frac{\lambda\left(5/2^-\right)}{\lambda\left(5/2^-\right)+\lambda\left(7/2^-\right)}=0.84$$

Note that the observed half-life of 433 yr. is again significantly longer than the predicted half-life of ~3 yr. This difference is attributed to the combined effects of the preformation factor and the hindrance effect of the odd proton in the americium parent (Z=95).

Beta Decay

The beta decay is a radioactive decay in which a proton in a nucleus is converted into a neutron (or vice-versa). In the process the nucleus emits a beta particle (either an electron or a positron) and quasi-massless particle, the neutrino.

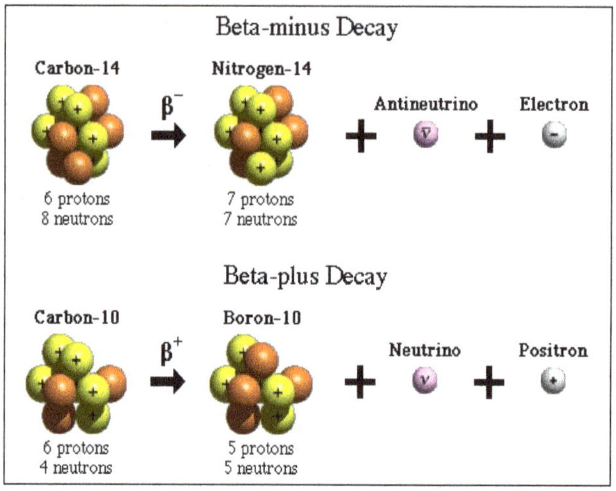

Beta decay schematics.

Nuclei with either too many protons or too many neutrons will undergo β decay via the weak nuclear force.

- A neutron-rich nucleus will undergo β^- emission.
- A proton-rich nucleus will either emit a β^+ particle or be transformed by capturing an atomic electron in a process called electron capture.

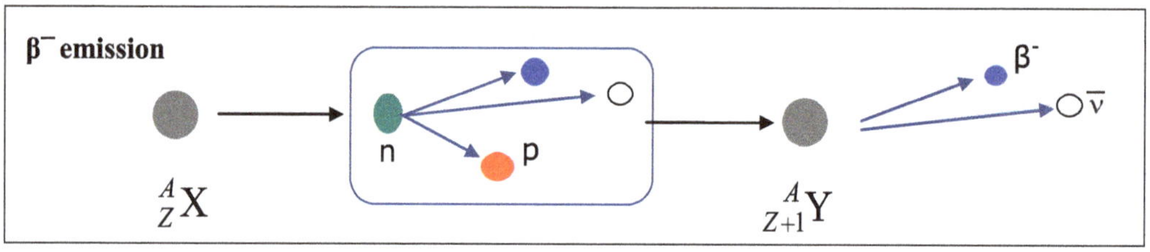

In β^- decay, the charge on the nucleus increases by one unit.

The β^- particle is an electron (mass 0.0005486 u = 511 keV and charge -e). It is considered to be created at the moment of decay. A second light particle, - antineutrino $(\bar{V})$- is also created and emitted. It has no charge and very small mass, which generally is assumed to be zero. It interacts extremely weakly with matter.

β^- decay is a three-body process - the kinetic energy released is shared between the β^- particle and the antineutrino. This results in emitted β^- particles having a distribution of energies, from 0 to the maximum allowed by the Q value (end point energy).

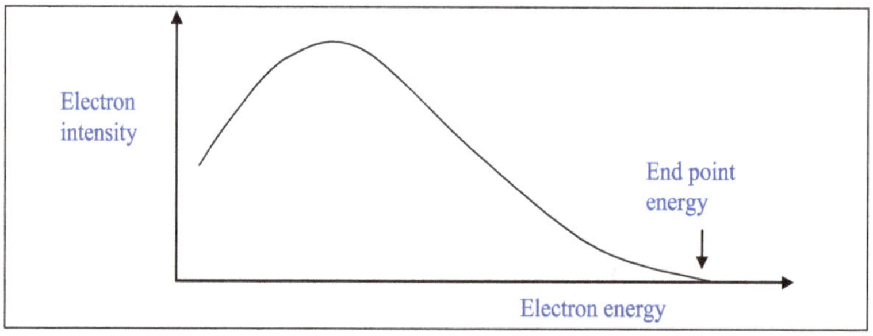

β^- decay involves the transformation of a neutron into a proton inside the nucleus:

$$n \rightarrow p + \beta^- + \overline{\nu}$$

Energy must be conserved and so the transformation given in Equation ($n \rightarrow p + \beta^- + \overline{\nu}$) can occur only if the daughter nucleus is lighter (more stable) than the parent. This means that the Q value:

$$Q_{\beta^-} = \left(m_p - m_D\right)c^2$$

must be > 0. Note that the masses m_P and m_D, for the parent and daughter, respectively, are atomic not nuclear masses and include the masses of the atomic electrons.

A beta particle is much lighter than an α particle. Its speed, therefore, for a given energy is much greater and it is much more penetrating. A few mm of material will stop a 1 MeV β particle.

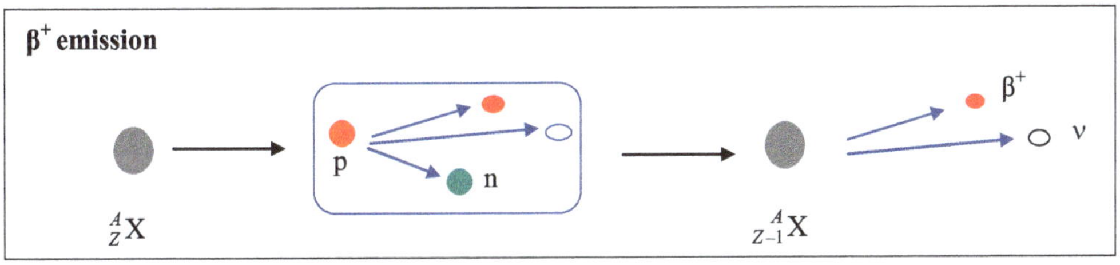

β^+ decay changes a proton rich nucleus into a more stable isobar. The nuclear charge is decreased by one unit.

A β^+ particle is an antielectron, called a positron. It is identical to an ordinary electron except that it is positively charged. Effectively, β^+ decay converts (via the weak nuclear force) a proton into a neutron and a positron. A neutrino is also created in the process. Note that this is an ordinary neutrino not an antineutrino.

$$p \rightarrow n + \beta^+ + \nu$$

There are 3 bodies in the final state and the positrons are emitted with a continuous range of energies, as are electrons in β^- decay.

Using energy conservation, the β^+ decay Q value, using atomic masses, is given by:

$$Q_{\beta^+} = \left(m_p - m_D - 2m\right)c^2$$

This must be greater than zero for the process to be able to occur.

Electron Capture

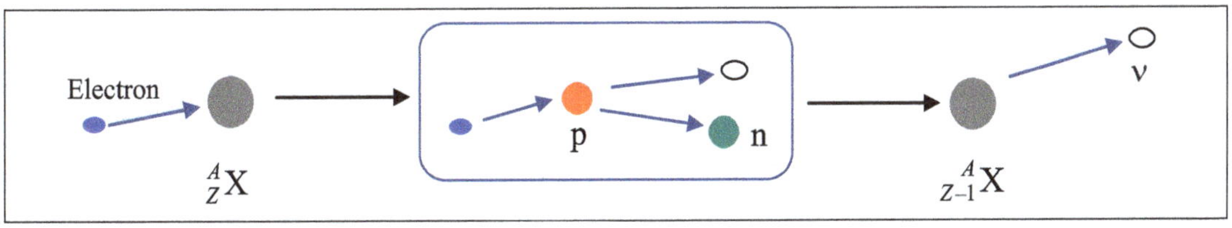

Electron capture EC is an alternative process to β^+ decay in that a proton is converted into a neutron. The parent absorbs an atomic electron, usually from the innermost orbit.

$$e^- + p \rightarrow n + \nu$$

Energy conservation gives the Q value as,

$$Q_{EC} = \left(m_p - m_D\right)c^2$$

Note that Q_{EC} is greater than $Q_{\beta+}$ by $2mc^2$. In electron capture, the mass of an atomic electron is converted into energy, whereas, in β^+ decay, energy is required to create a positron. This means that EC can occur when β^+ decay cannot. No β particle is emitted in electron capture and so, except for a very small amount of recoil energy of the daughter nucleus, the energy released escapes undetected.

Nuclear Transmutation

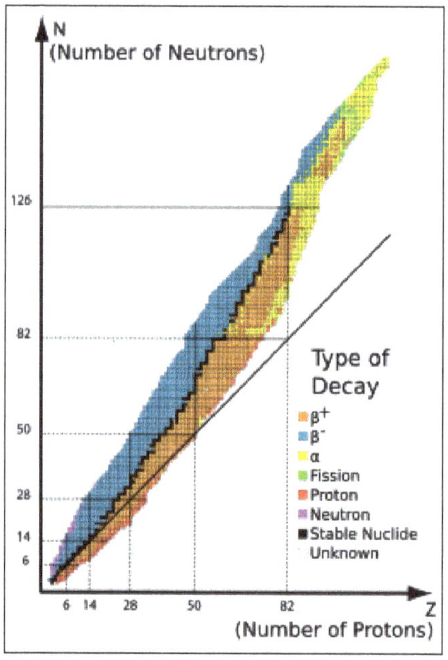

If the proton and neutron are part of an atomic nucleus, the above described decay processes transmute one chemical element into another. For example:

$$^{137}_{55}\text{Cs} \rightarrow {}^{137}_{56}\text{Ba} + e^- + \bar{\nu}_e \text{ (beta minus decay)}$$

$$^{22}_{11}\text{Na} \rightarrow {}^{22}_{10}\text{Ne} + e^+ + \nu_e \text{ (beta plus decay)}$$

$$^{22}_{11}\text{Na} + e^- \rightarrow {}^{22}_{10}\text{Ne} + \nu_e \text{ (electron capture)}$$

Beta decay does not change the number (A) of nucleons in the nucleus, but changes only its charge Z. Thus the set of all nuclides with the same A can be introduced; these *isobaric* nuclides may turn into each other via beta decay. For a given A there is one that is most stable. It is said to be beta stable, because it presents a local minima of the mass excess: if such a nucleus has (A, Z) numbers, the neighbour nuclei ($A, Z-1$) and ($A, Z+1$) have higher mass excess and can beta decay into (A, Z), but not vice versa. For all odd mass numbers A, there is only one known beta-stable isobar. For even A, there are up to three different beta-stable isobars experimentally known; for example, ${}^{124}_{50}Sn$, ${}^{124}_{52}Te$, and ${}^{124}_{54}Xe$ are all beta-stable. There are about 350 known beta-decay stable nuclides.

Competition of Beta Decay Types

Usually unstable nuclides are clearly either "neutron rich" or "proton rich", with the former undergoing beta decay and the latter undergoing electron capture (or more rarely, due to the higher energy requirements, positron decay). However, in a few cases of odd-proton, odd-neutron radionuclides, it may be energetically favorable for the radionuclide to decay to an even-proton, even-neutron isobar either by undergoing beta-positive or beta-negative decay. An often-cited example is the single isotope ${}^{64}_{29}Cu$ (29 protons, 35 neutrons), which illustrates three types of beta decay in competition. Copper-64 has a half-life of about 12.7 hours. This isotope has one unpaired proton and one unpaired neutron, so either the proton or the neutron can decay. This particular nuclide (though not all nuclides in this situation) is almost equally likely to decay through proton decay by positron emission (18%) or electron capture (43%) to ${}^{64}_{28}Ni$, as it is through neutron decay by electron emission (39%) to ${}^{64}_{30}Zn$.

Stability of Naturally Occurring Nuclides

Most naturally occurring nuclides on earth are beta stable. Those that are not have half-lives ranging from under a second to periods of time significantly greater than the age of the universe. One common example of a long-lived isotope is the odd-proton odd-neutron nuclide ${}^{40}_{19}K$ which undergoes all three types of beta decay (β^-, β^+ and electron capture) with a half-life of 1.277×10^9 years.

Conservation Rules for Beta Decay

Baryon Number is Conserved

$$B = \frac{n_q - n_{\bar{q}}}{3}$$

where,

n_q is the number of constituent quarks, and

$n_{\bar{q}}$ is the number of constituent antiquarks.

Beta decay just changes neutron to proton or, in the case of positive beta decay (electron capture) proton to neutron so the number of individual quarks doesn't change. It is only the baryon flavor that changes, here labelled as the isospin.

Up and down quarks have total isospin $I = \frac{1}{2}$ and isospin projections,

$$I_z = \begin{cases} \frac{1}{2} & \text{up quark} \\ -\frac{1}{2} & \text{down quark} \end{cases}$$

All other quarks have $I = 0$.

In general,

$$I_z = \frac{1}{2}(n_u - n_d)$$

Lepton Number is Conserved

$$L \equiv n_\ell - n_{\bar{\ell}}$$

so all leptons have assigned a value of +1, antileptons −1, and non-leptonic particles 0.

$$\begin{array}{ccccccccc} & n & \rightarrow & p & + & e^- & + & \bar{\nu}_e \\ L: & 0 & = & 0 & + & 1 & - & 1 \end{array}$$

Angular Momentum

For allowed decays, the net orbital angular momentum is zero, hence only spin quantum numbers are considered.

The electron and antineutrino are fermions, spin-1/2 objects, therefore they may couple to total *S=1* (parallel) or *S=0* (anti-parallel).

For forbidden decays, orbital angular momentum must also be taken into consideration.

Energy Release

The *Q* value is defined as the total energy released in a given nuclear decay. In beta decay, *Q* is therefore also the sum of the kinetic energies of the emitted beta particle, neutrino, and recoiling nucleus. (Because of the large mass of the nucleus compared to that of the beta particle and neutrino, the kinetic energy of the recoiling nucleus can generally be neglected.) Beta particles can therefore be emitted with any kinetic energy ranging from 0 to *Q*. A typical *Q* is around 1 MeV, but can range from a few keV to a few tens of MeV.

Since the rest mass of the electron is 511 keV, the most energetic beta particles are ultrarelativistic, with speeds very close to the speed of light.

β⁻ Decay

Consider the generic equation for beta decay,

$${}^{A}_{Z}X \rightarrow {}_{Z+1}^{\;\;A}X' + e^- + \nu_e.$$

The *Q* value for this decay is,

$$Q = [m_N({}^{A}_{Z}X) - m_N({}_{Z+1}^{\;\;A}X') - m_e - m_{\bar{\nu}_e}]c^2,$$

where $m_N({}^{A}_{Z}X)$ is the mass of the nucleus of the ${}^{A}_{Z}X$ atom, m_e is the mass of the electron, and $m_{\bar{\nu}_e}$ is the mass of the electron antineutrino. In other words, the total energy released is the mass energy of the initial nucleus, minus the mass energy of the final nucleus, electron, and antineutrino. The mass of the nucleus m_N is related to the standard atomic mass m by,

$$m({}^A_Z X)c^2 = m_N({}^A_Z X)c^2 + Zm_e c^2 - \sum_{i=1}^{Z} Bi.$$

That is, the total atomic mass is the mass of the nucleus, plus the mass of the electrons, minus the sum of all *electron* binding energies B_i for the atom. This equation is rearranged to find $m_N({}^A_Z X)$, and $m_N({}_{Z+1}^{A}X')$ is found similarly. Substituting these nuclear masses into the Q-value equation, while neglecting the nearly-zero antineutrino mass and the difference in electron binding energies, which is very small for high-Z atoms, we have,

$$Q = [m({}^A_Z X) - m({}_{Z+1}^{A}X')]c^2$$

This energy is carried away as kinetic energy by the electron and neutrino. Because the reaction will proceed only when the Q value is positive, β^- decay can occur when the mass of atom ${}^A_Z X$ is greater than the mass of atom ${}_{Z+1}^{A}X'$.

β^+ Decay

The equations for β^+ decay are similar, with the generic equation,

$${}^A_Z X \rightarrow {}_{Z-1}^{A}X' + e^+ + \nu_e$$

giving,

$$Q = [m_N({}^A_Z X) - m_N({}_{Z-1}^{A}X') - m_e - m_{\nu_e}]c^2.$$

However, in this equation, the electron masses do not cancel, and we are left with,

$$Q = [m({}^A_Z X) - m({}_{Z-1}^{A}X') - 2m_e]c^2$$

Because the reaction will proceed only when the Q value is positive, β^+ decay can occur when the mass of atom ${}^A_Z X$ exceeds that of ${}_{Z+1}^{A}X'$ by at least twice the mass of the electron.

Electron Capture

The analogous calculation for electron capture must take into account the binding energy of the electrons. This is because the atom will be left in an excited state after capturing the electron, and the binding energy of the captured innermost electron is significant. Using the generic equation for electron capture,

$${}^A_Z X + e^- \rightarrow {}_{Z-1}^{A}X' + \nu_e$$

we have,

$$Q = [m_N({}^A_Z X) + m_e - m_N({}_{Z-1}^{A}X') - m_{\nu_e}]c^2,$$

which simplifies to,

$$Q = [m({}^{A}_{Z}X) - m({}_{Z-1}^{A}X')]c^2 - B_n,$$

where B_n is the binding energy of the captured electron.

Because the binding energy of the electron is much less than the mass of the electron, nuclei that can undergo β^+ decay can always also undergo electron capture, but the reverse is not true.

Beta Emission Spectrum

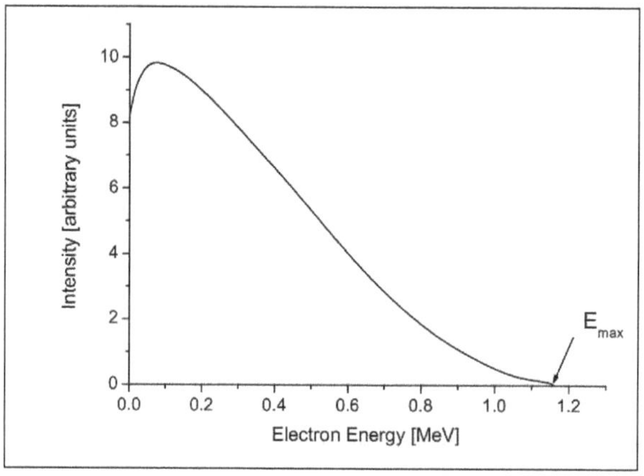

Beta spectrum of ^{210}Bi. $E_{max} = Q = 1.16$ MeV is the maximum energy.

Beta decay can be considered as a perturbation as described in quantum mechanics, and thus Fermi's Golden Rule can be applied. This leads to an expression for the kinetic energy spectrum $N(T)$ of emitted betas as follows:

$$N(T) = C_L(T)F(Z,T)pE(Q-T)^2$$

where T is the kinetic energy, C_L is a shape function that depends on the forbiddenness of the decay (it is constant for allowed decays), $F(Z, T)$ is the Fermi Function with Z the charge of the final-state nucleus, $E = T + mc^2$ is the total energy, $p = \sqrt{(E/c)^2 - (mc)^2}$ is the momentum, and Q is the Q value of the decay. The kinetic energy of the emitted neutrino is given approximately by Q minus the kinetic energy of the beta.

As an example, the beta decay spectrum of ^{210}Bi (originally called RaE) is shown to the right.

Fermi Function

The Fermi function that appears in the beta spectrum formula accounts for the Coulomb attraction/repulsion between the emitted beta and the final state nucleus. Approximating the associated wavefunctions to be spherically symmetric, the Fermi function can be analytically calculated to be:

$$F(Z,T) = \frac{2(1+S)}{\Gamma(1+2S)^2}(2p\rho)^{2S-2}e^{\pi\eta}\,|\Gamma(S+i\eta)|^2,$$

where p is the final momentum, Γ the Gamma function, and (if α is the fine-structure constant and r_N the radius of the final state nucleus) $S=\sqrt{1-\alpha^2Z^2}, \eta=\pm {Ze^2c}/{\hbar p}$ (+ for electrons, – for positrons), and $\rho = {r_N}/{\hbar}$.

For non-relativistic betas ($Q \ll m_e c^2$), this expression can be approximated by:

$$F(Z,T) \approx \frac{2\pi\eta}{1-e^{-2\pi\eta}}.$$

Other approximations can be found in the literature.

Kurie Plot

A Kurie plot (also known as a Fermi–Kurie plot) is a graph used in studying beta decay developed by Franz N. D. Kurie, in which the square root of the number of beta particles whose momenta (or energy) lie within a certain narrow range, divided by the Fermi function, is plotted against beta-particle energy. It is a straight line for allowed transitions and some forbidden transitions, in accord with the Fermi beta-decay theory. The energy-axis (x-axis) intercept of a Kurie plot corresponds to the maximum energy imparted to the electron/positron (the decay's Q value). With a Kurie plot one can find the limit on the effective mass of a neutrino.

Helicity of Neutrinos, Electrons and Positrons Emitted in Beta Decay

After the discovery of parity non-conservation, it was found that, in beta decay, electrons are emitted mostly with negative helicity, i.e., they move, naively speaking, like left-handed screws driven into a material (they have negative longitudinal polarization). Conversely, positrons have mostly positive helicity, i.e., they move like right-handed screws. Neutrinos (emitted in positron decay) have negative helicity, while antineutrinos (emitted in electron decay) have positive helicity. The higher the energy of the particles, the higher their polarization.

Types of Beta Decay Transitions

Beta decays can be classified according to the angular momentum (L value) and total spin (S value) of the emitted radiation. Since total angular momentum must be conserved, including orbital and spin angular momentum, beta decay occurs by a variety of quantum state transitions to various nuclear angular momentum or spin states, known as "Fermi" or "Gamow–Teller" transitions. When beta decay particles carry no angular momentum (L = 0), the decay is referred to as "allowed", otherwise it is "forbidden". Other decay modes, which are rare, are known as bound state decay and double beta decay.

Fermi Transitions

A Fermi transition is a beta decay in which the spins of the emitted electron (positron) and anti-neutrino (neutrino) couple to total spin S = 0, leading to an angular momentum change $\Delta J = 0$ between the initial and final states of the nucleus (assuming an allowed transition). In the non-relativistic limit, the nuclear part of the operator for a Fermi transition is given by:

$$\mathcal{O}_F = G_V \sum_a \hat{\tau}_{a\pm}$$

With G_v the weak vector coupling constant, $\tau_\pm$ the isospin raising and lowering operators, and a running over all protons and neutrons in the nucleus.

Gamow–Teller Transitions

A Gamow–Teller transition is a beta decay in which the spins of the emitted electron (positron) and anti-neutrino (neutrino) couple to total spin $S = 1$, leading to an angular momentum change $\Delta J = 0, \pm 1$ between the initial and final states of the nucleus (assuming an allowed transition). In this case, the nuclear part of the operator is given by,

$$\mathcal{O}_{GT} = G_A \sum_a \hat{\sigma}_a \hat{\tau}_{a\pm}$$

with G_A the weak axial-vector coupling constant, and σ he spin Pauli matrices, which can produce a spin-flip in the decaying nucleon.

Forbidden Transitions

When $L > 0$, the decay is referred to as "forbidden". Nuclear selection rules require high L values to be accompanied by changes in nuclear spin (J) and parity (π). The selection rules for the Lth forbidden transitions are:

$$\Delta J = L-1, L, L+1; \Delta\pi = (-1)^L,$$

where $\Delta\pi = 1$ or -1 corresponds to no parity change or parity change, respectively. The special case of a transition between isobaric analogue states, where the structure of the final state is very similar to the structure of the initial state, is referred to as "superallowed" for beta decay, and proceeds very quickly. The following table lists the ΔJ and $\Delta\pi$ values for the first few values of L:

Forbiddenness	ΔJ	Δπ
Superallowed	0	no
Allowed	0, 1	no
First forbidden	0, 1, 2	yes
Second forbidden	1, 2, 3	no
Third forbidden	2, 3, 4	yes

Rare Decay Modes

Bound-state β^- Decay

A very small minority of free neutron decays (about four per million) are so-called "two-body decays", in which the proton, electron and antineutrino are produced, but the electron fails to gain the 13.6 eV energy necessary to escape the proton, and therefore simply remains bound to it, as a neutral hydrogen atom. In this type of beta decay, in essence all of the neutron decay energy is carried off by the antineutrino.

For fully ionized atoms (bare nuclei), it is possible in likewise manner for electrons to fail to escape the atom, and to be emitted from the nucleus into low-lying atomic bound states (orbitals). This cannot occur for neutral atoms with low-lying bound states which are already filled by electrons.

Bound-state β decays were predicted by Daudel, Jean, and Lecoin in 1947, and the phenomenon in fully ionized atoms was first observed for $^{163}Dy^{66+}$ in 1992 by Jung et al. of the Darmstadt Heavy-Ion Research group. Although neutral ^{163}Dy is a stable isotope, the fully ionized $^{163}Dy^{66+}$ undergoes β decay into the K and L shells with a half-life of 47 days.

Another possibility is that a fully ionized atom undergoes greatly accelerated β decay, as observed for ^{187}Re by Bosch et al., also at Darmstadt. Neutral ^{187}Re does undergo β decay with a half-life of 42 × 10^9 years, but for fully ionized $^{187}Re^{75+}$ this is shortened by a factor of 10^9 to only 32.9 years. For comparison the variation of decay rates of other nuclear processes due to chemical environment is less than 1%.

Double Beta Decay

Some nuclei can undergo double beta decay (ββ decay) where the charge of the nucleus changes by two units. Double beta decay is difficult to study, as the process has an extremely long half-life. In nuclei for which both β decay and ββ decay are possible, the rarer ββ decay process is effectively impossible to observe. However, in nuclei where β decay is forbidden but ββ decay is allowed, the process can be seen and a half-life measured. Thus, ββ decay is usually studied only for beta stable nuclei. Like single beta decay, double beta decay does not change *A*; thus, at least one of the nuclides with some given *A* has to be stable with regard to both single and double beta decay.

"Ordinary" double beta decay results in the emission of two electrons and two antineutrinos. If neutrinos are Majorana particles (i.e., they are their own antiparticles), then a decay known as neutrinoless double beta decay will occur. Most neutrino physicists believe that neutrinoless double beta decay has never been observed.

Gamma Decay

Gamma ray decay occurs when a nucleus in an excited state releases its excess energy by emission of electromagnetic radiation, *i.e.*, a photon. Thus we have,

$$^{A}X^{*} \rightarrow {}^{A}X + \ldots$$

where the symbol * indicates an excited state of the nucleus. Note that there is no change in Z or A during this type of decay, only the release of energy. One can also get gamma ray emission from a high lying excited state to a lower lying state of the same nucleus. Thus gamma ray transitions do not have to go to the ground state of the nucleus. Figure depicts a typical situation in which a series of gamma rays de-excite the levels of a nucleus with so-called "crossover transitions" also occurring. Also note that the gammaray energy spectrum shows discrete lines corresponding to each transition. Note that the energies of the gamma rays can vary from a few keV to many MeV. Any nucleus from hydrogen to Element 112 can decay by gamma ray emission.

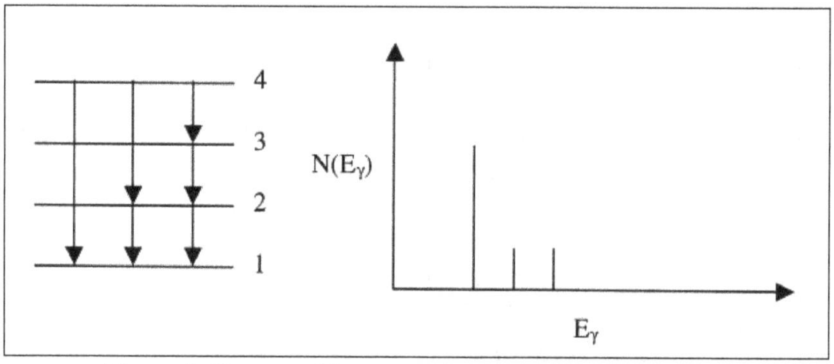

Schematic diagram of the gamma ray transitions for a nucleus and the resulting gamma ray energy spectrum.

In some unusual cases a nucleus can have two configurations of nucleons that have very similar low-lying energies but very different total angular momenta. One of these states will be lower in energy, of course, but the transition between the two states will be strongly hindered, due to the fact that the photon will have to balance the large change in angular momentum. This hindered decay is similar to the hindrance of triplet states in atomic and molecular systems. These long lived nuclear states are called *isomeric states* and their gamma ray decay is called an isomeric transition or simply IT decay. The ground state of the zinc nucleus is unstable with respect to b decay with a half-life of 56 minutes. The lowest excited state of this nucleus has an energy of only 439 keV but it has a much larger spin and the opposite parity compared to the ground state. The transition from the excited state to the ground state is hindered by the large change in angular momentum, $4\hbar$, combined with a change in parity which leads to an IT half-life of 14 hours.

Energetics of Gamma Decay

Imagine a gamma transition between two nuclear states. Applying the law of conservation of energy, we have,

$$M_0^*c^2 = M_0c^2 + E... + T_r$$

where $E\gamma$ is the photon energy, T_r the kinetic energy of the recoiling nucleus after γ-ray emission and $M_0{}^*$ and M_0 are the masses of the higher and lower nuclear states, respectively. Applying the law of conservation of momentum we have,

$$p... = p_r$$

where p_γ and p_r are the momenta of the photon and recoiling nucleus, respectively. The kinetic energy of the recoil is so small that non-relativistic mechanics can be used. Hence we have,

$$T_r = p_r^2/2M_0$$

Substituting,

$$T_r = p_\gamma^2/2M_0$$
$$T_r = E_\gamma^2/2M_0c^2$$

If E_γ = 2 MeV, and A=50, the recoil energy is about 40 eV, which is negligible except for Mossbauer studies.

Examples of Gamma Recoil Energy

(a) Calculate the recoil energy for the IT-decay of $^{69}Zn^m$ to the ground state of ^{69}Zn.

Using the energy of the excited state, we have:

$$\left(M_0^* - M_0\right)c^2 = E\gamma = 0.439\ MeV$$

Recall that $M^*_0c^2$ = amu 931.5 MeV/amu. Using the mass excess of 68.418 MeV found in the Wallet Cards, the mass of ^{69}Zn is 68.927 amu. This is obtained by 69 amu + (-68.418 MeV/931.5 MeV/amu) = 69 amu – 0.073 amu = 68.927 amu.

$$T_r = \frac{E_\gamma^2}{2M_0C^2} = \frac{\left(0.439\,MeV\right)^2}{2\left(68.927 \times 931.5\right)} = 1.50 \times 10^{-6}\ MeV = 1.5V$$

(b) Calculate the recoil energy from the emission of a 15.1 MeV photon by an excited ^{12}C nucleus. Recall that the mass excess of ^{12}C is 0 so that the mass of ^{12}C is 12.000 amu.

$$T_r = \frac{E_\gamma^2}{2M_0C^2} = \frac{\left(15.1MeV\right)^2}{2\left(12 \times 931.5\,MeV\right)}$$
$$T_r = 1.02\,x\,10^{-2}\ MeV = 10.2\,KeV$$

Classification of Decay Types

The conservation of angular momentum has provided an enormous amount of information on the structure of nuclei and plays a controlling role in the gamma ray decay process. From a schematic viewpoint, a stationary nucleus in a definite quantum mechanical state makes a transition to a lower energy state during gamma decay and emits a single photon. Both the initial and final states of the nucleus will have a definite angular momentum and parity and so the photon must connect the two states and conserve both parity and angular momentum. Photons each carry an exact integer number of angular momentum units ($\hbar$) and each has a definite parity. The conservation of angular momentum and parity are different, of course, and conservation of each has a different effect on the possible properties of the emitted photon.

The angular momenta of the initial and final states of final nucleus can be labeled as I_i h and I_f h and change in intrinsic angular momentum, ΔI ($\hbar$) is, of course, $\ell = \Delta l = \left(I_i - I_f\right)\hbar$ A photon must carry at least one unit of angular momentum so that ΔI=0 is forbidden for sngle photon emission. The emitted photon should have a minimum intrinsic spin of $\ell\hbar$ units to connect the two nuclear states. However, the standard coupling rules for angular momenta allow the photon to carry away up to a maximum of $\ell = (I_i - I_f)\hbar$ units. Therefore, given known values of the spins of initial and final states of the nucleus, the angular momentum carried by the photon can take any value in the range,

$$\left|(I_i - I_f)\right| \leq \ell \leq (I_i - I_f)\hbar.$$

The multipolarity of the photon is a quantification of the amount of angular momentum carried by the photon. The nomenclature is that a photon with ℓ units of angular momentum is called a 2 ℓ pole photon. (The nomenclature comes from the classical radiation patterns of electromagnetic radiation and the design of the antennas used to create those patterns.) For example, a photon with ℓ = 1 is called a dipole photon, ℓ = 2a quadrupole photon, and so on as indicated in table. The transition rate depends strongly on the angular momentum change so that the smallest value of $\ell = \left|(I_i - I_f)\right|\hbar$ is usually observed although conservation of parity plays a role. Transitions with the maximum change in the angular momentum of the nuclear states are called *stretched* transitions.

To understand the parity of electromagnetic transitions we need to recall that each of the initial and final states of the nucleus undergoing the transition can be viewed as having a definite distribution of matter and charge. When the excited nucleus makes a transition from the excited state to a lower energy state, the distribution of matter and charge will change in some way. For example, a nucleus that is spinning with a certain value of angular momentum will slow down as it de-excites and reaches the ground state. Thus, the emission of the photon can be associated with the change in the overall distribution of neutrons and protons but we can identify two different changes that are analogous to classical antennas. A shift in the distribution of charge (for example, the transition of a proton from one orbital to another) will give rise to an electric field but a shift in the distribution of current in the nucleus (for example, the shift of the direction of a proton-orbital) will give rise to a magnetic field. The parity of the photon depends on both the angular momentum and the type (electric or magnetic) of transition indicated in table. Notice that electric and magnetic radiation with a given multipole character have opposite parities.

Table: Gamma ray selection rules and multipolarities.

Radiation Type	Name	$\ell = \Delta I$	$\Delta\pi$
E1	electric dipole	1	(yes)
M1	magnetic dipole	1	(no)
E2	electric quadrupole	2	(no)
M2	magnetic quadrupole	2	(yes)
E3	electric octupole	3	(yes)
M3	magnetic octupole	3	(no)
E4	electric hexadecapole	4	(no)
M4	magnetic hexadecapole	4	(yes)

With the list of properties of photons we can generalize the procedure to identify the probable type of photon for a given transition between nuclear states. First the parity of the photon will be given by the difference in parities of the two nuclear states. Then the combination of allowed angular momenta and parity will determine the character of the electromagnetic radiation. For example, the first excited state in ^{7}Li has spin and parity 1/2- and the ground state is 3/2-. Possible electromagnetic transitions between the two states must have $\Delta\pi = no$ and $1 \leq \ell \leq 2$. Consulting table, we find that the only candidates are Ml $(\ell = l, \Delta\pi = no)$ and E2 $(\ell = 2, \Delta\pi = no)$ while the other combinations E1 and M2 are ruled out by parity. All of the allowed radiation types will be emitted but at substantially different rates so that the overall radiation usually has one predominant character.

Application of Selection Riles

Use the electromagnetic selection rules to identitythe character of the crossover transitions that could link the second excited state at 2.080 MeV (7/2+) in ^{23}Na with the ground state (3/2+). First, we should note that $\Delta\pi$ = no. Then,

$$\left|(I_i - I_f)\right| \leq \ell \leq I_i + I_f \Rightarrow |2| \leq \ell \leq 5$$

ℓ	Δπ	Type
2	no	E2
3	no	M3
4	no	E4
5	no	M5

As a final point on the topic of selection rules we noted that $\Delta\ell = 0$ is forbidden for the emission of a single photon. The electric monopole distribution (E0) corresponds to the static distribution of charge in the nucleus and is constant. Similarly the M0 distribution corresponds to the nonexistent magnetic monopole moment. Nonetheless there are a few examples of even-even nuclei that have first excited and ground states that are both 0+. Once populated, these states decay by internal conversion processes in which the atomic electrons, particularly s-electrons with significant penetration into the nucleus, are directly emitted from the atom.

Wave Length Calculation

Calculate the ratio of the wave length of the 439 keV photon emitted when the isomeric state of ^{69}Zn^m makes an IT to the diameter of this nucleus. For a photon,

$$E_\gamma = h\nu$$

$$\lambda\nu = c$$

$$\lambda = \frac{hc}{E_\gamma} = \frac{(6.626x10^{-34}\ js)(3.0x10^8\ m/s)}{(439x\ 10^3\ eV)(1.602x10^{-19}\ j/eV)}$$

$$\lambda = 2.8 \times 10^{-12}\ \text{m}$$

Recall that R = 1.2 x $A^{1/3}$ fm, so that:

$$\frac{\lambda}{2R}=\frac{2.82x10^{-12}m}{2x4.92x10^{-15}m}=287$$

Thus, the nucleus is not an effective antenna due to its small size compared to the wave length of the radiation. Gamma rays are in the long wave length limit and are not very sensitive to the detailed internal structure of the emitting nucleus.

Electromagnetic Transition Rates

Determining the rate at which an excited state will decay by the emission of a photon is a very general quantum mechanical problem that is not limited to the world of nuclei. The detailed derivation of the transition rate is beyond the scope of this text and we will only sketch out the results. The decay constant for the emission of a photon by a very well defined single state that has excess energy was shown in the appendix to be given by the general expression:

$$\lambda=\frac{2\pi}{\hbar}\left|\int \varphi^{*}_{final} V_{p} \varphi_{initail} dv\right|^{2}\rho\left(E_{f}\right)$$

which is also called Fermi's Golden Rule. The wave functions, j, represent the complete initial and final states of the entire system and V_p is a (very) small perturbative interaction between the nuclear and electric fields that stimulates the transition. The form and the strength of the perturbation will depend on the multipolarity of the transition. The last factor, $\rho(E_f)$ is the product of the density of nuclear and electromagnetic states that are available to the system after the transition. The initial wave function contains only the nuclear excited state, whereas the final wave function will have parts for the electromagnetic wave and the daughter nuclear state.

After some extensive calculus and input from the theory of electromagnetism we come to an expression for the electromagnetic decay rate:

$$\lambda=\left(\ell,l_{i},\pi\rightarrow l_{f},\pi\right)=\frac{8\pi\left(\ell+1\right)}{\ell\left[\left(2\ell+1\right)!!\right]^{2}}\frac{k^{2\ell+1}}{\hbar}B\left(\ell,l_{i},\pi\rightarrow l_{f},\pi\right)$$

where k is the photon wave number $\left(k=\frac{E_{\gamma}}{\hbar c}\right)$ The symbol !! calls for the double factorial of its' argument, which for the case of $\ell=2$ and $2\ell+1=5$ would be the product of the odd integers: 5!! = 5*3* 1 =15. The reduced transition probability $B\left(\ell,l_{i},\pi\rightarrow l_{j},\pi\right)$ is the matrix element for the reduced nuclear wave functions (*i.e.*, summed over magnetic orientations) and the multipole operator (either electric or magnetic in character),

$$B\left(\ell,l_{i},\pi\rightarrow l_{j},\pi\right)=\frac{1}{2l_{i}+1}\left|<l_{f}\xi\|O\ell\|l_{i}\xi>\right|^{2}.$$

In which the symbols ξ in the nuclear wave functions are meant to represent all the other relevant quantum numbers. (As an aside we should note that the two "types" of electric and magnetic radiation are only different in terms of their parity and in the orientation of their plane of polarization.) This expression is still somewhat complicated and is difficult to evaluate. Victor Weisskopf derived a general expression for the reduced transition probability with the assumption that the transition results from the change of a single particle inside a nucleus with a uniform density with the familiar radius function, $R = r_0 A^{1/3}$. His expression for electric multipole radiation, called the Weisskopf single-particle limit is:

$$B_{sp}(E,\ell)=\frac{1}{4\pi}\left[\frac{3}{(\ell+3)}\right]^2 (r_0)^{2\ell/3} e^2 \, f\, m^{2\ell}$$

The single particle limit for magnetic multipole radiation obtained by assuming that the change in current is due to a single nucleon is:

$$B_{sp}(M,\ell)=\frac{10}{\pi}\left[\frac{3}{(\ell+3)}\right]^2 (r_0)^{(2\ell-2)} A^{(2\ell-2)/3} \mu_n^{\,2} \, f\, m^{(2\ell-2)}$$

One of the nagging features of these expressions is that the radial integral from the multipole expansion introduces a factor of $r^{2\ell}$ and thus the dimensions of B (E,) and B_{sp} (E,) depend on.

Either of the single-particle limits for the reduced electric or magnetic transition probability can be substituted into the expression for the transition rate to obtain numerical estimates of the de-excitation rates under the assumption that one particle was responsible for the change in electric charge or electric current associated with the change in nuclear states. The transition rates vary over an enormous range depending most strongly on the value of ℓ. Electric transitions are faster than magnetic transitions by about two orders of magnitude. Looking back to the discussion of the fact that several different types of photons can be associated with a given nuclear transition we now see that we expect the rates of emission to favor the lowest multipolarity. This fact can be simply demonstrated by evaluating the expressions for the transition rate for electric dipole, $\ell = 1$, and electric quadrupole, $\ell = 2$, radiation with a typical nuclear radius parameter of $r_0 = 1.2$ fm. Combining the expressions for the transition rate and the reduced transition probability for an E1 transition we get:

$$\lambda_{sp}(E,\ell)=\frac{8\pi(\ell+1)}{\ell[(2\ell+3)!!]^2}\frac{k^{(2\ell-1)}}{\hbar}\frac{1}{4\pi}\left[\frac{3}{(\ell+3)}\right]^2\left(e^2 r_0 f\, m\right)^{2\ell} A^{2\ell/3}$$

Substituting in $\ell = 1$,

$$\lambda_{sp}(El)=\frac{8\pi(2)}{1[(2\ell+1)!!]^2}\left(\frac{E_\gamma}{\hbar c}\right)\frac{e^2}{4\pi\hbar}\left[\frac{3}{(1+3)}\right]^2(1.2 f\, m)^2 A^{2/3}\sec^{-1}$$

Recall that $\hbar c$ = 197.3 MeV-fm and e^2 = 1.440 MeV-fm so that,

$$\lambda_{sp}(El)=\frac{16\pi}{9}\left(\frac{E(MeV)}{197.3\, MeV-fm}\right)^3\frac{1.440 f\, m}{4\pi\hbar}\left[\frac{3}{4}\right]^2(1.2 f\, m)^2 A^{2/3}\sec^{-1}$$

$$\lambda_{sp}\left(El\right)=1.03x10^{14}\,E_{\gamma}^{3}\,A^{2/3}\,\sec^{-1}$$

Similar substitution into the expression for $\lambda_{sp}\left(E\ell\right)$ with $\ell = 2$ for electric quadrupole radiation will eventually yield:

$$\lambda_{sp}\,(\mathrm{E2})=7.28\,\mathrm{x}\,10^{7}\,E_{\gamma}\,{}^{5}\mathrm{A}^{4/3}$$

So we see that the rates depend very strongly on the energy of the photon and on the size (mass number) of the emitting nucleus. If we consider the specific hypothetical case of a 1 MeV transition in a medium mass nucleus, A=100, the ratio of transition rates is:

$$\frac{\lambda_{sp}\left(E1\right)}{\lambda_{sp}\left(E2\right)}=\frac{1.03x10^{14}\,E_{\gamma}^{3}\,A^{2/3}}{7.28xs10^{7}\,E_{\gamma}^{5}\,A^{4/3}}=1.41x10^{6}\,E_{\gamma}^{-2}\,A^{-2/3}$$

$$\frac{\lambda_{sp}\left(E1\right)}{\lambda_{sp}\left(E2\right)}=6.54x10^{4}$$

The formulas for the transition rates are summarized in table for the lowest five multipoles of each character. The transition rates always increase with a high power of the gamma ray energy so that low energy transitions, say below 100 keV, are much slower than high energy transitions, say above 1 MeV. The table also shows that in some cases, particularly in heavy nuclei, an $\ell+1$ electric transition can compete favorably with an ℓ magnetic transition.

Table: Weisskopf Single Particle transition rates (Eγ is in MeV).

Multipole	$E\ell$	$M\ell$
ℓ	$\lambda\left(\sec^{-1}\right)$	$\lambda\left(\sec^{-1}\right)$
1	$1.0x10^{14}\;A^{2/3}\,E_{\gamma}^{3}$	$3.1x10^{13}\,E_{\gamma}^{3}$
2	$7.4\,x10^{7}\,A^{4/3}\,E_{\gamma}^{5}$	$2.2\,x10^{7}\,A^{2/3}\,E_{\gamma}^{5}$
3	$3.5\,x10^{1}\,A^{2}\,E_{\gamma}^{7}$	$1.1\,x10^{1}\,A^{4/3}\,E_{\gamma}^{7}$
4	$1.1x10^{-5}\,A^{8/3}\,E_{\gamma}^{9}$	$3.3x10^{-6}\,A^{2}\,E_{\gamma}^{9}$
5	$2.4x10^{-12}\,A^{10/3}\,E_{\gamma}^{11}$	$7.4\,x10^{-13}\,A^{8/3}\,E_{\gamma}^{11}$

The Weisskopf estimates are usually good to within a factor of ten, which is remarkable given the large number of orders of magnitude that they span, and provide important references for comparison to the observed transition rates. Notice that if a transition occurs more rapidly than the single particle rate then the transition is more collective, that is more particles participate in the change. If the transition is significantly slower than the Weisskopf estimate then the nuclear matrix element must be smaller than the single particle limit, i.e., the overlap of the initial and final states is smaller. The ratio of the observed decay rate to the Weisskopf estimated rate is often quoted in the literature as the transition rate in Weisskopf units.

Example of Weisskopf Transition Rate

Use the electromagnetic selection rules to identity the character of the isomeric transition from the first excited state at 0.439 MeV (9/2+) in $^{69}Zn^{m}$ with the ground state (1/2-). Then calculate the Weisskopf single particle rates for the allowed transitions. First, we should note that $\Delta\pi$ = yes. Then,

$$\left|I_i - I_f\right| \le \ell \le I_i + I_f \Rightarrow |4| \le \ell \le 5$$

Thus, only M4 and E5 transitions are allowed. Using the expressions in table:

$$\lambda(M4) = 3.3\times10^{-6}A^2E_\gamma^9\left(\sec^{-1}\right)$$

$$\lambda(M4) = 9.5\times10^{-6}\left(\sec^{-1}\right)$$

and

$$\lambda(E5) = 2.4\times10^{-12}A^{10/3}E_\gamma^{11}\left(\sec^{-1}\right)$$

$$\lambda(E5) = 3.8\times10^{-10}\left(\sec^{-1}\right)$$

so we expect that the transition will be predominantly M4 in character. This example also shows that the observed transition occurs about twice as fast as the single particle estimate:

$$\lambda = \ln 2/\left(14hr*3600\sec/hr\right) = 1.37\times10^{-5}\left(\sec^{-1}\right)$$

which indicates that the "current" from more than one particle contributes to the magnetic transition.

It should be noted that E2 transitions are often enhanced by an order of magnitude compared to the single particle estimates. This enhancement of these specific transitions stems from collective nuclear motion and the enhancement is particularly strong for nuclei that lie in-between major shell closures. The excited nucleus cascades down from the 6+ level in a series of three E2 transitions with no crossover transitions. The lifetimes of the states, were used to calculate the transition rates in Weisskopf units. Notice that the rate of emission in this case ranges from 200 to 1100 times the single particle rate. If we take a closer look at the transition rate for electric quadrupole transitions then we would find that the reduced transition probability could be written in terms of the quadrupolc momcnt, Q_o, as:

$$B\left(E2, J_i \to J_f\right) = \frac{5}{16\pi}e^2Q_0^2\left\langle J_i, K, 2, 0 \mid J_f, K\right\rangle^2$$

in which the last term is a Clebsch-Gordon coefficient, which is a very general normalization coefficient for the values of the spins and the spin-projections. In the present case of transitions between the states of a rotational band, K = 0, J_i = J, and J_f = J −2, the normalization coefficient only contains values of J:

$$\left\langle J_i, K, 2, 0 \mid J_f, K\right\rangle^2 = \left\langle J_i, K, 2, 0 \mid (J-2), 0\right\rangle^2 = \frac{3J(J-1)}{2(2J+1)(2J-1)}$$

so that the reduced transition probability is:

$$B(E2, J \to J-2) = \frac{15}{32\pi} e^2 Q_0^2 \frac{J(J-1)}{(2J+1)(2J-1)}$$

Thus, the experimental transition rate provides a measurement of the quadrupole moment of the nucleus and we should not be surprised that a strongly deformed nucleus with a large quadrupole moment will have a larger E2 transition rate because the whole nucleus can participate in the transition compared to a single particle.

In the single particle estimates of γ-ray decay, one presumes a single nucleon interacts with a photon. This means there is an isospin selection rule,

$$\Delta T = 0 \ or \ 1$$

for γ-ray decay between two pure isospin states. Also E1 γ-transitions cannot occur when $\Delta T = 0$ in a self-conjugate nucleus (N = Z).

Internal Conversion

Internal conversion is a competing process to gamma ray decay and occurs when an excited nucleus interacts electromagnetically with an orbital electron and ejects it. This transfer of the nuclear excitation energy to the electron occurs radiationlessly without the emission of a photon. The energy of the internal conversion electron, E_{IC}, is given by,

$$E_{IC} = E_{transitions} - E_{electron\ binding\ energy}$$

Thus if a transition has $E_{transition}$ of 0.412 MeV, you would expect to see the spectrum of emitted internal conversion electrons. Note the different lines corresponding to the ejection of electrons from the K, L, and M shells. The nucleus will interact more readily with the K electrons than with the L electrons, than with the M electrons, etc. because the K electrons spend more time in the nucleus than the L electrons than the M electrons, etc.

To characterize this decay process and its competition with gamma ray emission, we define the *internal conversion coefficient*, α, by the relationship α = (number of internal conversion decays/ number of gamma ray decays) = $\Delta IC/\Delta$, where α can take on values from zero to infinity. Note further that,

$$\Delta = \Delta IC + \Delta_\gamma = \Delta_\gamma (1+\alpha)$$

One can define this ratio, the internal conversion coefficient, for electrons from the k shell only or for electrons from the m shell only, etc., giving rise to α_K, α_M, etc. Since the total probability of decay must equal the sum of the probabilities of decay via various paths we have,

$$\alpha \text{ total} = \alpha_K + \alpha_L + \alpha_M + \ldots$$

The internal conversion coefficient depends primarily on the density of the atomic electrons at the center of the nucleus, and thus it can be calculated using principles from atomic physics. Large tables and nomographs of internal conversion coefficients exist.

Rough approximate formulas for the internal conversion coefficients are:

$$\alpha(EL) = \frac{Z^3}{n^3}\left(\frac{L}{L+1}\right)\left(\frac{e^2}{4\pi\varepsilon_0\hbar c}\right)^4\left(\frac{2m_ec^2}{E}\right)^{L+5/2}$$

$$\alpha(ML) = \frac{Z^3}{n^3}\left(\frac{e^2}{4\pi\varepsilon_0\hbar c}\right)^4\left(\frac{2m_ec^2}{E}\right)^{L+3/2}$$

where Z is the atomic number of the atom in which the conversion is taking place, n is the principal quantum number of the bound electron being ejected and $\left(\frac{e^2}{4\pi\varepsilon_0\hbar c}\right)$, the fine structure constant is 1/137. Note that the internal conversion coefficient, α, increases approximately as Z^3, making internal conversion most important for heavy nuclei. The last factor in the equations gives the energy and multipolarity dependence with more internal conversion for low energies and higher transition multipolarities. The α_K/α_L ratio is approximately 8 due to the n^3 factor.

Example of Internal Conversion Coefficients

Determine the internal conversion coefficients for each shell for the transition from the first excited state at 0.08679 keV (2+) in ^{160}Dy to the ground state (0+). Then calculate the decay rates for internal conversion and for gamma ray emission.

We have to interpolate in a graph to find:

$$\alpha_K(E2, Z\approx 65, 0.090\,MeV) = 1.5$$

$$\alpha_{L1}(E2, Z\approx 65, 0.090\,MeV) = 0.1$$

$$\alpha_{L2}(E2, Z\approx 65, 0.090\,MeV) = 5$$

$$\alpha_{L3}(E2, Z\approx 65, 0.090\,MeV) = 2.5$$

$$\alpha = \alpha_K + \alpha_{L1} + \alpha_{L2} + \alpha_{L3} = 9.1$$

$$\lambda = (\ln 2)/2.02x10^{-9} = 3.34\times10^8 \sec^{-1} = \lambda(1+\alpha)$$

$$\lambda_\gamma = \frac{3.34\times10^8 \sec^{-1}}{(1-\alpha)} = 3.4\times10^7 \sec^{-1}$$

$$\lambda_{IC} = \lambda - \lambda_\gamma = 3\times10^8 \sec^{-1}$$

Note that internal conversion occurs approximately ten times faster than gamma ray emission for this transition in this nucleus.

References

- Rodejohann, W. (2012). "Neutrinoless double beta decay and neutrino physics". Journal of Physics G: Nuclear and Particle Physics. 39 (12): 124008. Arxiv:1206.2560. Doi:10.1088/0954-3899/39/12/124008.
- Basic-modes-of-radioactive-decay, online-first: intechopen.com, Retrieved 19 May, 2019

- L'Annunziata, Michael (2012). Handbook of Radioactivity Analysis (Third ed.). Elsevier Inc. P. 3. ISBN 9780123848741. Retrieved 4 October 2017.
- Radioactive-decay, entry: newworldencyclopedia.org, Retrieved 3 June, 2019
- Bijker, R.; Santopinto, E. (2015). "Valence and sea quarks in the nucleon". Journal of Physics: Conference Series. 578: 012015. Arxiv:1412.5559. Doi:10.1088/1742-6596/578/1/012015.
- Jevremovic, T. (2009). Nuclear Principles in Engineering. Springer Science + Business Media. P. 201. ISBN 978-0-387-85608-7.

5

Nuclear Power

Usage of nuclear reactions for the purpose of producing nuclear energy is termed as nuclear power. The thermal power station where heat is produced using nuclear power is called a nuclear power plant. This chapter closely examines nuclear power as well as the key components of nuclear power plants such as nuclear fuel, nuclear reactor and nuclear waste to provide an extensive understanding of the subject.

Nuclear power is the electricity generated by power plants that derive their heat from fission in a nuclear reactor. Except for the reactor, which plays the role of a boiler in a fossil-fuel power plant, a nuclear power plant is similar to a large coal-fired power plant, with pumps, valves, steam generators, turbines, electric generators, condensers, and associated equipment.

World Nuclear Power

Nuclear power provides almost 15 percent of the world's electricity. The first nuclear power plants, which were small demonstration facilities, were built in the 1960s. These prototypes provided "proof-of-concept" and laid the groundwork for the development of the higher-power reactors that followed.

The nuclear power industry went through a period of remarkable growth until about 1990, when the portion of electricity generated by nuclear power reached a high of 17 percent. That percentage remained stable through the 1990s and began to decline slowly around the turn of the 21st century, primarily because of the fact that total electricity generation grew faster than electricity from nuclear power while other sources of energy (particularly coal and natural gas) were able to grow more quickly to meet the rising demand. This trend appears likely to continue well into the 21st century. The Energy Information Administration (EIA), a statistical arm of the U.S. Department of Energy, has projected that world electricity generation between 2005 and 2035 will roughly double (from more than 15,000 terawatt-hours to 35,000 terawatt-hours) and that generation from all energy sources except petroleum will continue to grow.

In 2012 more than 400 nuclear reactors were in operation in 30 countries around the world, and more than 60 were under construction. The United States has the largest nuclear power industry, with more than 100 reactors; it is followed by France, which has more than 50. Of the top 15 electricity-producing countries in the world, all but two, Italy and Australia, utilize nuclear power to generate some of their electricity. The overwhelming majority of nuclear reactor generating capacity is concentrated in North America, Europe, and Asia. The early period of the nuclear power industry was dominated by North America (the United States and Canada), but in the 1980s that lead was overtaken by Europe. The EIA projects that Asia will have the largest nuclear capacity by 2035, mainly because of an ambitious building program in China.

A typical nuclear power plant has a generating capacity of approximately one gigawatt (GW; one billion watts) of electricity. At this capacity, a power plant that operates about 90 percent of the time (the U.S. industry average) will generate about eight terawatt-hours of electricity per year.

The predominant types of power reactors are pressurized water reactors (PWRs) and boiling water reactors (BWRs), both of which are categorized as light water reactors (LWRs) because they use ordinary (light) water as a moderator and coolant. LWRs make up more than 80 percent of the world's nuclear reactors, and more than three-quarters of the LWRs are PWRs.

Issues affecting Nuclear Power

Countries may have a number of motives for deploying nuclear power plants, including a lack of indigenous energy resources, a desire for energy independence, and a goal to limit greenhouse gas emissions by using a carbon-free source of electricity. The benefits of applying nuclear power to these needs are substantial, but they are tempered by a number of issues that need to be considered, including the safety of nuclear reactors, their cost, the disposal of radioactive waste, and a potential for the nuclear fuel cycle to be diverted to the development of nuclear weapons.

Safety

The safety of nuclear reactors has become paramount since the Fukushima accident of 2011. The lessons learned from that disaster included the need to: (1) adopt risk-informed regulation, (2) strengthen management systems so that decisions made in the event of a severe accident are based on safety and not cost or political repercussions, (3) periodically assess new information on risks posed by natural hazards such as earthquakes and associated tsunamis, and (4) take steps to mitigate the possible consequences of a station blackout.

The four reactors involved in the Fukushima accident were first-generation BWRs designed in the 1960s. Newer Generation III designs, on the other hand, incorporate improved safety systems and rely more on so-called passive safety designs (i.e., directing cooling water by gravity rather than moving it by pumps) in order to keep the plants safe in the event of a severe accident or station blackout. For instance, in the Westinghouse AP1000 design, residual heat would be removed from the reactor by water circulating under the influence of gravity from reservoirs located inside the reactor's containment structure. Active and passive safety systems are incorporated into the European Pressurized Water Reactor (EPR) as well.

Traditionally, enhanced safety systems have resulted in higher construction costs, but passive safety designs, by requiring the installation of far fewer pumps, valves, and associated piping, may actually yield a cost saving.

Economics

A convenient economic measure used in the power industry is known as the levelized cost of electricity, or LCOE, which is the cost of generating one kilowatt-hour (kWh) of electricity averaged over the lifetime of the power plant. The LCOE is also known as the "busbar cost," as it represents the cost of the electricity up to the power plant's busbar, a conducting apparatus that links the plant's generators and other components to the distribution and transmission equipment that delivers the electricity to the consumer.

The busbar cost of a power plant is determined by: 1) capital costs of construction, including finance costs, 2) fuel costs, 3) operation and maintenance (O&M) costs, and 4) decommissioning and waste-disposal costs. For nuclear power plants, busbar costs are dominated by capital costs, which can make up more than 70 percent of the LCOE. Fuel costs, on the other hand, are a relatively small factor in a nuclear plant's LCOE (less than 20 percent). As a result, the cost of electricity from a nuclear plant is very sensitive to construction costs and interest rates but relatively insensitive to the price of uranium. Indeed, the fuel costs for coal-fired plants tend to be substantially greater than those for nuclear plants. Even though fuel for a nuclear reactor has to be fabricated, the cost of nuclear fuel is substantially less than the cost of fossil fuel per kilowatt-hour of electricity generated. This fuel cost advantage is due to the enormous energy content of each unit of nuclear fuel compared to fossil fuel.

The O&M costs for nuclear plants tend to be higher than those for fossil-fuel plants because of the complexity of a nuclear plant and the regulatory issues that arise during the plant's operation. Costs for decommissioning and waste disposal are included in fees charged by electrical utilities. In the United States, nuclear-generated electricity was assessed a fee of $0.001 per kilowatt-hour to pay for a permanent repository of high-level nuclear waste. This seemingly modest fee yielded about $750 million per year for the Nuclear Waste Fund.

At the beginning of the 21st century, electricity from nuclear plants typically cost less than electricity from coal-fired plants, but this formula may not apply to the newer generation of nuclear power plants, given the sensitivity of busbar costs to construction costs and interest rates. Another major uncertainty is the possibility of carbon taxes or stricter regulations on carbon dioxide emissions. These measures would almost certainly raise the operating costs of coal plants and thus make nuclear power more competitive.

Radioactive-waste Disposal

Spent nuclear reactor fuel and the waste stream generated by fuel reprocessing contain radioactive materials and must be conditioned for permanent disposal. The amount of waste coming out of the nuclear fuel cycle is very small compared with the amount of waste generated by fossil fuel plants. However, nuclear waste is highly radioactive (hence its designation as high-level waste, or HLW), which makes it very dangerous to the public and the environment. Extreme care must be taken to ensure that it is stored safely and securely, preferably deep underground in permanent geologic repositories.

Despite years of research into the science and technology of geologic disposal, no permanent disposal site is in use anywhere in the world. In the last decades of the 20th century, the United States made preparations for constructing a repository for commercial HLW beneath Yucca Mountain, Nevada, but by the turn of the 21st century, this facility had been delayed by legal challenges and political decisions. Pending construction of a long-term repository, U.S. utilities have been storing HLW in so-called dry casks aboveground. Some other countries using nuclear power, such as Finland, Sweden, and France, have made more progress and expect to have HLW repositories operational in the period 2020–25.

Proliferation

The claim has long been made that the development and expansion of commercial nuclear power led to nuclear weapons proliferation, because elements of the nuclear fuel cycle (including uranium enrichment and spent-fuel reprocessing) can also serve as pathways to weapons development. However, the history of nuclear weapons development does not support the notion of a necessary connection between weapons proliferation and commercial nuclear power.

The first pathway to proliferation, uranium enrichment, can lead to a nuclear weapon based on highly enriched uranium. It is considered relatively straightforward for a country to fabricate a weapon with highly enriched uranium, but the impediment historically has been the difficulty of the enrichment process. Since nuclear reactor fuel for LWRs is only slightly enriched (less than 5 percent of the fissile isotope uranium-235) and weapons need a minimum of 20 percent enriched uranium, commercial nuclear power is not a viable pathway to obtaining highly enriched uranium.

The second pathway to proliferation, reprocessing, results in the separation of plutonium from the highly radioactive spent fuel. The plutonium can then be used in a nuclear weapon. However, reprocessing is heavily guarded in those countries where it is conducted, making commercial reprocessing an unlikely pathway for proliferation. Also, it is considered more difficult to construct a weapon with plutonium versus highly enriched uranium.

More than 20 countries have developed nuclear power industries without building nuclear weapons. On the other hand, countries that have built and tested nuclear weapons have followed other paths than purchasing commercial nuclear reactors, reprocessing the spent fuel, and obtaining plutonium. Some have built facilities for the express purpose of enriching uranium; some have built plutonium production reactors; and some have surreptitiously diverted research reactors to the production of plutonium. All these pathways to nuclear proliferation have been more effective, less expensive, and easier to hide from prying eyes than the commercial nuclear power route. Nevertheless, nuclear proliferation remains a highly sensitive issue, and any country that wishes to launch a commercial nuclear power industry will necessarily draw the close attention of oversight bodies such as the International Atomic Energy Agency.

Nuclear Power Plant

Nuclear power plants are a type of power plant that use the process of nuclear fission in order to generate electricity. They do this by using nuclear reactors in combination with the Rankine cycle, where the heat generated by the reactor converts water into steam, which spins a turbine and a generator. Nuclear power provides the world with around 11% of its total electricity, with the largest producers being the United States and France.

The Darlington nuclear power plant in Ontario produces power from four 878 MW CANDU reactors.

Aside from the source of heat, nuclear power plants are very similar to coal-fired power plants. However, they require different safety measures since the use of nuclear fuel has vastly different properties from coal or other fossil fuels. They get their thermal power from splitting the nuclei of atoms in their reactor core, with uranium being the dominant choice of fuel in the world today. Thorium also has potential use in nuclear power production, however it is not currently in use. Below is the basic operation of a boiling water power plant, which shows the many components of a power plant, along with the generation of electricity.

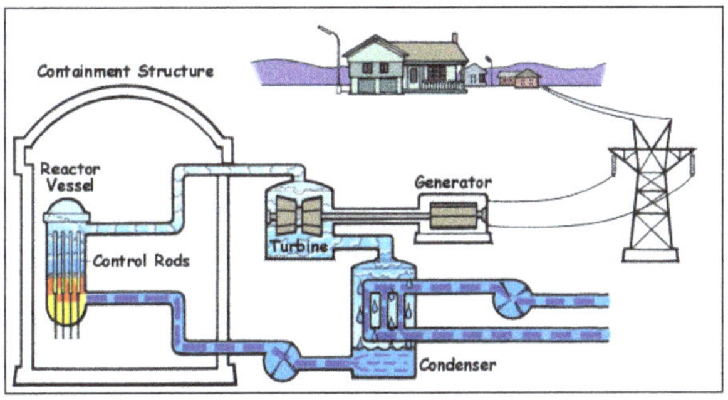

A boiling water nuclear reactor in combination with the Rankine cycle forms the basis of a nuclear power plant.

Components and Operation

Nuclear Reactor

The reactor is a key component of a power plant, as it contains the fuel and its nuclear chain reaction, along with all of the nuclear waste products. The reactor is the heat source for the power plant, just like the boiler is for a coal plant. Uranium is the dominant nuclear fuel used in nuclear reactors, and its fission reactions are what produce the heat within a reactor. This heat is then transferred to the reactor's coolant, which provides heat to other parts of the nuclear power plant.

Besides their use in power generation, there are other types of nuclear reactors that are used for plutonium manufacturing, the propulsion of ships, aircraft and satellites, along with research and medical purposes. The power plant encompasses not just the reactor, but also cooling towers, turbines, generators, and various safety systems. The reactor is what makes it differ from other external heat engines.

Steam Generation

The production of steam is common among all nuclear power plants, but the way this is done varies immensely.

Steam turbine in a power plant.

The most common power plants in the world use pressurized water reactors, which use two loops of circling water to produce steam. The first loop carries extremely hot liquid water to a heat exchanger, where water at a lower pressure is circulated. It then heats up and boils to steam, and can then be sent to the turbine section.

Boiling water reactors, the second most common reactor in power generation, heat the water in the core directly to steam.

Turbine and Generator

Two cooling towers of a nuclear power plant.

Once steam has been produced, it travels at high pressures and speeds through one or more turbines. These get up to extremely high speeds, causing the steam to loose energy, therefore, condensing back to a cooler liquid water. The rotation of the turbines is used to spin an electric generator, which produces electricity that is sent out the the electrical grid.

Cooling Towers

Perhaps the most iconic symbol of a nuclear power plant is the cooling towers. They work to reject waste heat to the atmosphere by the transfer of heat from hot water (from the turbine section) to the cooler outside air. Hot water cools in contact with the air and a small portion, around 2%, evaporates and raises up through the top. Moreover, these plants do not release any carbon dioxide—the primary greenhouse gas that contributes to climate change.

Many nuclear power plants simply put the waste heat into a river, lake or ocean instead of having cooling towers. Many other power plants like coal-fired power plants have cooling towers or these large bodies of water as well. This similarity exists because the process of turning heat into electricity is almost identical between nuclear power plants and coal-fired power plants.

Efficiency

The efficiency of a nuclear power plant is determined similarly to other heat engines—since technically the plant is a large heat engine. The amount of electric power produced for each unit of thermal power gives the plant its thermal efficiency, and due to the second law of thermodynamics there is an upper limit to how efficient these plants can be.

Typical nuclear power plants achieve efficiencies around 33-37%, comparable to fossil fueled power plants. Higher temperature and more modern designs like the Generation IV nuclear reactors could potentially reach above 45% efficiency.

Nuclear Fuel

Nuclear fuel is the fuel that is used in a nuclear reactor to sustain a nuclear chain reaction. These fuels are fissile, and the most common nuclear fuels are the radioactive metals uranium-235 and plutonium-239. All processes involved in obtaining, refining, and using this fuel make up a cycle known as the nuclear fuel cycle.

Uranium-235 is used as a fuel in different concentrations. Some reactors, such as the CANDU reactor, can use natural uranium with uranium-235 concentrations of only 0.7%, while other reactors require the uranium to be slightly enriched to levels of 3% to 5%. Plutonium-239 is produced and used in reactors (specifically fast breeder reactors) that contain significant amounts of uranium-238. It can also be recycled and used as a fuel in thermal reactors. Current research is being done to investigate how thorium-232 can be used as a fuel.

Production

Fuel fabrication plants are facilities that convert enriched uranium into fuel for nuclear reactors. For light water reactors, uranium is received from an enrichment plant in solid form. It is then converted into a gas and chemically converted into a uranium dioxide powder. This powder is then pressed into pellets and packed into fuel assemblies. A mixed oxide fuel can also be created when the uranium powder is packed along with plutonium oxide. The hazards present at fuel fabrication facilities—mainly chemical and radiological—are similar to the hazards at enrichment plants. These facilities generally pose a low risk to the public.

Use

When used in a reactor, the fuels used can have a variety of different forms a metal, an alloy, or some sort of oxide. Most nuclear reactors are fueled with a compound known as uranium dioxide. This uranium dioxide is put together in a fuel assembly and inserted into the nuclear reactor—where it can stay for several months or up to a few years. While in the reactor the fuel undergoes nuclear fission and releases energy. This released energy is used to generate electricity. Neutrons released during the fission process allow for a fission chain reaction to occur, allowing energy to be generated continually. The fuel is removed from the reactor after large amounts of the fuel—whether it is uranium-235 or plutonium-239—have undergone fission. The "used" nuclear fuel is known as *spent* or irradiated fuel. After use, the fuel must be cooled for a few years as it is extremely hot.

The spent fuel is placed in large, deep pools of water that act as a coolant and a radiation shield. The coolant property allows the water to remove the decay heat and the shielding abilities protect workers from the radioactivity of the fuel. After cooling, the fuel can be re-purposed or sent to storage depending on regulations.

Fuel Assembly

Nuclear reactors are powered by powdered uranium dioxide that has been compressed into small pellets, shown in figure. However, a power plant requires many of these pellets to run. Thus large numbers of these pellets are bundled into a fuel rod. A single uranium fuel pellet, only as big as a fingertip, contains as much energy as 481 cubic meters of natural gas, 807 kilograms of coal or 564 liters of oil. These rods are composed of numerous pellets of fissionable uranium fuel and can be several meters in length and about a centimeter in diameter. Then several of these rods, generally a dozen or more, are held together by strong metallic brackets in a fuel assembly. These rods are not bunched tightly together, rather there are several millimeters between each rod to allow coolant to flow between them. The tubes containing the pellets of uranium are generally composed of zirconium.

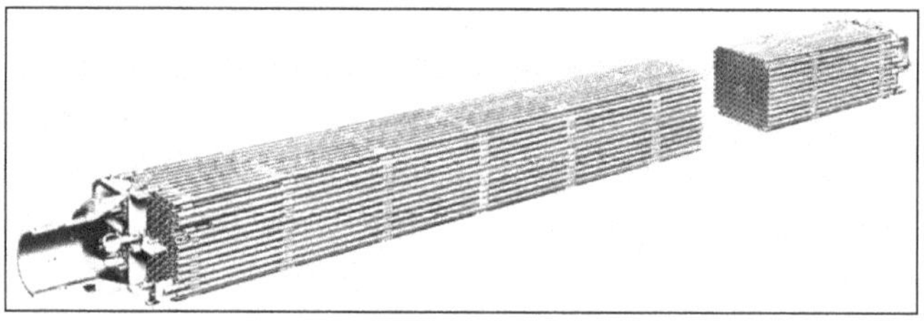

A nuclear fuel bundle.

Advantages and Disadvantages of Nuclear Fuels

Worldwide, there are extensive reserves of uranium left to mine. While nuclear fuel is not renewable, it is sustainable since there is so much of it. It will run out eventually, but not for centuries. Unlike fossil fuels, using nuclear fuels to produce energy does not directly produce carbon dioxide or sulfur dioxide. It should be mentioned that the processes of mining, transporting, and refining the fuel have carbon emissions associated with them, comparable to those of wind and solar power. Although the carbon footprint of using nuclear fuels is smaller, there are still disadvantages of using nuclear fuel. The waste, while a much lower volume must be handled very carefully because of its radioactivity. Nuclear fuels require far more complicated systems to extract their energy, which calls for greater regulation. These complex systems and regulation make for very long build times. In addition, public opinions on nuclear energy tend to be more negative than with other energy sources. The over-estimation of the dangers associated with releases of radioactive material is a significant issue, as large-scale nuclear incidents are rare.

Nuclear Reactor

Nuclear reactor refers to any of a class of devices that can initiate and control a self-sustaining series of nuclear fissions. Nuclear reactors are used as research tools, as systems for producing radioactive isotopes, and most prominently as energy sources for nuclear power plants.

Principles of Operation

Nuclear reactors operate on the principle of nuclear fission, the process in which a heavy atomic nucleus splits into two smaller fragments. The nuclear fragments are in very excited states and emit neutrons, other subatomic particles, and photons. The emitted neutrons may then cause new fissions, which in turn yield more neutrons, and so forth. Such a continuous self-sustaining series of fissions constitutes a fission chain reaction. A large amount of energy is released in this process, and this energy is the basis of nuclear power systems.

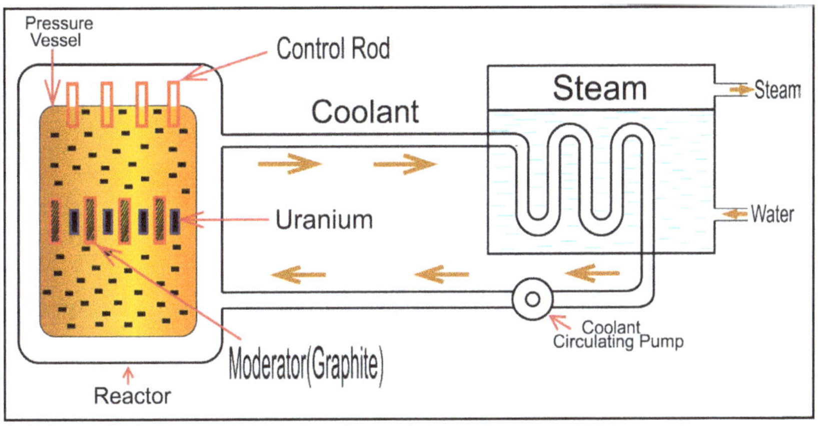

In an atomic bomb the chain reaction is designed to increase in intensity until much of the material has fissioned. This increase is very rapid and produces the extremely prompt, tremendously energetic explosions characteristic of such bombs. In a nuclear reactor the chain reaction is maintained at a controlled, nearly constant level. Nuclear reactors are so designed that they cannot explode like atomic bombs.

Most of the energy of fission—approximately 85 percent of it—is released within a very short time after the process has occurred. The remainder of the energy produced as a result of a fission event comes from the radioactive decay of fission products, which are fission fragments after they have emitted neutrons. Radioactive decay is the process by which an atom reaches a more stable state; the decay process continues even after fissioning has ceased, and its energy must be dealt with in any proper reactor design.

Chain Reaction and Criticality

The course of a chain reaction is determined by the probability that a neutron released in fission will cause a subsequent fission. If the neutron population in a reactor decreases over a given period of time, the rate of fission will decrease and ultimately drop to zero. In this case the reactor will be in what is known as a subcritical state. If over the course of time the neutron population is sustained at a constant rate, the fission rate will remain steady, and the reactor will be in what is called a critical state. Finally, if the neutron population increases over time, the fission rate and power will increase, and the reactor will be in a supercritical state.

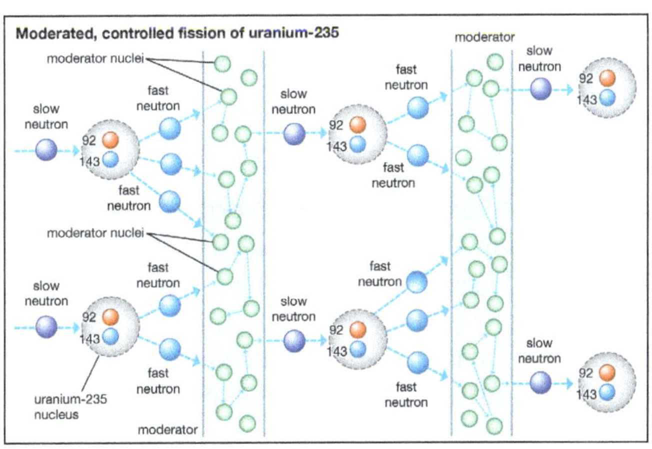

Chain reaction in a nuclear reactor at a critical state Slow neutrons strike nuclei of uranium-235, causing the nuclei to fission, or split, and release fast neutrons. The fast neutrons are absorbed or slowed by the nuclei of a graphite moderator, which allows just enough slow neutrons to continue the fission chain reaction at a constant rate.

Before a reactor is started up, the neutron population is near zero. During reactor start-up, operators remove control rods from the core in order to promote fissioning in the reactor core, effectively putting the reactor temporarily into a supercritical state. When the reactor approaches its nominal power level, the operators partially reinsert the control rods, balancing out the neutron population over time. At this point the reactor is maintained in a critical state, or what is known as steady-state operation. When a reactor is to be shut down, operators fully insert the control rods, inhibiting fission from occurring and forcing the reactor to go into a subcritical state.

Reactor Control

A commonly used parameter in the nuclear industry is reactivity, which is a measure of the state of a reactor in relation to where it would be if it were in a critical state. Reactivity is positive when a reactor is supercritical, zero at criticality, and negative when the reactor is subcritical. Reactivity may be controlled in various ways: by adding or removing fuel, by altering the ratio of neutrons that leak out of the system to those that are kept in the system, or by changing the amount of absorber that competes with the fuel for neutrons. In the latter method the neutron population in the reactor is controlled by varying the absorbers, which are commonly in the form of movable control rods (though in a less commonly used design, operators can change the concentration of absorber in the reactor coolant). Changes of neutron leakage, on the other hand, are often automatic. For example, an increase of power will cause a reactor's coolant to reduce in density and possibly boil. This decrease in coolant density will increase neutron leakage out of the system and thus reduce reactivity—a process known as negative-reactivity feedback. Neutron leakage and other mechanisms of negative-reactivity feedback are vital aspects of safe reactor design.

A typical fission interaction takes place on the order of one picosecond (10^{-12} second). This extremely fast rate does not allow enough time for a reactor operator to observe the system's state and respond appropriately. Fortunately, reactor control is aided by the presence of so-called delayed neutrons, which are neutrons emitted by fission products sometime after fission has occurred. The concentration of delayed neutrons at any one time (more commonly referred to as the effective delayed neutron fraction) is less than 1 percent of all neutrons in the reactor. However, even this small percentage is sufficient to facilitate the monitoring and control of changes in the system and to regulate an operating reactor safely.

Fissile and Fertile Materials

All heavy nuclides have the ability to fission when in an excited state, but only a few fission readily and consistently when struck by slow (low-energy) neutrons. Such species of atoms are called fissile. The most prominently utilized fissile nuclides in the nuclear industry are uranium-233 (^{233}U), uranium-235 (^{235}U), plutonium-239 (^{239}Pu), and plutonium-241 (^{241}Pu). Of these, only uranium-235 occurs in a usable amount in nature—though its presence in natural uranium is only some 0.7204 percent by weight, necessitating a lengthy and expensive enrichment process to generate a usable reactor fuel.

As an alternative to processing and enriching uranium-235, it is possible to go through the process of generating quantities of other fissile nuclides that are not as prevalent as uranium-235. Prominent sources of these nuclides are thorium-232 (^{232}Th), uranium-238 (^{238}U), and plutonium-240 (^{240}Pu), which are known as fertile materials owing to their ability to transform into fissile materials. For example, thorium-232, the predominant isotope of natural thorium, can be used to generate uranium-233 through a process known as neutron capture. When a nucleus of thorium-232 absorbs, or "captures," a neutron, it becomes thorium-233, whose half-life is approximately 21.83 minutes. After that time the nuclide decays through electron emission to protactinium-233, whose half-life is 26.967 days. The protactinium-233 nuclide in turn decays through electron emission to yield uranium-233.

Neutron capture may also be used to create quantities of plutonium-239 from uranium-238, the principal constituent of naturally occurring uranium. Absorption of a neutron in the uranium-238 nucleus yields uranium-239, which decays after 23.47 minutes through electron emission into neptunium-239 and ultimately, after 2.356 days, into plutonium-239.

If desired, plutonium-241 may be generated directly through neutron capture in plutonium-240, following the formula $^{240}Pu + {}^{1}n = {}^{241}Pu$.

A power reactor contains both fissile and fertile materials. The fertile materials partially replace fissile materials that are destroyed by fission, thus permitting the reactor to run longer before the amount of fissile material decreases to the point where criticality is no longer manageable. Plutonium-240 is particularly found to build up in reactors after long periods of operation, as it has a longer half-life than all its parent nuclides.

Heat Removal

A significant portion of the energy of fission is converted to heat the instant that the fission reaction breaks the initial target nucleus into fission fragments. The bulk of this energy is deposited in the fuel, and a coolant is required to remove the heat in order to maintain a balanced system (and also to transfer the heat energy to the power-generating plant). The most common coolant is water, though any fluid can be used. Heavy water (deuterium oxide), air, carbon dioxide, helium, liquid sodium, sodium-potassium alloy (called NaK), molten salts, and hydrocarbons have all been used in reactors or reactor experiments.

The Angra nuclear power plant, using pressurized-water reactors, at Angra dos Reis, near Rio de Janeiro, Brazil.

Some research reactors are operated at very low power and have no need for a dedicated cooling system; in such units the small amount of generated heat is removed by conduction and convection to the environment. Very high power reactors, on the other hand, must have extremely sophisticated cooling systems to remove heat quickly and reliably; otherwise, the heat will build up in the reactor fuel and melt it. Indeed, most reactors operate on the principle that their fuel cannot be allowed to melt; therefore, the systems designed to cool the fuel must operate sufficiently under both normal and abnormal conditions. Systems that enable sufficient cooling during all credible abnormal conditions in nuclear power plants are referred to as emergency core-cooling systems.

Shielding

An operating reactor is a powerful source of radiation, since fission and subsequent radioactive decay produce neutrons and gamma rays, both of which are highly penetrating radiations. A reactor must have specifically designed shielding around it to absorb and reflect this radiation in order to protect technicians and other reactor personnel from exposure. In a popular class of research reactors known as "swimming pools," this shielding is provided by placing the reactor in a large, deep pool of water. In other kinds of reactors, the shield consists of a thick concrete structure around the reactor system referred to as the biological shield. The shield also may contain heavy metals, such as lead or steel, for more effective absorption of gamma rays, and heavy aggregates may be used in the concrete itself for the same purpose.

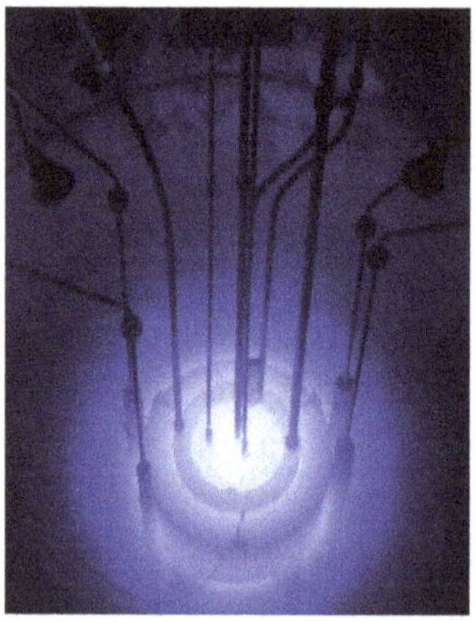

Cherenkov radiation emitted by the core of the Reed Research Reactor located at Reed College in Portland.

Critical Concentration and Size

Not every arrangement of material containing fissile fuel can be brought to criticality. Even if a reactor was designed such that no neutrons could leak out, a critical concentration of fissile material would have to be present in order to bring the reactor to a critical state. Otherwise, absorption of neutrons by other constituents of the reactor might dominate and inhibit a sustained chain fission reaction. Similarly, even where there is a high-enough concentration for criticality, the reactor must occupy an appropriate volume and be of a prescribed geometric form, or else more neutrons will leak out than are created through fission. This requirement imposes a limit on the minimum critical volume and critical mass within a reactor.

Although the only useful fissile material in nature, uranium-235, is found in natural uranium, there are only a few combinations and arrangements of this and other materials that enable a reactor to maintain a critical state for a period of time. To increase the range of feasible reactor designs, enriched uranium is often used. Most of today's power reactors employ enriched uranium fuel in which the percentage of uranium-235 has been increased to between 3 and 5 percent, approximately five and a half times the concentration in natural uranium. Large plants for enriching uranium exist in several countries; indeed, enrichment has become a commercial enterprise.

Thermal, Intermediate and Fast Reactors

Reactors are conveniently classified according to the typical energies of the neutrons that cause fission. Neutrons emanating in fission are very energetic; their average energy is around two million electron volts (MeV), nearly 80 million times the energy of atoms in ordinary matter at room temperature. As neutrons scatter or collide with nuclei in a reactor, they lose energy. This action is referred to as down-scattering. The choice of reactor materials and of fissile material concentrations determines the rate at which neutrons are slowed through down-scattering before causing fission.

In a thermal reactor, most neutrons down-scatter in the moderator material before interacting with a fissile material. Down-scattering events take place until the neutrons have reached thermal equilibrium with the reactor at energies of a few hundredths of an electron volt. Neutrons lose energy most efficiently by colliding with light atoms such as hydrogen (mass 1), deuterium (mass 2), beryllium (mass 9), and carbon (mass 12). For this reason, materials that contain atoms of these elements—water, heavy water, beryllium metal and oxide, and graphite—are deliberately incorporated into a thermal reactor and are known as moderators. Since water and heavy water also can function as coolants, they perform a dual purpose in thermal reactors.

One disadvantage of thermal reactors is that at low energies uranium-235 and plutonium-239 not only can be fissioned by thermal (or slow) neutrons but also can capture neutrons without undergoing fission. Neutron capture transforms these nuclides into, respectively, uranium-236 and plutonium-240, which are not fissile. The probability of neutron capture is much lower at higher energy levels than at thermal energies. To achieve higher energy levels and promote fission over neutron capture, a reactor can be built to operate without a moderator. Depending on the number of scattering events that take place with heavier atoms before fission occurs, the typical fission-causing neutrons may have energies in the range of 0.5 electron volt to thousands of electron volts (intermediate reactors) or several hundred thousand electron volts (fast reactors). Such reactors require higher concentrations of fissile material to reach criticality than do reactor designs that operate at thermal energy levels; however, they are more efficient at converting fertile material to fissile material. Fast reactors can be designed to produce more than one new fissile atom for each fissile atom destroyed. Such reactors are referred to as breeder reactors. Breeder reactors may become important if world demand for nuclear power turns out to be long-term and if access to naturally available sources of fissile material becomes limited.

Reactor Design and Components

There are a large number of ways in which a nuclear reactor may be designed and constructed; many types have been experimentally realized. Over the years, nuclear engineers have developed reactors with solid and liquid fuels, thick- and no-reflectors, forced cooling circuits and natural conduction or convection heat-removal systems, and so on. Most reactors, however, have certain basic components.

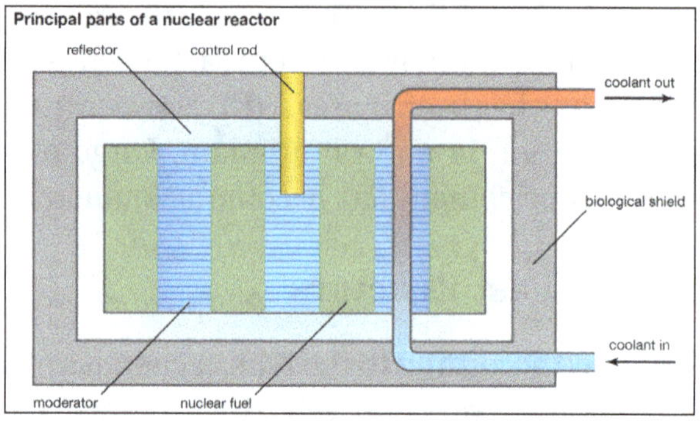

The basic components of a nuclear reactor.

Core

All reactors have a core, a central region that contains the fuel, fuel cladding, coolant, and (where separate from the latter) moderator. The fission energy in a nuclear reactor is produced in the core.

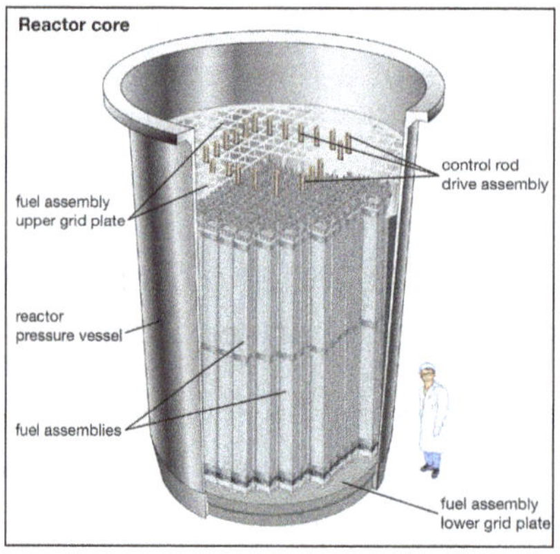

The core of a pressurized-water nuclear reactor.

The fuel is usually heterogeneous–i.e., it consists of elements containing fissile material along with a diluent. This diluting agent may be fertile material or simply material that has good mechanical and chemical properties and does not readily absorb neutrons. All diluents act as a matrix in which the fissile material can stably reside through its operable life. In solid fuels, the diluted fissile material is enclosed in a cladding–a substance that isolates the fuel from the coolant and minimizes the likelihood that radioactive fission products will be released. Cladding is often referred to as a reactor's first fission product barrier, as it is the first barrier that fissile material contacts after nuclear fission.

Fuel Types

A reactor's fuel must conform to the integral design of the reactor as well as the mechanisms that drive its operations. Following are brief descriptions of the fuel materials and configurations used in the most important types of nuclear reactors.

The light-water reactor (LWR), which is the most widely used variety for commercial power generation in the world, employs a fuel consisting of pellets of sintered uranium dioxide loaded into cladding tubes of zirconium alloy or some other advanced cladding material. The tubes, called pins or rods, measure approximately 1 cm (less than half an inch) in diameter and roughly 3 to 4 metres (10 to 13 feet) in length. The tubes are bundled together into a fuel assembly, with the pins arranged in a square lattice. The uranium used in the fuel is 3 to 5 percent enriched. Since light (ordinary) water, used in LWRs as both the coolant and the moderator, tends to absorb more neutrons than other moderators do, such enrichment is crucial.

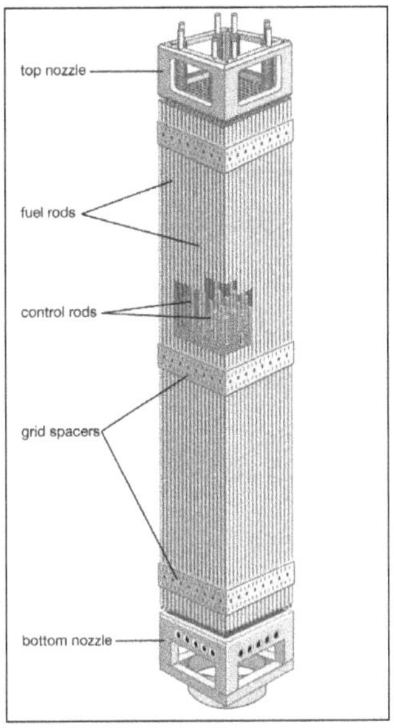

Nuclear fuel rods and control rods arranged by grid spacers into a fuel assembly for a pressurized-water reactor.

The CANDU (Canada Deuterium Uranium) reactor, which is the principal type of heavy-water reactor, uses natural uranium compacted into pellets. These pellets are inserted in long tubes and arranged in a lattice. A CANDU reactor fuel assembly measures approximately 1 metre (almost 40 inches) in length. Several assemblies are arranged end-to-end within a channel inside the reactor core. The use of heavy water rather than light water as the moderator enhances the scattering of neutrons rather than their capture, thereby increasing the probability of fission with the fuel material.

In one version of the high-temperature graphite reactor, the fuel is constructed of small spherical particles, or microspheres, containing uranium dioxide at the centre with concentric shells of carbon, silicon carbide, and carbon around them. These shells serve as localized cladding for each fuel sphere. The particles are then mixed with graphite and encased in a macroscopic graphite cladding.

In a sodium-cooled fast reactor, commonly called a liquid-metal reactor (LMR), the fuel consists of uranium dioxide or uranium-plutonium dioxide pellets (French design) or of uranium-plutonium-zirconium metal alloy pins (U.S. design) in steel cladding.

The most common type of fuel used in research reactors consists of plates of a uranium-aluminum alloy with an aluminum cladding. The uranium is enriched to slightly less than 20 percent, while silicon and aluminum are included in the "meat" of the plate to serve as the diluent and fuel matrix. Although aluminum has a lower melting point than other cladding materials, the flat plate design maintains a low fuel temperature, as the plates are often barely more than 1.25 mm thick. A common variety of research reactor known as TRIGA (from training, research, and isotope-production reactors–General Atomic) employs a fuel of mixed uranium and zirconium hydride, often doped with small concentrations of erbium and the whole clad in stainless steel.

Coolants and Moderators

A variety of substances, including light water, heavy water, air, carbon dioxide, helium, liquid sodium, liquid sodium-potassium alloy, and hydrocarbons (oils), have been used as coolants. Such substances are, in general, good conductors of heat, and they serve to carry the thermal energy produced by fission from the fuel and through the integral system, finally either venting the heat directly to the atmosphere (in the case of research reactors) or transporting it to the steam-generating equipment of the nuclear power plant (in the case of power reactors).

In many cases, the same substance functions as both coolant and moderator, as in the case of light and heavy water. The moderator slows the fast (high-energy) neutrons emitted during fission to energies at which they are more likely to induce fission. In doing so, the moderator helps initiate and sustain a fission chain reaction.

Reflectors

A reflector is a region of unfueled material surrounding the core. Its function is to scatter neutrons that leak from the core, thereby returning some of them back into the core. This design feature allows for a smaller core size. In addition, reflectors "smooth out" the power density by utilizing neutrons that would otherwise leak out through fissioning within fuel material located near the core's outer region.

The reflector is particularly important in research reactors, since it is the region in which much of the experimental apparatus is located. In some research reactor designs, reflectors are located inside the core as central islands in which high neutron intensities can be achieved for experimental purposes.

In most types of power reactors, a reflector is less important; this is due to the reactor's large size, which reduces the proportion of neutrons that may leak from the core region. The liquid-metal reactor represents a special case. Most sodium-cooled reactors are deliberately built to allow a large fraction of their neutrons—those not needed to maintain the chain reaction—to leak from the core. These neutrons are valuable because they can produce new fissile material if they are absorbed by fertile material. Thus, fertile material—generally depleted uranium or its dioxide—is placed around the core to catch the leaking neutrons. Such an absorbing reflector is referred to as a blanket or a breeding blanket.

Reactor Control Elements

All reactors need unique elements for control. Although control can be achieved by varying parameters within the coolant circuit or by varying the amount of absorber dissolved in the coolant or moderator, by far the most common method utilizes absorbing assemblies—namely, control rods or, in some cases, blades. Typically a reactor is equipped with three types of rods for different purposes: (1) safety rods for starting up and shutting down the reactor, (2) regulating rods for adjusting the reactor's power rate, and (3) shim rods for compensating for changes in reactivity as fuel is depleted by fission and neutron capture.

The most important function of the safety rods is to shut down the reactor, either when such a shutdown is scheduled or in case of a real or suspected emergency. These rods contain enough absorber to terminate a chain reaction under any conceivable condition. They are withdrawn before fuel is loaded and remain available in case a loading error requires their action. After the fuel is loaded, the rods are inserted, to be withdrawn again when the reactor is ready for operation. The mechanism by which they are moved is designed to be fail-safe in the sense that if there is a mechanical failure, the safety rods will fall by gravity or some other safe means into the core. In some cases, moreover, the safety rods have an automatic feature, such as a fuse, which releases them by virtue of physical effects independent of electronic signals.

Regulating rods are deliberately designed to affect reactivity only by a small degree. It is assumed that at some time the rods might be totally withdrawn by mistake, and the idea is to keep the added reactivity in such cases well within sensible limits. A well-designed regulating rod will add so little reactivity when it is removed that the delayed neutrons will continue to control the rate of power increase.

Shim rods are designed to compensate for the effects of burnup (i.e., energy production). Reactivity changes resulting from burnup can be large, but they occur slowly over periods of days to years, as compared with the seconds-to-minutes range over which safety actions and routine regulation take place. Therefore, shim rods may control a significant amount of reactivity, but they will work optimally only when constraints are imposed on their speed of movement. A common way in which shims are operated is by inserting or removing them as regulating rods reach the end of their most useful position range. When this happens, shim rods are moved so that the regulating rods can be reset.

The functions of shim and safety rods are sometimes combined in rods that have low rates of withdrawal but that can be rapidly inserted. This is usually done when the effect of burnup decreases reactivity. The rods are only partially inserted at the outset of operation. However, in the event that the system must be shut down as quickly as possible, the reactor operator may "scram" the reactor, fully dropping the control rods into the core and immediately sending the reactor into a subcritical state. The expression "scram" is said to stand for "safety control rod axe man," a reference to ad-hoc emergency preparations made for the earliest nuclear reactor.

The amount of shim control required can be reduced by the use of a burnable "poison." This is a neutron-absorbing material, such as boron or gadolinium, that burns off faster than the fissile material does over the lifetime of the core. At the beginning of operation, the inclusion of a burnable poison regulates the extra reactivity that has been built into the fuel to compensate for the amount of fuel consumed. At the end of an operating period, the absorbing material is often completely transformed through neutron capture.

Some boiling-water reactors utilize cruciform (T-shaped) control blades as the neutron-absorbing control mechanism. Because a number of these reactor vessels are designed with internal components above the core region, the control blades are inserted from below the core. Control blades operate on the same principle as control rods. However, since they are inserted upward into the core, they cannot use gravity to fall into place and put the reactor into a subcritical state in the event of a loss of power or some other abnormal condition. For this reason, control blades are connected to hydraulic drives that force compressed air into the mechanism upon initiation, injecting the control blades into the core.

Structural Components

The structural components of a reactor hold the system together and permit it to function as a useful energy source. The most important structural component in a nuclear power plant is usually the reactor vessel. In both the light-water reactor and the high-temperature gas-controlled reactor (HTGR), a reactor pressure vessel (RPV) is utilized so that the coolant is contained and operated under conditions appropriate for power generation—namely, elevated temperature and pressure. Within the reactor vessel are a number of structural elements: grids for holding the reactor core and solid reflectors, control-rod guide tubes, internal thermal hydraulic components (e.g., pumps or steam circulators) in some cases, instrument tubes, and components of safety systems.

Coolant System

The function of a power reactor installation is to extract as much heat of nuclear fission as possible and convert it to useful power, generally electricity. The coolant system plays a pivotal role in performing this function. A coolant fluid enters the core at low temperature and exits at a higher temperature after collecting the fission energy. This higher-temperature fluid is then directed to conventional thermodynamic components where the heat is converted into electric power. In most light-water, heavy-water, and gas-cooled power reactors, the coolant is maintained at high pressure. Sodium and organic coolants operate at atmospheric pressure.

Research reactors have very simple heat-removal systems, as their primary purpose is to perform research and not generate power. In research reactors, coolant is run through the reactor, and the heat that is removed is transferred to ambient air or to water without going through a power cycle. In research reactors of the lowest power, running at only a few kilowatts, this may involve simple heat exchange to tap water or to a pool of water cooled by ambient air. During operation at higher power levels, the heat is usually removed by means of a small natural-draft cooling tower.

Containment System

Reactors are designed with the expectation that they will operate safely without releasing radioactivity to their surroundings. It is, however, recognized that accidents can occur. An approach using multiple fission product barriers has been adopted to deal with such accidents. These barriers are, successively, the fuel cladding, the reactor vessel, and the shielding. As a final barrier, the reactor is housed in a containment structure, often simply referred to as the containment.

Installation of the dome of a containment structure at the Taishan nuclear power plant.

Containment Design Principles

The containment basically consists of the reactor building, which is designed and tested to prevent elevated levels of radioactivity that might be released from the fuel cladding, the reactor vessel, and the shielding from escaping to the environment. To meet this purpose, the containment structure must be at least nominally airtight. In practice, it must be able to maintain its integrity under circumstances of a drastic nature, such as accidents in which most of the contents of the reactor core are released to the building. It has to withstand pressure buildups and damage from debris propelled by an energy burst within the reactor, and it must pass appropriate tests to demonstrate that it will not leak more than a small fraction of its contents over a period of several days, even when its internal pressure is well above that of the surrounding air. The containment building also must protect components located inside it from external forces such as tsunamis, tornadoes, and airplane crashes.

The most common form of containment building is a cylindrical structure with a spherical dome, which is characteristic of LWR systems. This structure is much more typical of nuclear plants than the large cooling tower that is often used as a symbol for nuclear power. It should be noted that cooling towers are found at large modern coal- and oil-fired power stations as well.

Reactors other than those of the LWR type also have containment structures, though they vary in shape and construction. When it can be justified that major pressure buildups are not to be expected, the containment may be any functional form of airtight structure. In the United States and a number of other countries around the world, containment structures are required for all commercial power reactors and all high-power research reactors. In general, low-power research reactors are exempt, on the basis of the common assumption that an accident in such systems will not lead to a widespread release of radioactivity. In the United States, reactors operated by the Department of Energy and by the armed services are also exempt, a matter that has caused considerable controversy. Some of these have containment structures, whereas others do not.

Containment Systems and Major Nuclear Accidents

The concept of the containment originated in the United States during the 1950s and was generally accepted throughout much of the world. The Soviet-bloc countries, however, did not concur with this view, and when containment was added to Soviet reactor designs, it was generally not up to Western standards. For example, during the Chernobyl accident of 1986 in Ukraine, the power station's Unit 4, which suffered a catastrophic explosive accident and fire, had an internal structure that could withstand the loss of function of only a single pressure tube. Though the structure was called a containment, this was a misnomer by Western standards, and the structure would more suitably be referred to as a confinement.

Severe tests of Western-style containment systems occurred during the Three Mile Island accident in the United States in 1979 and the Fukushima accident in Japan in 2011. At Three Mile Island Unit 2, near Harrisburg, Pennsylvania, a stoppage of core cooling resulted in the destruction, including partial melting, of the entire core and the release of a large part of its radioactivity to the enclosure around the reactor—that is, the containment. In spite of a hydrogen deflagration that also occurred during the accident, the containment structure prevented all but a very small amount of radioactivity from entering the environment and is credited with having prevented a major radioactive release and its consequences.

At the Fukushima Daiichi ("Number One") plant in northeastern Honshu, Japan, a loss of main and backup power after an earthquake and tsunami led to a partial meltdown of fuel rods in three reactors. Melted material bored small holes in the lower head of two reactor pressure vessels; one of these was punctured again by an explosion. Radioactive water was soon discovered to have leaked into the ocean through cracks in the foundation of the containment. Within a few weeks, the cracks had been sealed, and within six months the reactors had been stabilized to the point where workers could begin to enter the containment. Despite a catastrophic sequence of natural events that led to the accident, the fission product barriers served their design purpose and kept all but a small amount of fission products from entering the environment.

Types of Reactors

Most of the world's existing reactors are power reactors, providing the heat needed to turn turbines that run electric-power generators. There are also numerous research reactors, and some navies of the world have submarines or surface ships driven by propulsion reactors. There are several types of power reactors, but only one, the light-water reactor, is widely used. Accordingly, this variety is discussed in considerable detail here. Other significant types are briefly described, as are research and propulsion reactors. Some attention is also given to the prospective uses of reactors for space travel and for certain industrial purposes.

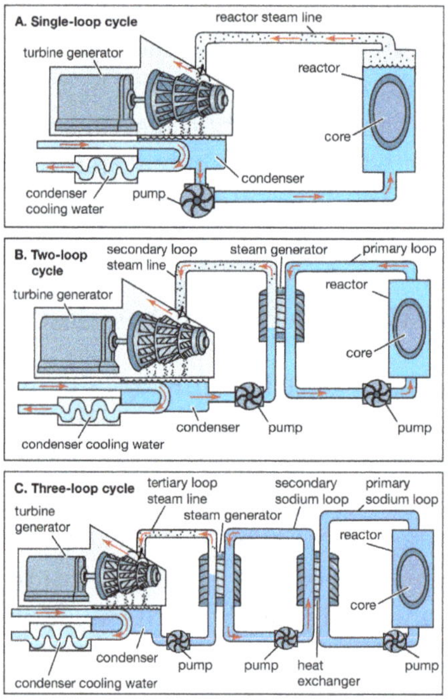

Basic power cycles in nuclear power plants (A) Single-loop cycle; as shown, it represents a boiling-water reactor (BWR), but it could also represent a direct-cycle, high-temperature gas-cooled reactor (HTGR) if helium were substituted for steam. (B) Two-loop cycle; the primary loop depicted here could constitute a pressurized-water reactor (PWR), a CANDU pressurized heavy-water reactor (PHWR), or a helium HTGR. (C) Three-loop cycle; this is found only in sodium-cooled reactors where an intermediate loop of nonradioactive sodium is provided between the radioactive primary loop and the steam generator.

Power Reactors

Light-water Reactors

Light-water reactors (LWRs) are power reactors that are cooled and moderated with ordinary water. There are two basic types: the pressurized-water reactor (PWR) and the boiling-water reactor (BWR). In the PWR, water at high pressure and temperature removes heat from the core and is transported to a steam generator. There the heat from the primary loop is transferred to a lower-pressure secondary loop also containing water. The water in the secondary loop enters the steam generator at a pressure and temperature slightly below that required to initiate boiling. Upon absorbing heat from the primary loop, however, it becomes saturated and ultimately slightly superheated. The steam thus generated ultimately serves as the working fluid in a steam-turbine cycle.

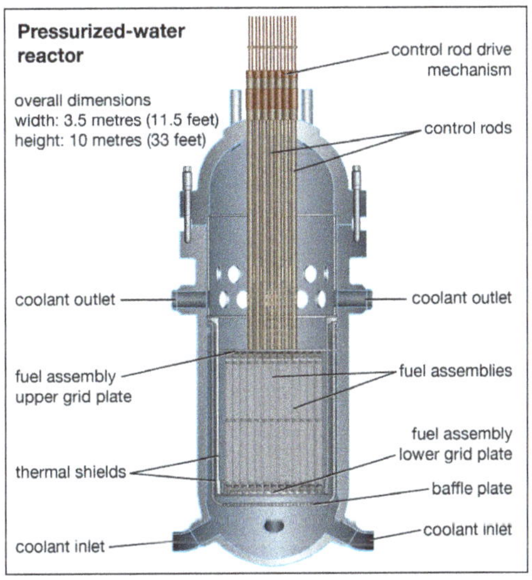

Section of a pressurized-water reactor, showing inlets and outlets for water coolant passing through the core.

A BWR operates on the principle of a direct power cycle. Water passing through the core is allowed to boil at an intermediate pressure level. The saturated steam that exits the core region is transported through a series of separators and dryers located within the reactor vessel that promote a superheated state. The superheated water vapour is then used as the working fluid to turn the steam turbine.

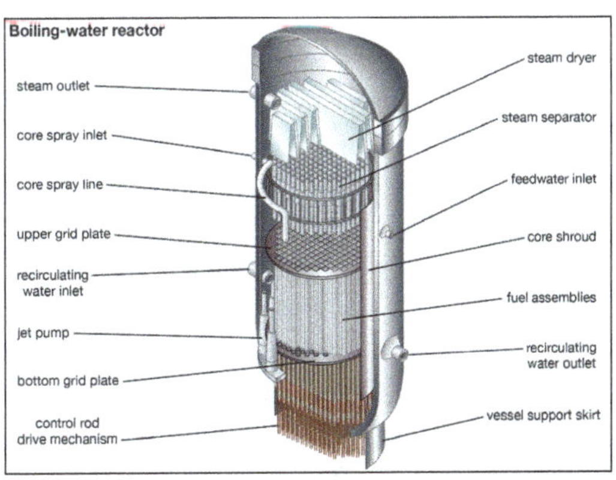

Cross section of a boiling-water reactor, showing the core, the steam separator, and the steam dryer.

Advantages and Disadvantages

Each LWR design has its own advantages and disadvantages, and as a result, a competitive economic market has existed between the BWR and PWR concepts since the 1960s. For instance, although there are fewer mechanical components in the steam cycle of a BWR design, additional components are required to support the reactor's emergency core-cooling system. Furthermore, the BWR vessel's internal system is more complex, since it includes internal recirculation pumps and complex steam separation and drying equipment that are not found in a PWR design. On the other hand, even though the internals of the PWR are simpler, a BWR power plant is smaller, because it has no steam generators. In fact, the steam generators of a PWR—there are typically four of them in a big plant—are larger than the reactor vessel itself.

The direct-cycle philosophy of a BWR design reduces heat loss between the core and the steam turbine, but the BWR operates at lower pressures and temperatures than the PWR, giving it less thermodynamic efficiency. Furthermore, because the BWR's power density is somewhat lower than that of the PWR, the pressure vessel must be built to a larger diameter for the same reactor power. On the other hand, because the BWR operates at lower pressure, its pressure vessel is thinner than the pressure vessel of a PWR.

Fueling and Refueling LWRs

Both standard light-water designs are fueled with uranium dioxide pellets in zirconium alloy cladding. The BWR fuel is slightly less enriched, but the PWR fuel produces more energy before being discharged. The control rods of a typical PWR are inserted from the top (through the reactor head), whereas those of a BWR are inserted from the bottom.

Light-water reactors are refueled by removing the reactor head—after lowering and unlatching the safety rods in the case of a PWR. This exposes the reactor to visual observation. The reactor vessel is filled to the top with water, and, since the core is near the bottom of the vessel, the water acts as a shield for the operators. Each fuel assembly to be removed is grasped by a fuel-handling tool and then lifted from its position in the core into a shielded cask, within which the assembly is transferred to a storage pool for cooling while it is still highly radioactive. This process is repeated for each element that requires removal.

In most LWRs a typical refueling cycle removes approximately one-third of the fuel assemblies. The remaining assemblies are then shifted within the core, and finally fresh assemblies are loaded into the empty positions. The purpose of shifting fuel at the time of reload is to achieve an optimal reactivity and power distribution for the next cycle of operation. Reloading is a time-consuming operation. In principle it could be accomplished in two weeks, but in practice nuclear power plants undergo maintenance during reload, which often takes considerably more time—up to several months. Utilities schedule maintenance and reload during the spring and fall, when electricity demand is lowest and the electrical grid usually has reserve capacity.

The removed fuel stored in the storage pool not only is highly radioactive but also continues to produce energy (referred to as decay heat). This energy is removed by natural circulation of the water in the storage pool. During the 1960s, when the nuclear industry was in its early stage, it was expected that spent fuel could be shipped out for reprocessing within two years. However, this option is currently practiced only in a few countries around the world—notably France, the United Kingdom, and Japan, where large-scale facilities employ a well-developed reprocessing technique known as PUREX. Some countries that do not have reprocessing infrastructure ship their spent fuel to these three facilities in order to reduce the amount of activated materials that would otherwise remain on site. In the United States, which does not reprocess, storage pools have continued to receive spent fuel, and some of the pools are filling up. Options available to nuclear plant operators are to store the spent fuel more densely than originally planned, to build new pools, or to store the oldest in aboveground silos (dry storage) locally on site. Ultimately, this fuel will be transferred to the U.S. Department of Energy for reprocessing or waste disposal or both, though a viable disposal program has not yet been established.

Global Status of LWRs

During the 1970s, light-water reactors represented the cheapest source of new electricity in most parts of the world, and it still is economical in many countries such as Japan, South Korea, Taiwan, France, and China (which, in the 1990s, embarked on an ambitious program of building nuclear power plants, almost all of them using light-water technology). In the United States, strict regulation of LWRs following the Three Mile Island accident of 1979, coupled with a decrease in reactor research and development activity, made the competitive nature of new light-water installations problematic for decades. In the early 21st century, signs began to emerge of a reviving nuclear power sector in the United States as demand for reliable energy sources continued to increase and it became evident that the fleet of U.S. nuclear power plants was aging.

Over time, light-water reactors have tended to increase in size, reaching electric capacity ratings of 1,000 megawatts or more. More recent designs—small modular reactors that produce 300 megawatts or less—may be able to provide energy on a more economic scale and thus appeal to a much broader market.

High-temperature Gas-cooled Reactors

The high-temperature gas-cooled reactor (HTGR), as mentioned above in Fuel types, is fueled by a mixture of graphite and fuel-bearing microspheres. There are two competitive designs of this reactor type: (1) a German "pebble bed" system that uses spherical fuel elements, nominally 60 mm (2.5 inches) in diameter, containing a graphite-and-fuel mixture coated in a graphite shell; and (2) an American version in which the fuel is loaded into precisely located graphite hexagonal prisms that interlock to create the core of the vessel. In both variants, the coolant consists of helium pressurized to approximately 100 bars, or roughly 100 standard atmospheres. In the German system, the helium passes through interstices in the bed of the spherical fuel elements, whereas in the American system, it passes through holes in the graphite prisms that align along the axis of the core region in the reactor vessel. Both are capable of operating at very high temperature, since graphite has an extremely high sublimation temperature and helium is completely inert chemically. The hot helium can be used directly as the working fluid in a high-temperature gas turbine, or its heat can be utilized to generate steam for a water cycle.

Experimental prototypes of both the American and German designs were built and operated from the 1960s to the 1980s, but by the early 1990s, they had resulted in no orders for commercial plants. In 2000 the High Temperature Reactor-10 (HTR-10), a Chinese reactor based on the

German pebble-bed design, began operating at a nominal steady-state power level of 10 megawatts. Intended to test the viability of a full-sized HTGR power plant, the HTR-10 provided power and heating for the campus of Tsinghua University near Beijing.

Liquid-metal Reactors

Sodium-cooled fast-neutron-spectrum liquid-metal reactors (LMRs) received much attention during the 1960s and '70s when it appeared that their breeding capabilities would soon be needed to supply fissile material to a rapidly expanding nuclear industry. When it became clear in the 1980s that this was not a realistic expectation, enthusiasm waned. The developmental work of the previous decades, however, resulted in the construction of a number of LMRs around the world—in the United States, Russia, France, Britain, Japan, and Germany. Most LMRs are fueled with uranium dioxide or mixed uranium-plutonium dioxides. In the United States, however, the greatest success has been with metal fuels.

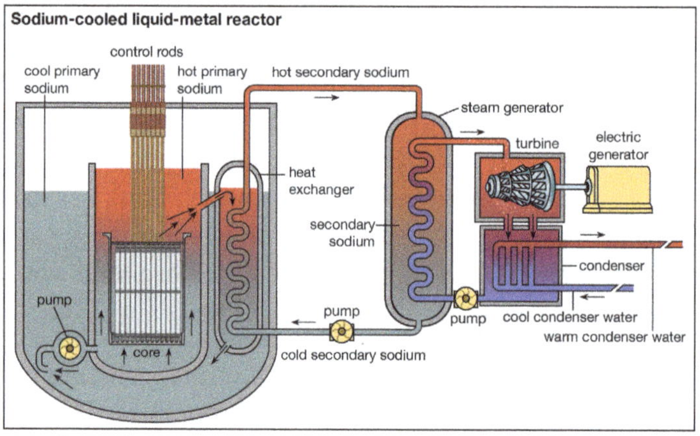

Schematic diagram of a nuclear power plant using a pool-type sodium-cooled liquid-metal reactor.

While some LMRs are of the loop type, equipped with heat exchangers and pumps outside the primary reactor vessel, others are of the pool variety, featuring a large volume of primary sodium in a pool that also contains the primary pumps and the primary-to-secondary heat exchanger. In all types the heat extracted from the core by primary sodium is transferred to a secondary, nonradioactive sodium loop, which serves as the heat source for a steam generator that heats the water in a tertiary loop to power a turbine. The pool type seems to have some safety advantage in that the large volume of primary sodium heats up only slowly even if no power is extracted; thus, the reactor is effectively isolated from upsets in the balance of the plant. The reactor core in all such systems is a tightly packed bundle of fuel in steel cladding through which the sodium coolant flows to extract the heat. Most LMRs are breeders or are capable of breeding, which is to say that they all produce more fissile material than they consume.

CANDU Reactors

Canada has focused its developmental efforts on reactors that utilize abundant domestic natural uranium as fuel without having to resort to enrichment services that would be supplied only by other countries. The result of this policy is the Canada Deuterium Uranium (CANDU) reactor—a line of natural uranium-fueled reactors moderated and cooled by heavy water. A reactor of this kind consists of a tank, or calandria vessel, containing a cold heavy water moderator at atmospheric pressure. The calandria is pierced by pressure tubes made of zirconium alloy in which the natural uranium fuel is placed and through which heavy water coolant is circulated. In contrast to the more common LWR reactor vessel, which is aligned in a vertical configuration, a CANDU reactor vessel is aligned horizontally. Electric-power generation is obtained through the transfer of the heat of fission into the heavy-water coolant, which is circulated to a steam generator. The light-water steam exiting the steam generator in the secondary loop is then transported through a conventional turbine cycle.

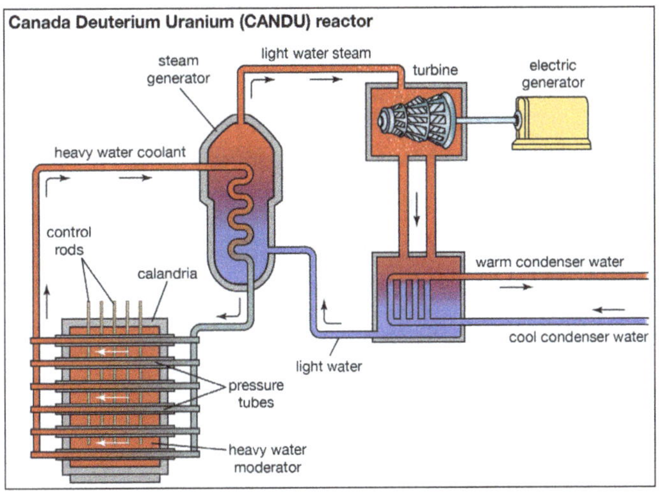

Schematic diagram of a nuclear power plant using a Canada Deuterium Uranium (CANDU) reactor.

The fuel assembly of a CANDU reactor, which consists of a bundle of short zirconium alloy-clad tubes containing natural uranium dioxide pellets, can be changed while the system is running. A new assembly is simply pushed into one end of a pressure tube, and the spent fuel that it replaces is collected as it is extruded at the opposite end. This feature has given the CANDU higher capacity factors than other reactor types. Capacity factor refers to the ratio of time that a reactor is operating at full power during a given period versus the total available time during that same period. Thus, a capacity factor of 1.0 refers to a reactor that operates at full power 24 hours a day over the entire period of time being considered. Several countries have purchased CANDU reactors for the same reason that they were developed by Canada—to be independent of foreign enrichment services.

Advanced Gas-cooled Reactor

The advanced gas-cooled reactor (AGR) was developed in the United Kingdom as the successor to reactors of the Calder Hall class, which combined plutonium production and power generation. Calder Hall, the first nuclear station to feed an appreciable amount of power into a civilian network opened in 1956. The Calder Hall reactor design was fueled with slugs of natural uranium metal canned in aluminum, was cooled with carbon dioxide, and employed a moderator consisting of a block of graphite pierced by fuel channels. In the AGR, fuel pins clad in Zircaloy (a trademark for alloys of zirconium having low percentages of chromium, nickel, iron, and tin) and loaded with approximately 2 percent enriched uranium dioxide are placed into zirconium-alloy channels that pierce a graphite moderator block. This design utilizes fast neutron energies and is therefore referred to as a fast reactor. The enriched fuel permits operation to economic levels of fuel burnup. A coolant of carbon dioxide absorbs and transports heat to a steam generator, where the heat is conducted to the secondary loop and ultimately to a traditional steam-turbine cycle.

Dungeness B, a nuclear power plant using an advanced gas-cooled reactor.

Although a number of AGRs have been built in the United Kingdom, maintenance and malfunction issues have proved to be more costly than expected, and no new AGRs are projected for construction.

Other Power Reactor Types

A large variety of reactor types have been built and operated on an experimental basis. A few examples include organic liquid-cooled and liquid-moderated reactors that can operate like a pressurized-water reactor but without requiring pressures in the primary circuit to be as high as those in the traditional PWR; sodium-cooled graphite-moderated reactors; and heavy-water reactors built in a pressure-vessel design. Each design has its own advantages and disadvantages.

Research Reactors

Water-cooled and Plate-fuel Reactors

These are the most common type of research reactor. Water-cooled, plate-fuel reactors use enriched uranium fuel in plate assemblies and are cooled and moderated with water. They operate over a wide range of thermal power levels, from a few kilowatts to hundreds of megawatts. As the primary mission of research reactors is not electricity production, they are characterized not by the amount of electric power that they produce but rather by thermal power, neutron density, and nominal neutron energy within the core region. It is these parameters that help quantify a research reactor's ability to perform specific research. The systems with the lowest licensed power levels are usually operated at universities and are used primarily for teaching, whereas those with the highest are used by research laboratories chiefly for materials testing and characterization as well as for general research.

A common form of the water-cooled, plate-fuel reactor is the pool reactor, in which the reactor core is positioned near the bottom of a large, deep pool of water. This has the advantage of simplifying both observation and the placement of channels, commonly referred to as beam ports, from which beams of neutrons can be directed and transported. At lower thermal power levels, no pumping is required, as the natural convection of the coolant past the fuel plates provides sufficient heat removal to maintain a safe operating state. A heat exchanger is usually located at or near the top of the pool, where the hottest water is stratified. At higher operating power levels, pumping becomes necessary to augment the natural circulation.

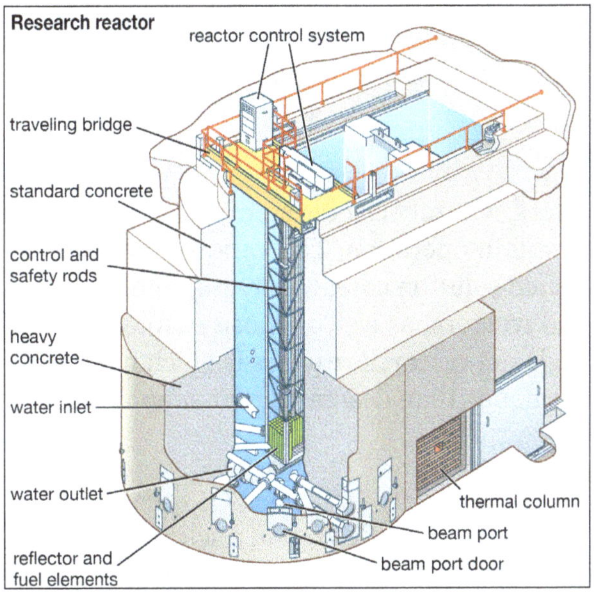

A water-cooled research reactor, with graphite blocks serving as an internal reflector to moderate the reaction.

Most pool reactors use the water of the pool as a reflector, but some have blocks of a solid moderator (canned graphite or beryllium metal) around the core that serves as an inner reflector. The inclusion of graphite and beryllium provides a means for reflecting neutrons back into the core (or a useful location adjacent to the core) that would otherwise leak out of the core. Following interactions between the reflector components, neutrons tend to down-scatter quickly, producing relatively large local regions of thermal neutrons. These regions are often exploited for experimental use by transporting the neutrons down the beam ports and to the experimenter as required.

At higher power levels, it becomes more convenient to employ a tank-type reactor, because it is simpler to control the flow path of pumped water in such a system. Low-power educational reactors also are available in the tank form. The core and reflector arrangement and the position of these components within the tank are similar in both tank-type and pool-type systems. However, solid concrete shielding is increased in robustness around the sides of tank-type reactors, whereas pool-type reactors rely primarily on water as a biological shield.

TRIGA Reactors

The training, research, and isotope-production reactors–general atomic (TRIGA) system is a popular variety of research reactor. It is another tank-type water-cooled system, but its fuel differs from that employed by the plate-fuel research reactors. The fuel element of the TRIGA reactor consists of stainless steel- or aluminum-clad rods containing mixed uranium and zirconium hydrides that are often doped with small concentrations of erbium. In contrast to thin plate-type fuel, TRIGA fuel elements are nominally 3.8 cm (1.5 inches) in diameter and in general approximately 67 cm

(26 inches) in total length. A unique characteristic of this fuel is that it exhibits an extremely large negative power-reactivity coefficient—so large that the TRIGA reactor can be placed in an extremely supercritical state for an instant, causing its power to rise very rapidly, after which it quickly shuts itself down on the basis of the fuel's inherent material composition and characteristics. The resulting power transient is referred to as a pulse and is useful for a number of dynamic experiments that require large bursts of neutrons over a short period of time. The total energy released in a pulse does not draw concern toward the safety of the reactor, since the automatic shutdown occurs very quickly and the energy release is proportional to both peak power and pulse duration.

Other Research Reactors

As in the case of power reactors, a number of different reactor types have seen service as research reactors, and a number are still in operation. The variety is so great as to defy cataloging. There have been homogeneous (fueled solution cores), fast, graphite-moderated, heavy-water-moderated, and beryllium-moderated reactors, as well as those adapted to use fuels left over from power reactor experiments. The design of research reactors is much more fluid and sensitive to a greater variety of unique research demands than designs for other applications are.

Ship-propulsion Reactors

The original, and still the major, naval application of nuclear reactors is the propulsion of submarines. The chief advantage of using nuclear reactors for submarine propulsion is that they, unlike fossil-fuel combustion systems, require no air for power generation. Consequently, a nuclear-powered submarine can remain underwater for prolonged periods, whereas a conventional diesel-electric submarine must surface periodically to run its engines in air. Nuclear power confers a strategic advantage on naval surface vessels as well, because it eliminates their dependence on refueling from vulnerable tankers or in foreign ports.

The nuclear-powered submarine USS Nautilus entering.

The design of military nuclear power plants is classified for defense security purposes, and so only general information pertaining to them has been published. It is known that U.S. naval power plants are fueled with highly enriched uranium and moderated and cooled with light water. The design of the first nuclear submarine reactor, that of the USS Nautilus, was heavily influenced by high-power research reactor design. Unique features include the incorporation of a very large reactivity margin to accommodate long burnups without refueling and to permit restart after shutdown. For submarine use, the power plant also must be extremely quiet to avoid sonic detection.

Various reactor core models have been developed to address the specific requirements of different classes of submarines.

The nuclear power plants for U.S. aircraft carriers are believed to have been derived from the power plant designs for the largest submarines, but again the particulars of their design have not been published.

USS Enterprise, commissioned in 1961, the first nuclear-powered aircraft carrier of the U.S. Navy.

Besides the United States, nuclear submarines are deployed by the United Kingdom, France, Russia, China, and India. In each case, the design was developed in secret, but it is generally believed that they are all rather similar; the demands of the application usually lead to similar solutions. Russia also has a small fleet of nuclear-powered icebreakers, whose reactors are thought to be essentially the same as those of the earliest Soviet submarines. As with naval vessels, the ability to operate without refueling is an enormous advantage for Arctic icebreakers.

The Soviet/Russian nuclear-powered icebreaker Lenin.

Prototypes of nuclear-powered commercial cargo ships were built and operated by a handful of countries in the latter half of the 20th century, but they were soon decommissioned. These vessels did not operate very economically, and opposition to their docking in a number of major ports was also a factor in their decommissioning. The prototypes were powered by reactors of the pressurized-water type.

Production Reactors

The very first nuclear reactors were built for the express purpose of manufacturing plutonium for nuclear weapons, and the euphemism of calling them production reactors has persisted to this day. At present, most of the material produced by such systems is tritium (^3H, or T), the fuel for hydrogen bombs. Plutonium has a long half-life of approximately 24,100 years (specific to plutonium-239), so countries with arsenals of nuclear weapons using plutonium as fissile material generally have more than they expect to need. In contrast, tritium has a half-life of approximately 12 years; thus, stocks of this radioactive hydrogen isotope have to be continuously produced to maintain the required stockpiles. The United States, for example, operates several reactors moderated and cooled by heavy water that produce tritium at the Savannah River facility in South Carolina.

The plutonium isotope that is most desirable for sophisticated nuclear weapons is plutonium-239. If plutonium-239 is left in a reactor for a long time after production, plutonium-240 builds up as an undesirable contaminant. Accordingly, a significant feature of a production reactor is its capability for quick throughput of fuel at a low energy-production level. Any reactor that can be operated this way is a potential production reactor.

The world's first plutonium production reactors, built by the United States in Hanford, Washington, during World War II, were fueled with natural uranium, moderated by graphite, and cooled by light water. It is believed that the early Soviet production reactors were the same sort, and the French and British versions differed only in that they were cooled with gas. The first significant power reactor, the Calder Hall reactor in Cumbria, northwestern England, was actually a dual-purpose production reactor.

Specialized Reactors

Nuclear reactors have been developed to provide electric power and steam heat in far-removed isolated areas. Russia, for instance, operates smaller power reactors specially designed to supply both electricity and steam for heating to accommodate the needs of a number of remote Arctic communities, and in China, the 10-megawatt HTR-10 reactor supplies both heat and electricity for the research institution at which it is located. Independent developmental work on small automatically operated reactors with similar capabilities has been undertaken by Sweden and Canada. Between 1960 and 1972, the U.S. Army used small pressurized-water reactors to provide power for remote bases in Greenland and Antarctica. They were replaced with oil-fired power plants, but it is still technically feasible to employ nuclear power for such applications, as nuclear reactors require less fuel maintenance than a traditional fossil fuel plant and, in general, run at a consistently high capacity. Small modular reactors being designed in the United States might offer unique capabilities such as remote operations.

Reactors have been developed to supply power and propulsion in space. Between 1967 and 1988, the Soviet Union deployed small intermediate reactors in Earth-orbiting satellites (mostly in the Cosmos series) for powering equipment and telemetry, but this policy became a target for criticism. At least one of the Soviet Union's reactor-powered spacecraft reentered the atmosphere and distributed radioactive contamination in remote areas of Canada. The United States launched only one reactor-powered satellite, in 1965, but developmental activity continues for such possible deep-space missions as manned exploration of other planets or the establishment of a permanent lunar base. Reactors for these applications would necessarily be high-temperature systems based on either the HTGR or the LMR design but would use enriched fuel to last the entire life of a prolonged space mission. A power cycle in space must be run at a very high temperature to minimize the size of the radiator from which heat is to be rejected. In addition, a reactor for space applications has to be compact so that it can be shielded with a minimum amount of material and reduce the weight during launch and space flight.

Reactor Safety

Nuclear reactors contain very large amounts of radioactive isotopes—mostly fission products but also such heavy elements as plutonium. If this radioactivity were to escape the reactor, its impact on the people in the vicinity would be severe. The deleterious effects of exposure to high levels of ionizing radiation include increased probability of cancer, cellular damage, an increased number of developmental abnormalities in children exposed in the womb, and even death within a period of several days to months when irradiation is extreme. For this reason, the primary consideration in the design of a reactor is ensuring that a significant release of radioactivity does not occur as a result of any plausible accident scenario. This is accomplished by a combination of preventive measures and mitigating measures. Preventive measures are those that are taken to avoid accidents, and mitigating measures are those that decrease the adverse consequences. In spite of the most stringent preventive and mitigating measures, however, it is still possible that accidents will reach an emergency scale, and in these cases, the nuclear industry and regulators have prepared a set of emergency responses.

Preventive Measures

Design and Operating Standards

Essentially, preventive measures are the set of design and operating rules that are intended to make certain that a reactor is operated safely. The nuclear industry in the United States created a design philosophy referred to as “defense in depth” that numerous other countries have also adopted. In a nuclear power plant following the defense-in-depth model, all safety systems are required to be functionally independent, inherently redundant, and diverse in design.

Among the most well-known preventive measures are the reports and inspections for double-checking that a plant is properly constructed, rules of operation, and qualification tests for operating personnel to ensure that they know their jobs. Nuclear reactors must operate under a very high standard of quality assurance, requiring staff members to audit, evaluate, survey, and verify that all procedures and maintenance are being performed as they should be.

An important part of a safety system is strict adherence to design requirements: the reactor must have a negative power-reactivity coefficient; the safety rods must be injectable under all circumstances; and no single regulating rod should be able to add substantial reactivity rapidly. Another important design requirement is that the structural materials used in the reactor must retain acceptable physical properties over their expected service life. Finally, construction is to be covered by stringent quality-assurance rules, and both design and construction must be in accordance with standards set by major engineering societies and accepted by regulatory bodies.

Since no human activity can be shown to be absolutely safe, all these measures cannot reduce the risks to zero. However, it is the aim of the regulations and safety systems to minimize risks to the point where a reasonable individual would conclude that they are trivial. What this de minimis risk value is, and whether it has been achieved by the nuclear industry, is a subject of bitter controversy, but it is generally accepted that independent regulatory agencies—the United States' Nuclear Regulatory Commission (NRC), the United Kingdom's Office for Nuclear Regulation (ONR), the International Atomic Energy Agency (IAEA), and similar agencies around the world—are the proper judges of such matters.

Windscale Accident

Prior to the development of current preventive-design philosophies, the world's first large-scale nuclear reactor accident took place in October 1957 at Windscale, Cumberland (now part of Cumbria), northwestern England. The Windscale plant was powered by a pair of identical reactors, known as Piles 1 and 2, that were an air-cooled, graphite-moderated design. Initially constructed between 1946 and 1950 to produce plutonium for nuclear weapons, they also provided energy for electricity production. However, with Britain fully engaged in the nuclear arms race, the reactors' operations were subject to direct political influence, and Windscale engineers were driven to modify the fuel design in order to increase plutonium and tritium production. These modifications increased heat generation in the fuel and therefore temperature levels within the reactor cores.

After approximately seven years of successful operation, the increased production rate became too much for Pile 1 to handle. Possibly as a result of imperfections in one of the fuel elements, the core began to overheat, and fire broke out within the lithium-magnesium-clad uranium fuel slug. In an attempt to remove heat from the core, operators turned on all the reactor's air-cooling fans. Unfortunately, this only fanned the flames and spread the fire through the core region. In addition, a significant amount of radioactive contamination was released through the ventilation stacks when the fans were turned on. Operators ultimately put out the fire by turning the fans off and forcing water through the core.

The Windscale event caused much less damage than the Three Mile Island, Chernobyl, and Fukushima accidents of later years. Nevertheless, it provided an explicit demonstration that political agendas (in this case, participation in the nuclear arms race) must be separated from those of energy and safety.

Reactor Safety Study

In 1972, as part of an effort to evaluate the risks from nuclear power plants, the U.S. Atomic Energy Commission (a predecessor of the NRC) authorized a major safety study. Conducted with major assistance from a number of laboratories, the AEC's study involved the application of probabilistic risk assessment (PRA) techniques for the first time on a system as complex as a large nuclear power reactor. Also for the first time, the study compared the risk of a nuclear power plant accident with other events such as natural disasters and human-caused events. This work resulted in the publication in 1975 of a report titled Reactor Safety Study, also known as WASH-1400. The most useful aspect of the study was its delineation of components and accident sequences (scenarios) that were determined to be the most significant contributors to severe accidents.

The Reactor Safety Study concluded that the risks of an accident that would injure a large number of people were extremely low for the light-water reactor (LWR) systems being analyzed. This conclusion, however, was subject to very large quantitative uncertainties and was challenged. One basic problem with PRA techniques is that it cannot easily be confirmed by experience when the level of risk has been reduced to low values. That is to say, if PRA predicts that a reactor is subject to, say, one failure in 10,000 years, there is no way to prove that statement with only a few, or even with 10,000, years of experience. Thus, the results of the Reactor Safety Study as to risk levels were not confirmable. Nevertheless, updated versions of the report still provide the framework and reference for nuclear-related probability risk assessment.

Three Mile Island and Chernobyl

The principles established by the Reactor Safety Guide were given an unexpected test in 1979 when Three Mile Island Unit 2 near Harrisburg, Pennsylvania, suffered a severe accident. Through the failure of an important valve to operate correctly, cooling water to the core was lost, parts of the core were melted and the rest of it destroyed, and a large quantity of fission products was released from the primary reactor system to the interior of the containment structure. The equipment failure was exacerbated by reactor operator error, as the emergency core cooling system was deactivated by operator action because of a misinterpretation of the type of accident that was occurring. Fortunately, the containment vessel of the reactor building fulfilled its function, and only a small amount of radioactivity was released, demonstrating the wisdom of incorporating this structure into a rigorous design. All the same, a severe accident had occurred.

Aerial view of Three Mile Island near Harrisburg, Pennsylvania.

Many investigations of the Three Mile Island accident followed. Recommendations differed among them, but a common thread was that the human element was a much more important factor and higher risk to the safe operations of a nuclear power plant than had been recognized. The human element pertained not only to the operating staff but also to the management of nuclear plants and even the NRC itself. As a result, following the accident many changes in operator training and in technical and inspectorate staffing were implemented, just as a number of hardware enhancements were introduced. It is generally believed that these changes have been effective in reducing the likelihood of the occurrence of accidents as severe as that at Three Mile Island. As a side issue to this, however, the operating costs of nuclear power plants have escalated sharply as more and more highly trained people have been added to the operating staffs.

The significance of the human element, particularly as it relates to plant management and rigorous high-level regulatory decision making, was borne out again by the Chernobyl disaster of 1986. One of the four reactors in a nuclear power station approximately 100 km (60 miles) north of Kiev, Ukraine (then part of the Soviet Union), exploded and caught fire as the result of an ill-conceived experiment (a test to quantify how long the steam turbines would run while coasting to a stop if the reactor was abruptly shut down). Before the event was brought under control, an estimated 25 percent of the radioactive contents of the reactor had been released in a high cloud plume. Approximately 116,000 people had to be evacuated, and a large area surrounding the plant received fallout so great that it could not be farmed or pastured as a result of this accident. Significant levels of radiation were detected as far north as Scandinavia and as far west as Switzerland.

In September 2005 the Chernobyl Forum, comprising seven United Nations organizations and programs, the World Bank, and the governments of Belarus, Russia, and Ukraine, published a three-volume, 600-page report assessing the impact of the accident on public health. Approximately 50 emergency workers had died of acute radiation sickness shortly after the accident, and 9 children had died from thyroid cancer because of radiation exposure. From among the 200,000 emergency workers who were present at the site in the first year following the accident, the people who were evacuated, and the 270,000 residents of the most heavily contaminated areas, an additional 3,940 people were likely to die from cancer during a prolonged period after the accident.

Investigation of the Chernobyl accident placed the largest blame, as with the Three Mile Island mishap, on poor management both at the plant and within the government bureaucracy. Because these accidents primarily resulted from human failings rather than from some intrinsic factor, most experts have continued to believe that nuclear energy can be a safe source of power. There is, however, a condition on the conclusion that nuclear power is by and large a safe form of power. The facilities for generating this power must be designed, built, and operated to high standards by knowledgeable, well-trained professionals, and a regulatory mechanism capable of enforcing these standards must be in place.

Mitigating Measures

Systems and Structures

Mitigating measures, also referred to as safety systems, are systems and structures that prevent accidents from proceeding to a catastrophic outcome in the event they do occur. Two of the principal mitigating measures are: (1) the safety rod systems that quickly put the reactor into a subcritical state and prevent a supercritical accident and (2) the containment structure that prevents radioactive materials from being released into the atmosphere. Other significant mitigating measures include the emergency core-cooling system, whose purpose is to provide sufficient cooling of the core and fuel region within the vessel upon a loss of reactor coolant, and the emergency power system, which is designed to supply electrical power to support systems in the event that the normal supply is disrupted. Emergency power systems are necessary so that detectors, circulating pumps, valves, and other critical components continue to operate as necessary to remove decay heat. An extreme mitigating measure is the evacuation of personnel who might otherwise be heavily exposed in a reactor installation.

According to PRA studies, three categories of events are primarily responsible for the risks associated with LWRs—namely, station blackout, so-called transient without scram, and loss of cooling. In station blackout, a failure in the power line to which the station is connected is postulated. The proposed emergency defense is a secondary electrical system, typically a combination of diesel generators big enough to drive the pumps and a battery supply sufficient to run the instruments. In transient without scram, the assumed event is an insertion of positive reactivity—for example, through an undesired withdrawal of the shim rods. The protective safety system response in this case is the rapid and automatic insertion of the safety rods. In loss of coolant, the event is assumed to be caused by a mechanical failure of the normal cooling system such that a certain amount of the coolant is lost. The emergency response is activation of an emergency core-cooling system. In all such measures, proper operator action and proper functioning of the appropriate backup system are paramount aspects of emergency response.

Other reactor designs pose different types of risk. For example, neither the pool-type liquid-metal reactor (LMR) nor the high-temperature gas-cooled reactor (HTGR) is at major risk with regard to loss of coolant flow and perhaps not with regard to station blackout. However, the LMR and perhaps the HTGR are at some risk from events that might cause air or water to enter the coolant system. The hazard is that reactor materials, sodium or graphite, could chemically react with air and water, causing what is known as an exothermic reaction that releases large amounts of heat in addition to the decay heat already existing within the core region. The hazard is greater with sodium in the LMR than it is with graphite in the HTGR.

Fukushima

A failure of the main power line and a loss of backup power were at the heart of the second worst nuclear accident in the history of nuclear power generation (after Chernobyl)—a partial meltdown in 2011 at the Fukushima Daiichi ("Number One") plant in Japan. That facility, located on Japan's Pacific coast in northeastern Fukushima prefecture, was made up of six boiling-water reactors (BWRs) constructed between 1971 and 1979, three of which were operational and one of which was under maintenance, its fuel having been stored out of the core in the reactor's spent fuel storage pool. A powerful earthquake shook all units at the plant, initiating an automatic shutdown, or scram. Immediately after the earthquake, all safety systems in each unit were operable, though a few were slightly damaged. However, less than one hour after the earthquake, a tsunami struck the shoreline where the reactor units were built. The tsunami reached heights much greater than the reactors were designed to withstand, and ultimately it cut off the main power supply to the facility and damaged the backup generators by flooding their housing structures. Although the reactors withstood both an earthquake and a tsunami beyond their design requirement, the prolonged power outage drained backup batteries incorporated into the emergency core-cooling system, which led to a loss of capability to remove decay heat. Despite the best efforts of the reactor operators and emergency responders, rising temperatures within each reactor's core eventually caused a partial meltdown of the fuel rods, a fire in the storage reactor, explosions in the outer containment buildings (caused by a buildup of hydrogen gas), the release of radioactive steam into the air, and the leakage of radioactive water into the ocean. As workers struggled to cool and stabilize the three cores by pumping seawater and boric acid into them, government officials established a 30-km (18-mile) evacuation zone around the plant. Approximately one month after the initiating event, the reactor cores were stabilized, cracks in the foundations of the containment vessels were sealed, and irradiated cooling water began to be pumped to a storage building until it could be properly treated.

Two of the damaged containment buildings at the Fukushima Daiichi nuclear power plant, northeastern Fukushima prefecture, earthquake and tsunami that crippled the installation.

The Fukushima accident made it all too clear that another type of risk can arise from external events: earthquakes and tsunamis may not be two separate events but rather be two successive events in which an earthquake will cause structural damage to a reactor and will also initiate a tsunami. The risk associated with an earthquake of plausible magnitude is minimized by building plants away from faults and by making use of earthquake-resistant mechanical design and construction features. Furthermore, the addition of dikes and water barriers reduces the risk of damage by a tsunami. Added construction features such as water barriers must be able to withstand both an earthquake and a tsunami, as these are likely to be coupled events.

In contrast to the Three Mile Island and Chernobyl accidents, which were largely blamed on staffing issues, the "weak link" in the Fukushima accident seemed upon immediate observation to be the physical plant itself rather than human error. However, because the plants were not designed to handle the natural disaster that took place, fault can be found with the design process, in a sense pointing out human error once again as the most failure-prone component in the nuclear industry.

Emergency Response

Each regulating body that oversees the operations of a country's nuclear power has its own methods for identifying and responding to emergency conditions. In the United States, the NRC has an emergency classification system that identifies four levels of severity in conditions at a nuclear power plant:

1. Notification of unusual events: Potential degradation in the level of safety of the plant, but no release of radioactive material requiring off-site response or monitoring.
2. Alert: Actual or potential substantial degradation in the level of safety of the plant, with a release of radioactive material from the plant expected.

3. Site area emergency: Actual or likely major failures of plant functions needed for protection of the public, with radioactivity levels potentially above acceptable thresholds at the boundary of the power plant.

4. General emergency: Actual or imminent substantial core damage or melting of reactor fuel with the potential for loss of containment integrity; radioactive material is released and may be above acceptable thresholds beyond the boundary of the power plant.

A man is checked for radiation exposure after having been evacuated from the quarantine area around a nuclear power station in Fukushima prefecture, Japan, that was damaged in the March 11, 2011, earthquake and tsunami.

On a worldwide scale, the IAEA has developed the International Nuclear and Radiological Event Scale (INES), to be applied to any event occurring in the agency's signatory states that is associated with nuclear facilities and with the transport or storage of nuclear materials and radiation sources. The INES offers a common event scale for all parties that interact with nuclear power or radiological sources in any part of the world. The scale includes seven independent event levels; the lower three are referred to as "incidents" and the upper four as "accidents." A declaration of a specific level is determined by identifying specific criteria that have an impact on defense-in-depth of the nuclear power plant, radiological barriers and controls, and people and the environment. The seven levels and some of the important criteria are as follows:

1. Anomaly: Minor problems with safety components, with significant defense-in-depth remaining.

2. Incident: Significant contamination within the facility into an area not expected by design, with exposure of a worker in excess of the statutory annual limits.

3. Serious incident: Severe contamination in an area not expected by design, with a nonlethal health effect such as a burn on a worker from radiation.

4. Accident with local consequences: Fuel melt or damage to fuel resulting in more than 0.1 percent release of core inventory; release of significant quantities of radioactive material within an installation, with a high probability of significant public exposure and at least one death from radiation.

5. Accident with wider consequences: Severe damage to reactor core; release of large quantities of radioactive material within an installation, with a high probability of significant public exposure and several deaths from radiation.

6. Serious accident: Significant release of radioactive material likely to require implementation of planned countermeasures.
7. Major accident: Major release of radioactive material with widespread health and environmental effects requiring implementation of planned and extended countermeasures.

Under the INES, Three Mile Island is classified as Level 5, an accident with wider consequences, whereas both Fukushima and Chernobyl are Level 7, major accidents.

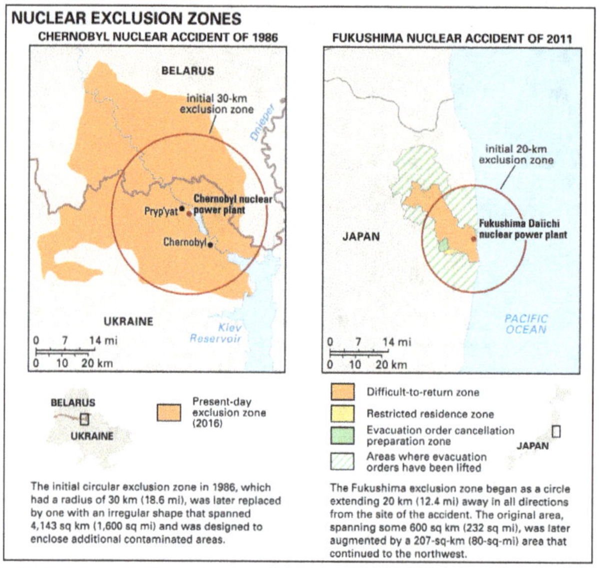

Exclusion zone: Chernobyl disaster; Fukushima accident Map of the exclusion zones following the nuclear accidents at Chernobyl, Soviet Union (now in Ukraine), and Fukushima, Japan.

Nuclear Fuel Cycle

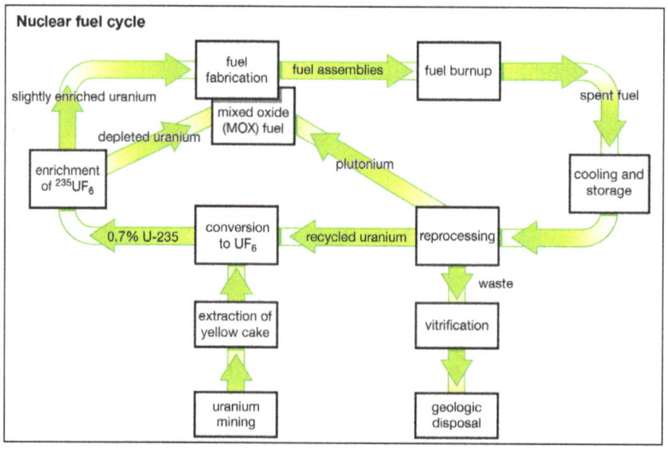

A "closed loop" nuclear fuel cycle, showing the reprocessing of uranium-235 and plutonium from spent fuel for use in new fuel assemblies.

No discussion of nuclear power is complete without a brief exposition of the nuclear fuel cycle. The whole point of a reactor is, after all, to initiate and control the process of fission on a very large scale in nuclear fuel, and the low cost of fueling is the chief reason for the economic competitiveness of nuclear power. The principal steps of the fuel cycle include uranium mining and extraction from its ore (processing), uranium enrichment, fuel fabrication, loading and irradiation in the reactor (fuel management), unloading and cooling, reprocessing, waste packaging, and waste disposal.

The nuclear fuel cycle also is an integral step in the production of plutonium for nuclear weapons, and the technologies of enrichment and reprocessing in particular have been key factors in the proliferation of these weapons around the world. For this reason and also for a host of other political, environmental, and economic reasons, the various steps in the nuclear fuel cycle are closely regulated and frequently observed under terms of international treaties. Conflicts between some countries' nuclear ambitions and various international conventions have sometimes generated great controversy.

Uranium Mining and Processing

Uranium is extracted from ores whose uranium content is often less than 0.1 percent (one part per thousand). Most ore deposits occur at or near the surface; whether they are mined through open-pit or underground techniques depends on the depth of the deposit and its slope. The mined ore is crushed and the uranium chemically extracted from it at the mouth of the mine. The residue remains naturally radioactive, as it contains long-lived radioactive daughter nuclei of uranium and has to be carefully managed to minimize the release of radioactive contaminants into the environment. The uranium concentrate, which is known as yellow cake, consists of uranium compounds (typically 75 to 95 percent). It is shipped to a chemical plant for further purification and chemical conversion.

Enrichment

Several enrichment techniques have been developed, though only two of these methods are used on a large scale; these are gaseous diffusion and gas centrifuging. In gaseous diffusion, natural uranium in the form of uranium hexafluoride gas (UF_6), a product of chemical conversion, is encouraged (through a mechanical process) to seep through a porous barrier. The molecules of $^{235}UF_6$ penetrate the barrier slightly faster than those of $^{238}UF_6$. Since the percentage of ^{235}U increases by only a very small amount after traversal of the barrier, the process must be repeated over and over in thousands of stages to obtain the necessary enrichment for commercial nuclear power use.

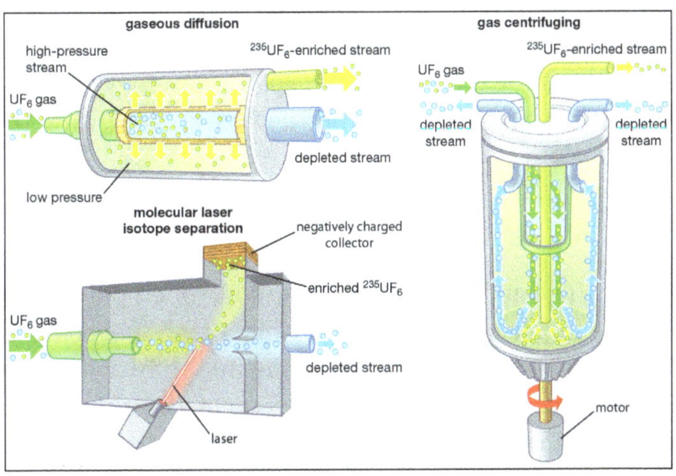

Three uranium-enrichment processes.

In gas centrifuging, the UF_6 gas is fed into a high-speed centrifuge. The centrifuge is balanced very well at the top bottom and spins at an extremely high rate. Because of the relative centripetal forces that each atom experiences, the lighter species of this mixture of gaseous molecules, including ^{235}U, tend to concentrate near the centre of the spinning centrifuge, while the heavier ones accumulate along the wall. These mixtures are then siphoned off. The degree of enrichment per stage in a centrifuge is greater than that obtained in a gaseous diffusion chamber, and the process uses less energy than gaseous diffusion does, but centrifuges are more expensive pieces of equipment.

An experimental enrichment method with much commercial potential is laser separation. This process is based on the principle that isotopes of different molecular weight absorb light of different frequencies. Once a specific isotope has absorbed radiation and has reached an excited state, its properties may become quite different from the other isotopes; it is then separated on the basis of this difference. In one method known generically as MLIS (molecular laser isotope separation)—or commercially as SILEX (separation of isotopes by laser excitation)—gaseous UF_6 is exposed to high-powered lasers tuned to the correct frequencies to cause the molecules containing ^{235}U (but not ^{238}U) to lose electrons. In this (ionized) form, the ^{235}U-containing molecules are separated from the stream on the basis of their different electric charge. Proponents of laser separation claim that the method consumes less energy and wastes less starting material than, for example, gaseous diffusion.

Fabrication

This step involves the conversion of the suitably enriched product material to the chemical form desired for reactor fuel. The only fuel fabricated on a large scale is for light-water reactors (LWRs).

The chemical form prepared for the LWR is uranium dioxide. Produced in the form of a ceramic powder, this compound is ground to a very fine flourlike consistency and inserted into a die, where it is pressed into a pellet shape—in the case of some LWR fuels, approximately 6 mm in diameter and 10 mm in length (that is, about 0.25 × 0.4 inch). Next the pellet is sintered in a furnace at 1,500–1,800 °C (approximately 2,700–3,300 °F). This sintering, similar to the firing of other ceramic ware, produces a dense ceramic pellet. The pellets are loaded into prefabricated zirconium alloy cladding tubes, which are then filled with an inert gas and welded shut. Once the zirconium alloy tubes have been sealed, they go through significant testing to verify that there are no leaks. These tubes, called rods or pins, are then bundled together with proper spacing ensured by top and bottom manifolds through which the ends of the pins pass as well as spacer grids distributed along the middle portion of the pins. Together with other necessary hardware, the bundle constitutes a fuel assembly.

Fuel Management

Fuel is loaded into a reactor in a very specific and well-controlled pattern so as to obtain the most energy production before the material becomes unusable. Fresh fuel is more reactive than old fuel. Typically, a reactor is fueled in cycles, each cycle lasting one to two years, and a fuel batch is kept in the reactor for three or four cycles. At the end of each cycle, the oldest fuel is removed—normally this consists of about one-third the fuel content in the core—and fresh fuel loaded. The partially burned fuel that remains, however, is shuffled before the fresh fuel is installed. The objective of this procedure is to achieve a fuel assembly arrangement of maximum reactivity while keeping the power distribution among the different fuel assemblies as even as possible and within technical specifications.

Fuel burnup—that is, energy production—is limited by two factors. After significant burnup has occurred, the physical properties of the fuel become degraded, and it is not prudent to continue to keep it in the reactor. Also, after some burnup, the old fuel no longer contributes useful reactivity to the reactor. The fuel design, including its initial enrichment, is such that these two limits are made to coincide approximately.

Unloading and Cooling

Spent reactor fuel is extremely radioactive, and its radioactivity also makes it a source of heat. When the spent fuel is removed from the reactor, it must continue to be both shielded and cooled. This is accomplished by placing the spent fuel in a water-storage pool, or spent-fuel cooling pool, located next to the reactor. The water in the pool contains a large amount of dissolved boric acid, which is a strong absorber of neutrons; this ensures that the fuel assemblies in the pool do not go critical. Pool water is also a common source of emergency cooling water for the reactor.

Spent nuclear fuel rods stored underwater at the Bruce Power site near Tiverton, Ontario, Canada.

Pools vary in size; older spent-fuel cooling pools are able to accommodate only about 10 years' worth of spent fuel. As the pools fill up, more spent fuel storage is needed. Additional storage space can be gained by loading spent fuel into the pool more densely than originally planned, by building a new pool, or by removing the oldest fuel assemblies from the existing pool and storing them in air-cooled concrete and steel silos—called spent-fuel storage casks—located aboveground. This last method becomes feasible after fuel has been stored in cooling pools for two or three years, because radioactivity and the rate of heat generation decrease rapidly over this period. Dense storage in existing pools and casks tends to be less expensive and more economical for utilities than building new pools.

Reprocessing

Both the converted plutonium and residual uranium-235 in spent fuel can be recycled by chemically reprocessing the fuel and extracting the specific elements of interest. Reprocessing not only provides a means to recycle nuclear fuel, but it also can reduce the volume and radioactivity of the waste material that must ultimately be eliminated by some method of permanent disposal.

One motivation for reprocessing is ultimately to provide a "closed-loop" fuel cycle within the nuclear industry. Closed-loop refers to recycling with 100-percent efficiency of all materials that are fabricated for use as nuclear fuel (including the most commonly used fuel, uranium dioxide pellets). Though the goal of 100-percent efficiency has yet to be attained by any country's nuclear industry, a closed-loop fuel cycle is not an unrealistic ambition, based on current progress in reprocessing technology. Many benefits would result from fuel recycling, including lower cost for fuel (once the recycling infrastructure was in place) and reduced quantities of spent fuel to be stored on reactor sites around the world.

Reprocessing Methods

The most common method for reprocessing, known as the PUREX (for plutonium-uranium extraction) process, begins with dissolving the spent fuel in nitric acid and contacting the acid solution with oil in which tributyl phosphate (TBP) has been dissolved. TBP is a complexing agent for uranium and plutonium, forming compounds with them that bring them into the oil solution. A physical separation of the (immiscible) oil and acid serves to remove the desired products from the nitric acid solution (which still contains all the fission products). The uranium and plutonium are then washed out of the TBP back into a water solution and separated from each other by various means to the degree desired. Thus, reprocessing produces three product streams: (1) a purified uranium product, (2) a plutonium product that may be either pure or mixed with uranium, and (3) a waste stream of fission products dissolved in nitric acid.

Reprocessing Policies

During the period of ambitious nuclear power plant construction in the United States in the 1950s and '60s, it was generally assumed that after two to five years, spent fuel would be delivered to a reprocessing plant. Some commercial reprocessing plants were built or planned, but by the mid-1970s the cost of reprocessing had escalated to a point where its economics became questionable. Also, in 1977 Pres. Jimmy Carter, in order to take a public, symbolic stand against nuclear proliferation, declared that the federal government would permanently defer all permits for the commercial reprocessing and recycling of plutonium. Carter's directive was rescinded by his successor, Ronald Reagan, and it has not been reinstated by any subsequent president. Even so, reprocessing is still not done commercially in the United States, partly because of the huge costs of building a reprocessing plant in a period when the supply of uranium ore has been sufficient to satisfy demand relatively cheaply.

Policy and institutional arrangements have been different in France and the United Kingdom, where commercial plants reprocess spent fuel not only from nuclear plants in the host countries but also from plants in other countries. The reprocessed plutonium can be used not only as fuel for proposed future liquid-metal reactors (LMRs) but also to help fuel existing LWRs. In the latter application, the plutonium is utilized in mixed oxide (MOX) form—a combination of uranium and plutonium dioxides having 3 to 6 percent plutonium.

A number of other countries, including Russia, India, Japan, and China, reprocess their spent fuel or plan to do so. The proactive reprocessing efforts of these countries have reduced the waste scheduled for long-term disposal to amounts well below those that are accumulating in countries that do not reprocess.

Waste Disposal

In the absence of reprocessing, spent fuel is considered to be waste and must be prepared for permanent disposal in a separate facility. In addition, the waste stream from spent-fuel reprocessing must also be disposed of. Many nuclear countries, from the United States to China to Finland, have researched the technologies and geologic locations for disposal sites, but no permanent disposal site is in use anywhere in the world. Pending approval and construction of disposal sites, all spent fuel and processed waste are being kept either in cooling pools or in aboveground storage casks.

Waste Conditioning

Spent fuel must be sealed in containers that are expected to remain viable in stable (and presumably underground) disposal sites over centuries and even millennia. Though no permanent disposal site is currently operational, the preparing, or conditioning, of spent fuel for disposal is expected to follow the same basic process. After storage aboveground for one to five years, the fuel pins are to be removed from their assemblies. End manifolds and components within the fuel assembly that contain no fuel are to be removed and the pins repacked into a dense lattice emplaced in a corrosion-resistant stainless-steel canister. A cover is to be welded on and the canister covered with an overpack.

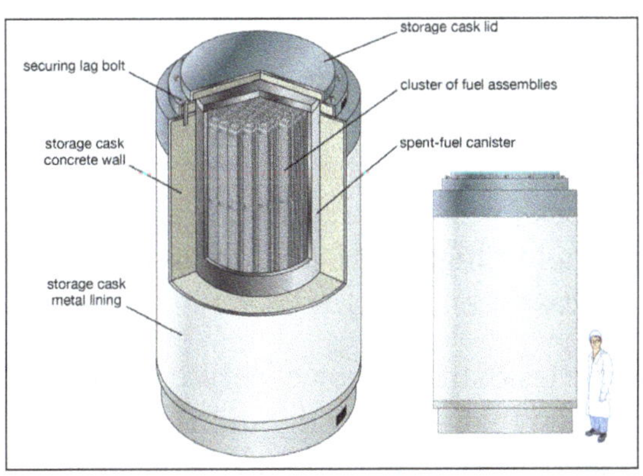

Cutaway diagram of a dry storage cask for spent nuclear fuel, showing fuel assemblies packed into a metal canister that is encased in a concrete cask.

Some waste is generated every year in the form of the fission-product solution that arises from reprocessing. Reprocessed fuel from nuclear-weapons production reactors also generates this type of waste. One glassmaking process for conditioning this waste is operational on an industrial scale in France, the United Kingdom, and Japan and has been tested in many other countries. The waste solution is completely evaporated, leaving behind the fission products in the solid residue, which is heated until all the constituent nitrate salts have been converted to oxides. These oxides are then put into a glass-forming oven and mixed with materials that will produce a borosilicate glass (similar to the commercial glass known as Pyrex). The fission-product oxides dissolve in the glass as it forms. The glass melt is subsequently poured into a steel canister, 200–400 mm (8–16 inches) in diameter and approximately 1 metre (40 inches) high, where it solidifies. Once covered with an overpack of bentonite clay (for shielding, molecular diffusion, and chemical isolation), the solid canister-like block is ready for disposal.

Geologic Disposal

The waste-disposal method currently being planned by all countries with nuclear power plants is called geologic disposal. This means that all conditioned nuclear wastes are to be deposited in mined cavities deep underground. Shafts are to be sunk into a solid rock stratum, with tunnel corridors extending horizontally from the central shaft region and tunnel "rooms" laterally from the corridors. The waste would be emplaced (by remotely controlled or robotic devices) in holes drilled into the floors of these rooms, after which the boreholes would be sealed and the rooms and corridors backfilled. When the entire operation has been completed (perhaps after approximately 30 years of operation), the shafts too would be backfilled and sealed.

Risks of Nuclear Waste Disposal

When a holistic view is taken of the nuclear waste-disposal process, the risks seem extremely small, yet among the general public, these risks are one of the most-feared aspects of the nuclear fuel cycle. Nuclear waste retains its very intense level of radioactivity for several hundred years, but after a thousand years have passed, the remaining radioactivity, while persistent, is at a level comparable to (though still greater than) that of an equivalent quantity of natural uranium ore. This separates the safety problem into two time periods: a first millennium during which it is crucial to ensure tight retention of the wastes in the repository, and a subsequent period during which it is necessary only to ensure that any release that occurs is small and slow.

One possible route for the emergence of radioactive waste to the surface would be the impingement of groundwater into the underground disposal site, followed by corrosion of the waste canisters, dissolving of waste material, and discharge of the resulting solution to the environment. However, water migrates slowly in most rock formations, and, contrary to the popular belief that any dissolution of waste would quickly lead to high-level contamination, only low levels of contamination are projected in even the worst-case scenarios.

The migration of radioactive species that has been observed at shallow burial sites for low-level radioactive waste is not an indication that similar migration can be expected for high-level waste located in a repository deep underground. In addition to the near insolubility of the waste material, waste form engineering, particularly of corrosion-resistant containers, provides extra protection against such dispersal. Moreover, most of the dispersal problem in shallow disposal sites is caused by biochemical products that would not exist in deep formations.

Finally, a great deal of care is to be expended in selecting the site of the repository. Various conditions are mandatory: the repository must not be near a populated area; the rock stratum selected must be deep (300 metres [1,000 feet] or more) and, as much as possible, naturally sealed from aquifers; and any discharge of the water table into the surface waters should be slow. Furthermore, the site must be in a tectonically inactive zone so that earthquakes will not break the seal.

The risk of high-level waste burial is almost certainly smaller than the risks of reactor accidents and even smaller than the risks arising from improperly managed mine tailings. Nonetheless, the siting of a repository must be handled with political sensitivity, and the confirmation of acceptable hydrologic and geologic conditions must have a high degree of validity. There are many acceptable sites in principle, but confirming acceptability for any one of them is a large and expensive technical undertaking. Both politically and technically, site selection is probably the biggest problem in the nuclear fuel cycle, causing the postponement, revision, and even termination of many siting efforts around the world.

Nuclear Waste

The term nuclear waste is often used to refer to spent nuclear fuel. This term is general, as there are many different types of nuclear waste, but refers to any radioactive waste substance that is produced from industrial processes, including nuclear power plants. Nuclear waste is also produced at hospitals and other medical facilities.

This spent nuclear fuel is placed in a cold pool to cool.

Spent nuclear fuel is specifically high level waste, but it is the waste that many think of. There are many different grades of nuclear waste. Generally, the more radioactive the waste, the less there is of it. The categories of waste are:

- Exempt waste: This type of waste is also referred to as very low level waste. This type of waste is any material that has so little radioactivity, and thus poses no threat. Waste of this kind includes concrete from decommissioned nuclear power plant buildings.
- Low level waste: This type of waste is generated by facilities that use radioactive isotopes. This includes the nuclear power industry, but also includes medical facilities (such as hospitals or radiology clinics) that use medical isotopes to assess and treat diseases like cancer.

The radioactivity from low level waste is far less intense than from spent fuel, but the volume is much greater.

- Intermediate level waste: This waste has sufficient radiation that it requires shielding and needs to be handled more carefully than low level waste. There is much less intermediate level waste than low level waste. This type of waste includes the metal around the fuel pellets if the fuel is reprocessed.
- High level waste: This waste is the used fuel and the material that has been drawn out of the used nuclear fuel.

Exempt waste, as the name implies, isn't monitored as radioactive waste, but the rest of the types are. Different levels of waste have different laws proscribing how each level is to be handled.

References

- Nuclear-power, technology: britannica.com, Retrieved 3 June, 2019
- Nuclear-power-plant, encyclopedia: energyeducation.ca, Retrieved 17 May, 2019
- Nuclear-fuel, encyclopedia: energyeducation.ca, Retrieved 13 February, 2019
- Nuclear-reactor, technology: britannica.com, Retrieved 29 March, 2019
- Nuclear-waste, encyclopedia: energyeducation.ca, Retrieved 26 July, 2019

Permissions

All chapters in this book are published with permission under the Creative Commons Attribution Share Alike License or equivalent. Every chapter published in this book has been scrutinized by our experts. Their significance has been extensively debated. The topics covered herein carry significant information for a comprehensive understanding. They may even be implemented as practical applications or may be referred to as a beginning point for further studies.

We would like to thank the editorial team for lending their expertise to make the book truly unique. They have played a crucial role in the development of this book. Without their invaluable contributions this book wouldn't have been possible. They have made vital efforts to compile up to date information on the varied aspects of this subject to make this book a valuable addition to the collection of many professionals and students.

This book was conceptualized with the vision of imparting up-to-date and integrated information in this field. To ensure the same, a matchless editorial board was set up. Every individual on the board went through rigorous rounds of assessment to prove their worth. After which they invested a large part of their time researching and compiling the most relevant data for our readers.

The editorial board has been involved in producing this book since its inception. They have spent rigorous hours researching and exploring the diverse topics which have resulted in the successful publishing of this book. They have passed on their knowledge of decades through this book. To expedite this challenging task, the publisher supported the team at every step. A small team of assistant editors was also appointed to further simplify the editing procedure and attain best results for the readers.

Apart from the editorial board, the designing team has also invested a significant amount of their time in understanding the subject and creating the most relevant covers. They scrutinized every image to scout for the most suitable representation of the subject and create an appropriate cover for the book.

The publishing team has been an ardent support to the editorial, designing and production team. Their endless efforts to recruit the best for this project, has resulted in the accomplishment of this book. They are a veteran in the field of academics and their pool of knowledge is as vast as their experience in printing. Their expertise and guidance has proved useful at every step. Their uncompromising quality standards have made this book an exceptional effort. Their encouragement from time to time has been an inspiration for everyone.

The publisher and the editorial board hope that this book will prove to be a valuable piece of knowledge for students, practitioners and scholars across the globe.

Index

www.ingramcontent.com/pod-product-compliance
Lightning Source LLC
LaVergne TN
LVHW080911210926
848485LV00004B/629